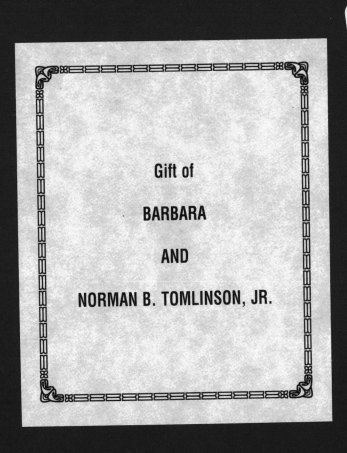

RAND McNALLY

ATLAS OF THE OCEANS

ATLAS OF THE
OCEANS

JOHN PERNETTA
FOREWORD BY COLIN SUMMERHAYES,
DIRECTOR OF THE INSTITUTE OF OCEANOGRAPHIC SCIENCES

RAND McNALLY

CONTENTS

Rand McNally
ATLAS OF THE OCEANS

Published by Rand McNally in 1994 in the U.S.A.

General editor
Dr John Pernetta

Commissioning editor
Stephen Luck

Art Director
Andrew Sutterby

Art Editor
Karen Stewart

Designer
Patrick Tate

Cartographic editor
Andrew Thompson

Picture research
Caroline Hensman

Production
Simon Shelmerdine
Katy Sawyer

Executive editor
Robin Rees

Contributors
Dr. William A. Nierenberg
Dr. Henry Charnock
Sir George Deacon
Dr. A. S. Laughton
H. A. G. Lewis
Robert Allen
Dayton Lee Alverson
Alan Archer

Dr. Tim Barnett
John Bevan
Patricia Birnie
Dr. Quentin Bone
Dr. W. R. P. Bourne
Dr. David Cartwight
Dr. David Cushing
Sir George Deacon
Magaret Deacon
Dr. Brian Durrans
Dr. N. C. Fleming
Dr. B. N. Fletcher
Prof. B. M. Funnell
Dr. Edward D. Goldberg
Dr. James Greenslate
Dr. Roy Harden-Jones
Prof. John D. Isaacs
Dr. E. J. W. Jones
Dr. Niel H. Kenyon
Dr. Robert B. Kidd
Dr. Alan R. Longhurst
Dr. Peter Lonsdale
Dr. Jerome Namias
Joyce Pope
G. H. Rhys
Dr. David A. Ross
Ann Sayer
Reg L. Vallintine
Prof. F. J. Vine
Dr. Peter Wadhams
Cdr David Waters
Dr. David Webb

First published in Great Britain in 1977
by Mitchell Beazley
an imprint of Reed Consumer Books Limited
Michelin House, 81 Fulham Road, London SW3 6RB
and Auckland, Melbourne, Singapore and Toronto

Revised and updated in 1994

Rand McNally and Company.
    [Atlas of the oceans]
    Atlas of the oceans / [editor], John Pernetta.
        p.   cm.
    Rev. ed. of: Rand McNally atlas of the oceans. c1977.
    Includes biographical references and index.
    ISBN 0-528-83703-6
    1. Ocean--Maps.   2. Marine resources--Maps. I. Rand McNally
atlas of the oceans. II. Title.
    G2800.R3 1994 <G&M>
    551.46'022'3--dc20   94-22176
                                              CIP
                                              MAP

Typeset in Times, Gill Sans and Univers

Reproduction by Mandarin Offset, Singapore

Printed and bound by Butler and Tanner Limited,
Great Britain

Half-title page: *Krill*

Opposite title page:
Pacific breaker

FOREWORD

The vast majority of people in the world depend on the oceans in one way or another. Some for food, others for transport, and yet more for mineral resources, such as oil and gas. The ocean controls our climate. It damages our coastlines. Yet, at the same time, we like to swim in it, look at it, live by it, and sail on it. Unfortunately, we also use it as a trash can, polluting our coastal waters through rivers and sewers. To use the ocean wisely means that we must know how it works, and what we are doing to it, so that we can manage it better in the future.

Despite our interest in and use of the ocean, we are surprisingly ignorant about it. This is partly because the oceans are so vast, covering 72 per cent of the Earth's surface, and partly because oceanography is such a young science. Although oceanography started early, with Major James Rennell (1742–1830), the English geographer, mapping currents as "rivers in the sea", and continued with the first global oceanographic expedition, aboard HMS *Challenger* in 1872–6, it was not until after World War II that oceanography really grew as a global science.

Today, most oceanography is carried out remotely, using complex, scientific instruments. Advanced mapping systems, for example, using sound rather than light, produce the undersea equivalent of aerial photographs of the seabed, covering an area the size of Massachusetts in a day. Even with these tools, the area of the ocean is so vast it will take many years to map.

Although there is still much to do, we do know a fair amount about the shape of the ocean floor. The ocean hides the world's longest, highest mountain range, the Mid-Ocean Ridge, whose volcanic Everest peaks above the water in Iceland, the home of rifted valleys, hot springs, lava fields and volcanoes. The Mid-Ocean Ridge lies along the crack formed by the rupture of the Earth's rigid outer shell by the forces involved in plate tectonics. It was these forces that pulled Europe and Africa away from the Americas at about the same rate a fingernail grows to create the Atlantic Ocean. That 30- to 60-mile wide fissure is just as active beneath the sea as it is above the waves in Iceland. Searches of this rift valley are revealing as many as 100 volcanoes every 400 square miles, and hot vents spewing out clouds of metal-rich sulfides at temperatures of up to 350°C. Around the vents thrive bizarre organisms such as giant tube worms up to 8 feet long, and blind shrimp with a third eye capable of detecting heat. They feed on unusual bacteria which live on the energy derived by converting sulfide to sulfate. This new world of vents, first discovered in 1975, is reminiscent of Jules Verne's science fiction world beneath the sea which he imagined over 100 years ago. Although we still know of less than 50 vent sites, we estimate that there may be as many as 5000, a striking indication of what there is to find out about our planet.

Some of our new knowledge about these vent sites may be exploitable. Through biotechnology, scientists may find some application for bacteria living at high temperatures. Understanding how the sulfide deposits form may help us to discover where they are on land.

What other sorts of things don't we know? Why do thousands of seabirds die around our coasts from time to time? Why have wave heights increased by 25 per cent in some coastal areas of Europe in the past 25 years? What triggered a major submarine landslide which created a tidal wave that wiped out coastal settlements in Scotland 10,000 years ago? Will events like these recur, and if so when and where?

Are these sorts of things predictable, or are the oceans chaotic and so unpredictable in detail? We need to know if we are to manage and exploit them more effectively. In the past few decades massive growth in computer power has revolutionized the field of mathematical modeling of environmental systems, enabling us to determine better the role of ocean currents and eddies in transporting heat. Few people know that there is as much heat stored in the top 3 feet of the ocean as there is in the entire atmosphere. Eddies and currents move the vast store of heat freely about the world. The Gulf Stream and its branches carry heat all the way from Florida to northwest Europe. We are also beginning to model the role the ocean ecosystem plays in controling the content of greenhouse gases like carbon dioxide in the atmosphere. Will we soon be able to use our models to predict climate change?

You will find in this atlas the answers to this and many other questions about the manifold resources of the oceans and the problems of managing them sustainably for the benefit of our descendants. The Atlas of the Oceans covers every aspect of the oceans, from their geological evolution through the way in which they work, to their chemistry and the life that it sustains. I hope it will lead you to realize how much there is yet to discover, and how worthwhile and exciting the journey of discovery may be for all of us. The oceans are our last great unexplored frontier.

◄ **Coral reefs** are the most diverse of all marine habitats, with over 1000 species of coral recorded to date. Sadly, however, today many reef environments are being threatened by the growing density of people in coastal areas.

DR. COLIN SUMMERHAYES
DIRECTOR OF THE INSTITUTE OF OCEANOGRAPHIC SCIENCES, ENGLAND

THE
OCEANS

◄ *The Maldive island*
archipelago is found in the
Indian Ocean, southwest of
Sri Lanka. Comprising a
total of 1190 islands, of
which 202 are inhabited, the
warm and sandy Maldives
have become increasingly
popular with tourists.

Genesis of the Oceans

In Greek mythology the great river or sea that surrounded the flat disk of the Earth was personified by Oceanus, the god of the primeval waters. Now, of course, we know that the world is spherical and that just over 70 per cent of its surface is covered by the oceans.

From space, the blue of the Earth reflects the vast expanse of water covering its surface. From a point in outer space above the centre of the Pacific Ocean, the entire planet seems to be composed of water, since no land is visible.

The oceans are by far the largest store of water on the planet. In comparison, the freshwater contained in ice caps, rivers, lakes and groundwater is much smaller, while the clouds and water vapour in the atmosphere make up only a minute fraction of the total. Water leaves the oceans through evaporation, and enters via rainfall, rivers and groundwater seepage. The total volume of water involved in this continuous cycle is approximately 450,000 cubic kilometres (108,000 cubic miles).

Surprisingly, the Earth's crust contains significant amounts of water. The water either permeated into the underlying rocks, or was incorporated into sediments during their formation. In areas where volcanic activity occurs, the molten rock (magma) absorbs water as it rises towards the surface of the crust where it is released as superheated steam during volcanic eruptions. It seems that the water in the crust is critical to the formation of magmas since it lowers the melting point of the rock at depth.

Ocean formation

Most of the water now lying in the oceans has been on Earth since very early in geological history, although initially this would have been mainly in the form of water vapour in the atmosphere. As the Earth cooled, the condensing water vapour fell as rain, forming rivers and fill-ing the lower-lying areas to form the oceans. Although we cannot be certain as to when this process took place, geological evidence suggests that surface waters existed 3800 million years ago. The evidence takes the form of the oldest-known sedimentary (water-formed) rock, discovered in west Greenland. While evidence of simple organisms has been found in types of rock 3400 to 2000 million years old, the earliest traces of multi-celled animals are a mere 600 million years old.

It is in the oceans that life began. Initially it took the form of simple organic molecules, followed later by self-replicating, large organic molecules – the nucleic acids. These would eventually provide the mechanism of reproduction enabling cells to divide and give rise to new cells like themselves.

For almost 2000 million years, simple, single-celled micro-organisms were the only form of life on the planet. Their existence depended on the presence of water, and it was their life processes that altered the composition of the atmosphere. At first, Earth's atmosphere was dominated by water vapour, methane and ammonia, but the mere existence of photosynthesizing micro-organisms led to increasing concentrations of oxygen in the atmosphere. This oxygen led in turn to the formation of the ozone shield which now protects life on Earth from the harmful effects of the Sun's ultraviolet radiation.

The changing oceans

The ocean basins have not always had their present form. The movement of the continents and ocean floor spreading have resulted in periods when the oceans were quite differently arranged. Some 225 million years ago the land formed a single, large continental landmass called Pangaea, which was surrounded by a super-ocean, Panthalassa. When this landmass broke up some 180 mil-

▶ *The oceans* are the largest store of water on the planet. Four thousand million years ago the atmosphere was dominated by water vapour. As the Earth cooled, condensing water vapour fell as rain, forming rivers and filling the lower-lying areas to form the oceans.

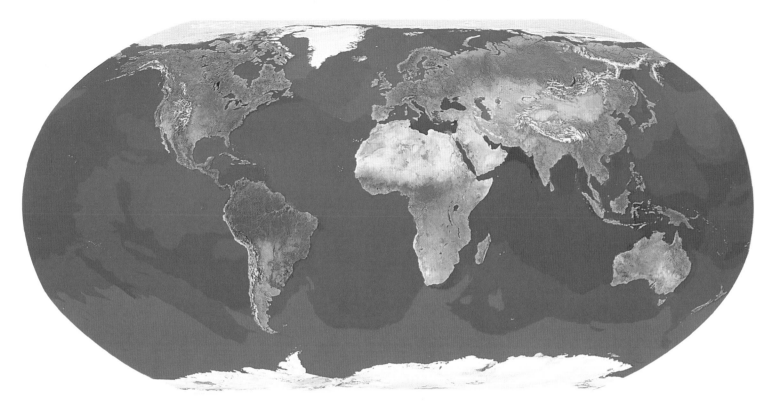

▲ **This geosphere** image of the Earth based on the Robinson projection shows land areas in natural tones, while the ocean colour provides an impression of temperature – cooler areas in light blue, warmer areas in darker tones. This picture is constructed from thousands of cloud-free images taken by the NOAA weather satellites and blended using a graphics supercomputer.

lion years ago to form two super continents (Laurasia to the north and Gondwanaland to the south), the Tethys Sea was formed between them. This pan-tropical sea was subsequently closed by the collision of what is now Africa with the Eurasian mainland. The present ocean basins are thus relatively young, being less than 80 million years old.

The ocean environment

The oceans are in constant motion. They are influenced by the rotation of the Earth, the gravitational pull of the Sun, the Moon and nearby planets, and by the movements of the atmosphere. This motion has important consequences since the oceans circulate heat from the warmer tropics to the colder, higher latitudes, so influencing the climate of neighbouring landmasses. Ocean circulation is not only an important component of the global climate system, but in addition it results in quite different physical and chemical conditions for life in different parts of the ocean basins.

The environment of the ocean varies drastically. Conditions range from the warmer, sunlit surface waters to the dark and colder ocean deeps, and from the high-salinity ocean basins such as the Red Sea to the low-salinity conditions of the Bay of Bengal.

The surface waters of the ocean are generally low in nutrients, particularly in tropical areas, and this limits the rate of primary production by phytoplankton. Phytoplankton are single-celled plants that use sunlight to form more complex chemicals and which serve as the basis for marine life. The deep ocean, on the other hand, contains vast stores of nutrients, and where these are brought to the surface by vertical water movement, known as upwelling, there is a high production of plankton and fish. High concentrations of nutrients also occur in coastal regions where surface run-off carries nitrogen, phosphorus and organic material to the coastal waters.

This spatial variation in physical and chemical conditions influences the types of plant and animal which can survive in different sections of the world's ocean basins. It is this variation in physical features that accounts for the diversity of life in the marine environment.

The Earth's Moving Crust

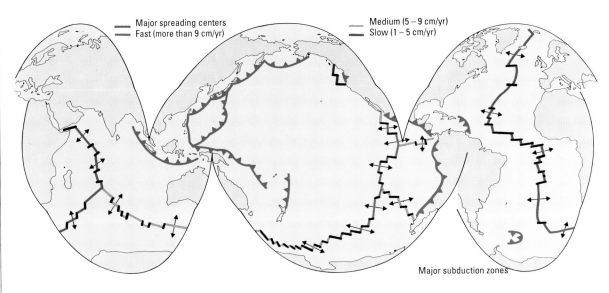

Major spreading centers
Fast (more than 9 cm/yr)

Medium (5 – 9 cm/yr)
Slow (1 – 5 cm/yr)

Major subduction zones

▼ *The Earth* *consists of several distinct layers: the core of nickel and iron, existing below about 2880 km (1800 miles); the stony mantle; and the crust, which averages only 16–40 km (10–25 miles) in thickness. The mantle is solid except for a layer near the top called the asthenosphere, which is thought to be fairly mobile. Above this lies a layer about 96 km (60 miles) thick, consisting of the uppermost mantle and the crust. This rigid outer layer, called the lithosphere, represents the plates that move over the Earth's surface. Each plate, such as the one centred on Antarctica is ringed by active plate margins.*

Planet Earth is made up of several concentric spheres of slightly different composition. The crust, or outer shell, consists of relatively light materials and is separated from the denser, underlying mantle by a region known as the Mohorovičić discontinuity. Beneath the continents, the Earth's crust is much thicker, 30–40 kilometres (19–25 miles), than beneath the ocean basins where it is some 6 kilometres (4 miles) thick; and where its physical properties suggest that it is much more closely related to the underlying mantle. In addition, the thickness of the ocean crust is essentially constant throughout the ocean basins. In contrast, the thickness of the continental crust varies considerably reflecting surface relief. The greatest depth of crust being found beneath mountain ranges.

From mantle to core

The mantle extends to a depth of approximately 2900 kilometres (1800 miles) and is thought to consist of ferromag-

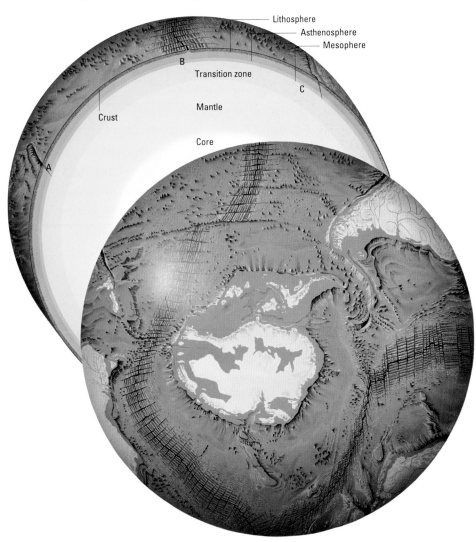

Lithosphere
Asthenosphere
Mesophere

Transition zone

Mantle

Crust

Core

B

C

A

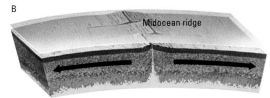

Island arc   Trench

A

Crust   Mantle   Lithosphere   Asthenosphere

▲ *The Tonga Trench is a destructive plate margin, involving two oceanic plates. The margin is characterized by a deep trench and an arc of volcanic islands formed by melting within and above descending plates.*

B

Midocean ridge

▲ *The East Pacific Rise is a constructive margin, where two oceanic plates are being continuously generated. Material is constantly rising to the surface, to be added to the edge of the plates as they move apart.*

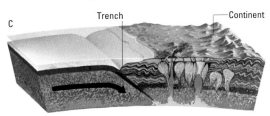

Trench   Continent

C

▲ *The Peru-Chile Trench is a destructive margin, where one oceanic and one continental plate meet. The oceanic plate forms a trench as it is carried down. The lighter continental material rises into a mountain range.*

nesian silicate minerals such as olivene and pyroxene. The division between the upper and lower, denser sections of the mantle occurs at a depth of around 700 kilometres (435 miles). The upper mantle is in turn divided into the upper lithosphere, which extends down for between 50 and 250 kilometres (30 and 155 miles). Below this lies the asthenosphere, a partially molten zone some 200 kilometres (125 miles) thick. The asthenosphere defines the lower boundary of the relatively rigid outer lithosphere, which includes the crust. The plates of the lithosphere move over the semi-molten asthenosphere, giving rise to the phenomenon of continental drift.

Below the asthenosphere lies the mesosphere, which is separated, at a depth of around 400 kilometres (250 miles) by a transition zone, from the lower mantle which lies between 1000 and 2900 kilometres (620 and 1800 miles) in depth. The lower boundary of the mantle is marked by the Gutenberg discontinuity, which separates the lower mantle from the underlying dense core. The core itself is divided into an inner and outer layer at a depth of between 4980 and 5120 kilometres (3090 and 3170 miles). The core most probably consists of an iron-nickel alloy, with a liquid outer core and an inner solid core with a radius of 1300 kilometres (810 miles).

Seafloor spreading

Large-scale convection currents occur in the asthenosphere, and where these currents rise towards the Earth's surface mid-ocean ridges are formed. Alternatively, where the currents flow towards the Earth's core, deep ocean trenches are created. The thin, oceanic crust forms at mid-ocean ridges and is absorbed back into the mantle beneath the trenches.

Since the mid-ocean ridges are, as their name suggests, found along the mid-line of the ocean basins, new oceanic crust is formed on each side of the ridge. As new crust is formed it pushes the older crust further away from the ridge. The youngest crust is, therefore, found near the centre of the mid-oceans, and the oldest in the vicinity of the deep ocean trenches. It is now thought that the oldest ocean crust is no more than 200 million years old.

The mid-ocean ridges are crossed by great transform faults which result in the ridges being offset, reflecting the irregularities of the edges of the plates. Along these transform faults, earthquakes are generated where newly formed parts of adjacent crustal plates slide past one another in opposite directions.

The outer skin of the Earth is divided into a relatively small number of rigid plates which are approximately 100 kilometres (60 miles) thick and bounded by mid-ocean ridges and deep-sea trenches. There are six major plates and several small ones that are moving relative to each other. The direction of the motion between any two plates is determined by the nature of the plate boundary. Plates separated by a mid-ocean ridge are clearly moving apart, while plates separated by deep-ocean trenches and active mountain chains, or both, are moving towards each other. Where plates are moving past each other without the creation or destruction of crustal material they are are separated by major fault zones, as in the Caribbean Basin.

The movement of plates is also indicated by volcanic island chains, such as the Hawaiian islands, which result from the movement of the ocean plate over a fixed "hotspot" in the mantle. Passing northeastwards along the island chain the islands become progressively older, with the oldest volcanoes no longer reaching the surface of the sea, but continuing the chain in the form of a line of submerged guyots, or sea mounts.

◀ **The size of the plates** is not constant. The African plate is bounded on three sides by mid-ocean ridges that indicate the presence of constructive margins. Since the plate is constantly enlarging from these ridges, and there is no intervening destructive margin, it follows that the ridges cannot be stationary relative to each other but are moving apart as the plate enlarges.

The Magnetic Record

We know new rocks are formed at ocean ridge crests because magnetic particles in them are immediately aligned parallel to the Earth's magnetic field. This magnetic "signature" is locked into the rocks as they solidify and is retained as they continue to spread.

Magnetic north

The Earth's magnetic field has changed its polarity in the past few million years. Any newly formed rocks are magnetized in the new direction, which over time creates a pattern of alternately polarized strips.

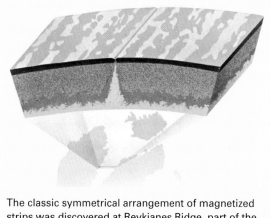

The classic symmetrical arrangement of magnetized strips was discovered at Reykjanes Ridge, part of the Mid-Atlantic Ridge south of Iceland. The pattern of polarity reversals correlates with those found in the Pacific and with the known sequence of reversals over the past ten million years.

Continental Drift

The idea that the continents were once joined together is not a new one. Even as early as 1858, maps were published that depicted the continents welded into a single, large landmass surrounded by a vast ocean. Subsequently, the evidence of the similarity in both fossil type and age of geological formations on both sides of the Atlantic led Alfred Wegener, in 1915, to publish a series of reconstructions of the continents.

However, it was not until the 1960s, when the evidence of palaeomagnetism (the record of the Earth's magnetic history) in the ocean crust became available and mechanisms to explain the process of continental drift were postulated, that the theory became accepted.

Pangaea and Panthalassa

During the Cambrian period (590–505 million years ago), most of today's Southern Hemisphere landmasses were fused into a supercontinent known as Gondwanaland. North America, northern Europe and most of Asia (later to become Laurasia) remained separate landmasses. By the Carboniferous period (350 million years ago), North America and northern Europe had collided. Where the two continental plates met, the Caledonian mountain range was formed. For a time, the newly formed landmass was separated from the larger continent of Gondwanaland, but by the Permian period (280–225 million years ago), the continents joined together to create the supercontinent of Pangaea. Pangaea was surrounded by a single superocean known as Panthalassa.

The movement of southern Europe towards northern Europe raised the Hercynian mountains, while the collision of Asia and Europe created the Urals. Subsequently, between 200–180 million years ago, Pangaea began to split into the northern continent of Laurasia and a southern

▲ *Antonio Snider-Pelligrini* was the first to reassemble the continents according to geological as well as geometrical evidence. He published the above maps in his book, La Création et ses Mystères Dévoilés, *in 1858. His ideas, however, were deemed too farfetched, and were forgotten for over 50 years.*

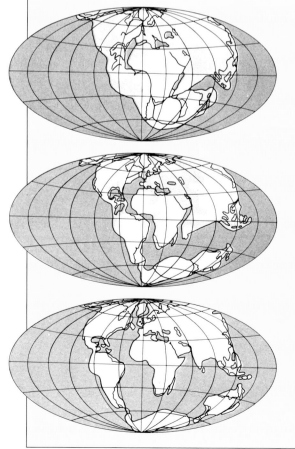

Continental History
Alfred Wegener, a German geophysicist, astronomer and meteorologist, published a series of paleogeographical reconstructions of the continents in 1915 (translated in 1924 as *Origin of Continents and Oceans*). His evidence was based on the similarity in fossil type and age of geological formations on both sides of the Atlantic. His work was not universally recognized until the 1960s when the evidence of palaeomagnetism in the ocean crust, seafloor spreading and the theory of plate tectonics were established. At the time, he was at a loss to account for the mechanism of continental drift and proposed such unlikely processes as tidal force and variations in gravity at different points on the globe. Although Wegener saw the continents as flexible masses rather than rigid plates, and although he was mistaken about the speed of drift, his maps are still acceptable today.

continent of Gondwanaland. By 200 million years ago Laurasia began to break up, forming the North Atlantic. Thirty million years later Gondwanaland began to separate, forming the Indian and South Atlantic oceans, the latter being 100 million years old.

Ocean basins

The ocean basins have been formed, been lost and been reformed many times. Six hundred million years ago, for example, the North Atlantic region was covered by an ocean, not unlike that of the present day. Around 500 million years ago, this proto-Atlantic had started to close and was flanked by Andean-type mountain ranges. About 100 million years later it had completely closed and the movement of the continents towards one another formed the Caledonian Mountains. The present-day North Atlantic has only been in existence for the last 200 million years, and the split which formed this ocean basin occurred roughly along the line of closure of its predecessor, the proto-Atlantic.

Today, the Red Sea represents an early stage in the formation of a new sea. If a new area of upwelling in the asthenosphere starts beneath a continental area then this sets up tensions in the continental crust. The tension causes the crust to split and form the classic rift valleys such as those of eastern Africa. As the two parts of the continent move away from one another a new sea is formed, along the centre line of which a mid-ocean ridge develops and new ocean crust is formed.

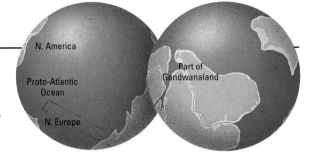

◄ The configuration of continents today reflects the movements of the continental plates over the past few hundred million years. Similar geological features on separate continents allow geologists to reconstruct the past positions of the land and oceans.

Mountains and volcanic arcs

When two continental plates collide, the continental crust buckles and is forced both upwards and deeper into the underlying lithosphere. The result is often the formation of mountain ranges, such as the Alps and Himalayas. However, not all plate margins are of these simple types. Where, for example, the margin of an oceanic plate meets a plate margin of continental crust, the oceanic crust is forced beneath the continental crust in a subduction zone.

An example of a subduction zone can be found off the west coast of Latin America. It takes the form of a deep ocean trench, where the denser ocean crust is being forced downwards. In such situations the less dense continental crust is folded and uplifted to form extensive mountain ranges such as the Andes in South America, the Rocky Mountains in North America, and during a geologically earlier phase of continental drift, the Appalachians.

In some ocean basins where two plate margins, both formed of ocean crust, move towards each other, one is forced beneath the other. Such destructive plate margins are characterized by a deep ocean trench where one plate is driving down into the crust and an arc of volcanic islands forms on the other plate.

▲ The Cambrian world would be unrecognizable today. Gondwanaland existed as a single landmass that would not split into Australia, India, South America, Africa and Arabia for another 400 million years. North America, northern Europe and most of Asia were separate continents.

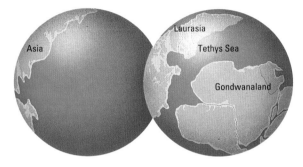

▲ In the Carboniferous, at the time of the coal forests and giant amphibians, North America and Europe had closed the proto-Atlantic Ocean forming the landmass of Laurasia with the Caledonian Mountains between. Laurasia was separated from the southern landmass of Gondwanaland by the Tethys Sea and the South Pole lay somewhere in South America.

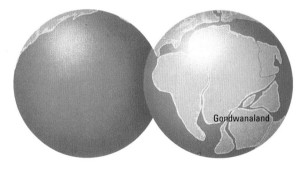

▲ The Permian was the time of Pangaea and Panthalassa, with the land forming a single supercontinent. The movement together of Laurasia and Gondwanaland pushed up the Ural and Hercynian mountain ranges. At this time much of the northern landmass was desert, reptiles roamed the land and icthyosaurs and ammonites dominated the oceans.

▲ In the early Cenozoic, flowering plants had appeared and mammals had replaced the reptiles. Pangaea had broken up and the Atlantic had formed. Antarctica and Australia were still together, but India was drifting northward to collide with Asia and produce the Himalayas.

Eroding Shores

Shorelines are highly varied in their form since they are shaped by the action of both wind and waves. The movement of material by currents in near-shore waters may result in the deposition of sands, gravels or larger material, such as boulders, depending on the power of the waves. In other areas, where sedimentary material is scarce and the waves move material away from the shoreline, erosion dominates the coast. Erosional shorelines tend to occur in areas of high wave energy, while depositional shorelines are characteristic of sheltered coasts.

▲ *An upland area* sloping gently into the sea is not a stable configuration. Waves attack the slope and erode it at the base.

▲ *The first-formed* feature of coastal erosion is usually a notch in the slope, forming a small cliff at high-tide level.

▲ *The cliff* is worn back by the action of waves cutting into the base, forming an overhang which eventually collapses.

▲ *A wave-cut platform* is left as the cliff is worn back and an offshore terrace is built of eroded material.

Erosional shores

Eroding shorelines are characterized by cliffs and dramatic rocky features such as the Giant's Causeway off the Irish coast. Although the height of cliffs facing the sea is largely determined by the height of the land behind them, their form is dependent on a wide variety of factors, including the properties of the rocks and sediments of which they are composed, the topography of the landscape into which they are cut, and the geological history, including sea level and tectonic changes. Great variations are possible but features such as promontories, blow holes, arches and sea stacks are frequently formed in well-jointed rocks and give rise to some of the most distinctive coastal scenery.

In general the nature of eroding shorelines reflects the ability of the rock to resist the erosive forces of the waves and the suspended materials thrown by the water against the shore. It is not the water itself which erodes the land, but the sand, pebbles and gravel which are hurled against the cliff face by the force of the waves. Most erosion takes place along weaknesses in the rock such as faults, joints, bedding planes and layers of softer materials interspersed between harder bands.

Erosion occurs in a relatively narrow band centred on mean sea level. The width of the band is narrow in areas with a small tidal range and wider where the range is greater. The restricted width of the zone of active erosion means that the main removal of material occurs at the foot of the cliff and results in slumping or collapse of materials as they are undercut by wave action. In areas of soft rock cliffs the rate of erosion may be limited only by the rate at which longshore currents remove the slumped material from the base of the cliff where it forms a protective beach. Along hard rock formations the undercutting may result in only infrequent collapse of the cliff face above, and shallow sea caves or overhangs may develop.

In rocks such as sandstones, which have even bedding planes (surfaces that separate different layers of rock), erosion occurs along the planes, softer strata are worn away faster, leaving a ragged profile to the cliff face. In fine grained chalks where bedding planes are not well defined, smooth, sheer cliff faces develop, while in basalt erosion takes place along the joints. Basalt cliffs, therefore, appear to have been built of individual blocks.

Beaches on eroding shorelines

Along any eroding shoreline a beach of eroded materials will be formed as the cliff face collapses into the sea. These erosional products may be either removed by currents and deposited elsewhere, or deposited as an offshore terrace or a wave-cut platform beneath the erosional face of the cliff.

When a wave approaches the shore it is slowed down in shallow areas and on meeting an irregular coastline is turned along the shore. Most of the wave's energy is thus expended against projecting headlands and as a result these are eroded more rapidly than the bayheads and beaches where sand and gravel accumulate. Eventually the headlands may be completely eroded and a straight shoreline will evolve.

Coastline Evolution

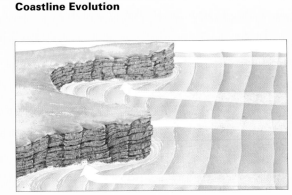

A headland or cliff erodes first along joints or bedding planes, which enlarge to form deep clefts.

The beaches found at the foot of cliffs are transitory, often changing their form with the season as changing wind and wave patterns alter the rate of removal of the material. In general, the finer materials are removed more rapidly so that pebbles, cobbles and even boulders tend to dominate such beaches. Most of the erosional materials that form these beaches are removed offshore to deeper waters, or moved along the coastline to more sheltered areas where depositional coastal formations occur.

Along some coastlines, such as those of the eastern Mediterranean, pocket beaches are found at the head of small inlets. Such beaches are generally surrounded by high cliffs and these coastal landforms are highly susceptible to changes in sea level or wave patterns, which result in the sediments being lost offshore.

In the case of many Mediterranean islands which have been formed by uplift along fault lines, cliffs descend vertically into the sea and erosion rates may be quite slow since the tidal range is small. In contrast, along low-lying coastlines where relative sea level is rising and the materials are soft and unconsolidated, erosion rates may be extremely high, as much as several metres a year. Often the erosion occurs in short periods associated with storms, resulting in the development of an erosional scree slope, before the materials are gradually removed.

Life on eroding shores
Erosional shores are generally inhospitable environments for animals and plants. Since they are characteristically high wave energy environments, the animals and plants must be able to either attach themselves firmly to the rocks or hide away in the cracks. Where the rates of erosion are high the diversity of the community of animals and plants tends to be limited since the shore does not provide a suitably stable base for attachment. On hard rock, erosion-resistant shores, the diversity of organisms tends to be greater and the zonation of the animal and plant communities is well marked.

Above the high-tide level in the splash zone are found species which are resistant to desiccation, while soft-bodied animals generally occur lower down in the intertidal zone where they are active only when covered by the tide. Above the high-tide line occur a wide variety of lichens and some blue-green algae which are replaced lower down by the typical sequence of green, brown, and red seaweeds. On rocky shores where the surface is smooth, animals tend to exist lower in the intertidal zone than on shores where cracks and fissures are present which can provide moist hiding places. Thus the nature of the material forming the shoreline not only affects the rate of erosion but also the distribution of the organisms that live there.

◄ **These old red** sandstone cliffs on Hoy in the Orkneys show the effects of erosion along the bedding planes. Small embayments have been formed, and the eroded material is moved along the shore before being deposited on the sheltered beaches at the head of these bays.

▼ **Erosion of this** shoreline at Etratet in Normandy, France, has resulted in the formation of pinnacle-like stacks and natural arches. Eventually the arch in this photograph will collapse, forming a second stack, and a new arch will be formed in the headland behind.

A vertical cleft subjected to constant erosion will eventually enlarge to form a cave.

When air in a cave is repeatedly compressed by wave action, the pressure forces a hole in the roof.

Continued deepening of caves on opposite sides of a headland causes them to meet, forming a rock arch.

A natural arch will continue to widen until its lintel collapses, leaving an isolated sea stack.

Depositional Shores

▶ *The swirls of blue and beige in this landsat picture of the Yangtse River delta in China show that the delta is continuing to grow seawards as sediments are brought by the river from the inland areas of the catchment.*

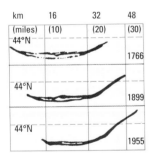

Path of oblique waves

▲ *Longshore drift is caused by waves meeting the beach at an angle.*

km	16	32	48
(miles)	(10)	(20)	(30)
44°N			
			1766
44°N			
			1899
44°N			
			1955

▲ *Sand-formed features may be moved great distances by longshore drift. In 200 years Sable Island, off Nova Scotia, travelled 14 km (9 miles) eastward.*

Depositional shorelines are soft shoreline types formed by the build-up of sediments. Most shoreline sediments are ultimately derived from the weathering of exposed rock surfaces and erosion from land. Eroded materials, including soil, are carried from catchment basins into the river systems of the world. They may remain for some time in flood plains and river levees before being discharged to the waters of coastal areas.

Deltas and estuaries

Most of the major rivers of the world enter the sea via large, deltaic systems. These systems are constructed of sand and fine muds deposited at the delta margin as a result of the reduced speed of water movement where the river meets the ocean. As a consequence of the reduced velocity, coarser sediments settle out and the movement of the tides reworks the sediments into flats which may be exposed at low tide.

The relative strength of the river flow, the volumes of sediment, the offshore bottom topography and the tidal influence all affect the shape and size of deltas and estuaries. Where large volumes of sediment are carried out to sea by a river discharging into a shallow coastal region, deltas may accrete seawards. In contrast, where the sediment load is small, estuaries are formed and salt water may penetrate the river channels far inland if the river flow is weak.

The land-based products of erosion are often moved along the coast by the process of longshore drift. Thus beaches may be dependent on a source of sediment that is a considerable distance away. When the supply of sediment from a river is cut off (by dam construction, for example), beach erosion can occur down current.

Sediments along the coast may be moved offshore depending on the local patterns of waves and currents. Alternatively, in some areas the major source of sediments that form beaches may be offshore deposits, particularly where the seafloor is covered by loose glacial material. Currents and storms may then move this material towards the shore, forming storm beaches or ridges that may continue to grow through further, wind-blown deposition of sand.

Nutrients

The productivity of coastal waters is much higher than that of the open oceans, due in part to the input of nutrients being washed from land to coastal waters. Nitrogen and phosphorus are carried by river and groundwater flows into the coastal zone where they provide the basis for phytoplankton production. Human influences have greatly enhanced the rates of nutrient transfer to the coastal ocean through the discharge of sewage, agricultural and industrial wastes. As a consequence phytoplankton "blooms" are now more common and more widespread than in the past.

Even in areas where human influence is minimal, natural inputs of nutrients from land provide the basis for the high productivity of coastal waters. Some of the most productive areas of the coastal ocean are the deltas, estuaries and wetlands associated with major river outflow into near-shore waters.

Sediment types

The nature of the sediments found along shorelines reflects their source, in terms of the rock from which the materials have been derived. At high latitudes along the shores of Mediterranean and temperate regions, beaches are normally formed of materials of mineral rather than organic origin. In some areas extensive shell sand beaches line the coasts in the vicinity of offshore shell reefs.

In contrast, along many tropical shorelines entire beaches are comprised of bioclastic sand – sand formed by the weathering of the skeletons of corals, molluscs and other marine organisms. In many atoll environments much of the calcium carbonate found on beaches was originally laid down in the living tissues of corals and coralline algae. These white beaches contrast sharply with the black sand beaches formed of weathered volcanic materials on oceanic islands and island arcs.

◄ **Sand dunes** such as these in South Wales are stabilized by the marram grass underground rhizome and root system. The leaves reduce wind erosion and trap sand blown from the beach, which continually adds to the height of the dune.

The action of the waves and currents not only reduces the size of individual particles through mechanical abrasion, but also sorts the material into different sizes. Towards the landward side of a beach the particle size is larger than lower in the intertidal zone, while the finest sediments are found offshore as sub-tidal muds.

Life in soft shores

The sediments on the sea bottom are constantly moved by tidal and other currents – ripple marks and other small-scale features of the surface reflect this movement. This disturbance is greatest in the tidal and sub-tidal reaches of the shore, and as a result the surface of soft shorelines is home to relatively few animals and plants. The shifting surface provides no permanent anchorage and animals must retreat beneath the surface if they are to survive.

Most organisms on soft shores, therefore, have adaptations for burrowing and spend much of their life below the surface, emerging to feed and mate when water conditions are suitable, or merely extending the feeding apparatus above the surface while the body of the animal remains in its protective burrow.

At low tide a mud flat may appear lifeless and barren, but this appearance is misleading. Buried beneath the sediments lie a multitude of animals ranging from small microscopic forms which move between individual sand grains, to large burrowing clams and lugworms eagerly sought by sea birds that forage in such areas at low tide. The distribution of these animals reflects not only the influence of the tidal range but also the size of sediment particles with different species inhabiting areas with different types of sand, mud and finer particles.

◄ **This picture** shows the effects of road construction running parallel to the shoreline in San Francisco Bay. Water exchange between the sea and the pools on the right no longer occurs, resulting in the formation of hypersaline lagoons.

Man-made Coastlines

At the present time, around 60 per cent of the world's population lives within 60 kilometres (40 miles) of the coast, and in many countries this proportion is much higher. In addition, the population of coastal areas is, in many tropical countries, growing at twice the rate of national population growth, due to migration from inland areas to the coast. This problem is not confined to developing countries. Historically, large-population centres tended to develop in coastal locations, deriving much of their economic viability from international trade and commerce. Today, over two-thirds of the world's cities with populations greater than 1 million people are located on the coast, often in highly productive estuarine areas where the productivity of both the coastal land and inshore waters are lost as a consequence of the urbanization of the land surface and pollution of the coastal waters through urban and industrial discharges.

Coastal space

As a consequence of these trends, the problems of conflict between different uses of coastal space and inshore waters have grown considerably in recent years. The demand for highly priced coastal land near major ports and harbours has resulted in expensive reclamation schemes. In some cases, where the absolute area of land is limited, such as on islands, the need for living space has necessitated the creation of new land on the surface of neighbouring coral reefs or shallow water areas. This is exemplified in the case of Male', capital city of the Republic of the Maldives. Male' occupies the entire surface of an island 1700 metres (5610 feet) long and 700 metres (2310 feet) wide, on which 56,000 people live. Around half the island has been artificially constructed by pumping lagoon sand onto the coral reef flat to create new land. As a result, the island occupies virtually the entire surface of the coral reef platform on which it stands and the seaward edge of the island is only some 30 metres (100 feet) from the edge of the reef platform. To protect the island, a sea wall was constructed around the entire perimeter to a height of about 2 metres (7 feet) above sea level. This sea wall was, however, inadequate to protect against long-distance swells from a severe storm in the Indian Ocean in 1987, which resulted in extensive flooding of the capital city. A breakwater has now been constructed on the seaward side of the island at a cost of US$12,000 per metre. Land reclamation and the subsequent need to protect the investment inevitably results in further investment in protection. This example is replicated in different forms all around the world.

Coastal investment and protection

As the investment in coastal areas has risen, so has the need to protect that investment against flooding, storm surges, tsunamis and the general erosive action of winds and waves. Shorelines are armoured, sea walls constructed, groynes extended seawards and beaches replenished with sand to cater for the increasing influx of tourist visitors to coastal areas. Under natural conditions shorelines change. Sediments are moved onshore and offshore according to the relative influence of different wind and wave patterns. In many areas, sediments are moved along a coastline (longshore drift) and the construction of hard structures at right angles to the shore interrupts this flow of sediment, resulting in accretion on the up-drift side of the groyne and enhanced erosion on the down-drift side.

The dynamic nature of natural shorelines is an inconvenience to many human activities, and, therefore, considerable effort is expended in trying to hold the shoreline in a constant position. A retreating coast which threatens structures such as tourist hotels or roads may be armoured. Although this may protect the individual site, it frequently results in increased erosion elsewhere along the coast as the pattern of waves and local currents is altered by these structures. In many areas, where tourism is important, hotels are constructed close to the shore, often without any information on the dynamic changes which might be expected in the area. When erosion commences, sea walls are built to protect the hotel and groynes constructed in an attempt to trap sand on the beach area in front of the hotel.

▶ *Expensive* engineering structures such as the Thames Flood Barrier are built to protect low-lying cities such as London from flooding. Flooding may be caused by a combination of high tides and onshore winds which temporarily raise the sea level.

▶ **Beach erosion** such as that illustrated in this picture of the Florida coastline often results in action to try and stabilize the shore. Such actions frequently result in increased erosion elsewhere along the coast, and are often expensive to maintain.

Loss of sand has resulted in expensive beach replenishment schemes. Sand is pumped from deeper waters offshore onto the beach to maintain its form. Like dredging of harbours, this activity is an on-going expense.

The nature of many coastlines worldwide is being dramatically altered, such that in many areas the entire coastline is artificial. In some countries, such as the Netherlands, the whole shoreline is artificial, and what was once a dynamic boundary between land and sea, subject to erosion and/or accretion under different conditions, has been fixed at one point in space. Land reclamation inevitably requires protection. Much of the Netherlands is actually below sea level, having been reclaimed through construction of dikes and pumping of water out of the enclosed polder. Maintaining this dry land requires constant and continuous pumping of water into the sea from behind the protective dikes which line the coast. Many of these dikes are based on old dune systems, but unlike the dunes, which would have changed position over time, these must now be constantly maintained if the sea is not to encroach on what is now intensively used farmland and densely populated areas.

Maintaining artificial coastlines will become increasingly expensive in the future. Sea level is already rising worldwide at a rate which could increase in the next century under the influence of global warming.

◀ **To prevent** coastal erosion, groynes are often built that are designed to prevent the loss of sediments through the action of longshore drift. The construction of groynes results in sediment build-up on the up-current side and enhanced erosion on the down-current side.

Changing Levels of the Sea

Anyone who has lived near the sea or spent time on the seashore is aware of the daily rhythm of the tides. The tides can change the meeting point of sea and land by as much as 15 metres (50 feet) vertically and thousands of metres horizontally. In addition to the daily variation in sea level resulting from the tidal cycle, the position of high and low water varies between spring and neap tides depending on the phase of the lunar cycle. Unusually high water levels may occur when onshore winds occur at the same time as high tides. During such events seawater may penetrate far inland along freshwater courses and inundate coastal land which is normally above sea level.

The more subtle and gradual changes in sea level are less obvious. Some of these changes represent long-term trends such as the apparent rise in global sea level of around 1.5 millimetres ($^1/_{16}$ inch) per year which has occurred over the last one hundred years. In contrast, other changes represent shorter term responses to major shifts in ocean currents such as the lowering of sea level in the western Pacific by as much as 14 centimetres (6 inches) during El Niño years (see p.162). Far greater changes in sea level have occurred in the recent geological past. The glacial and interglacial periods were marked by changes in sea level of up to 120 metres (400 feet).

Global changes in sea level

Global sea level is affected by a variety of factors, including the volume of water in the ocean basins, thermal expansion of the surface layers of the ocean, and changes in depth of the ocean basins. The great changes which occurred during the Pleistocene period reflect the large volumes of water alternately locked into ice caps on land or released as water to the ocean basins. During glacial epochs sea level was lower than today, while during the warmer interglacials the melting ice sheets resulted in higher sea levels.

Local changes

On a local scale, relative sea level, that is the vertical position of the land relative to the sea, may be affected by a wide variety of phenomena, both natural and man-made. The movements of the Earth's crust may result in tectonic uplift of coastal areas, as in the case of the Huon Peninsula of northern New Guinea where terraces of raised coral reefs provide a visual record of changes in sea level and land movements in the recent past. In the Mediterranean, tectonic movements have resulted in the submergence of some Roman and Bronze Age ports and the uplift of others which are no longer at sea level.

Tectonic changes along coastlines may be abrupt, with sudden uplift events occurring during earthquake activity. In other areas, coasts are sinking or rising more gradually as a consequence of post-glacial rebound, the gradual rising of the land when the weight of the glacial ice sheets is removed. The southern coast of the United Kingdom, for example, is slowly sinking, while that of northern Scotland is rising. During the glacial periods the northern half of these islands was covered by extensive ice sheets, the weight of which depressed the Earth's crust, causing the north to sink and the south to rise. Once the ice melted the crust readjusts to the reduced weight above, causing the south to sink and the north to rise. Erosion of the shoreline

▲ **The Huon Peninsula** in northern New Guinea is lined with raised coral reefs. This coast is being raised by uplift of the crust, while the sea level has risen and fallen during the Pleistocene period, alternatively exposing and inundating the reef systems.

▶ **The combined effects** of high tides and erosion are clearly shown by this palm tree on Grenada in the Caribbean, which has at some time in the past fallen towards the sea. Subsequently the tree has resumed a much more upright pattern of growth.

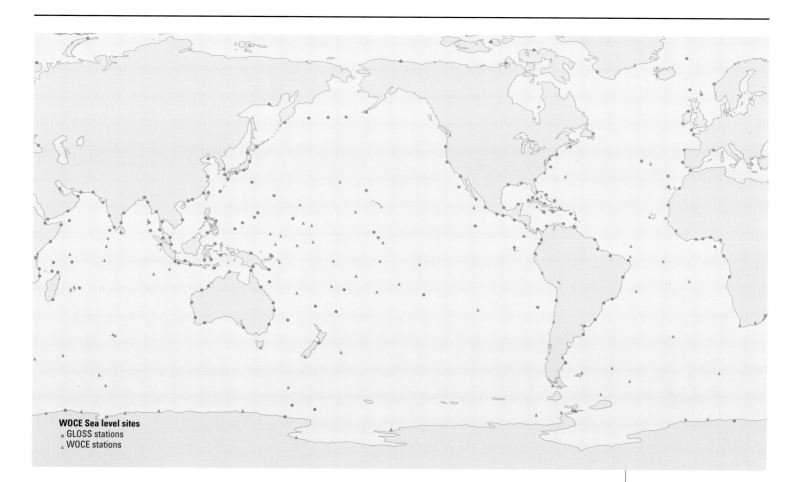

WOCE Sea level sites
● GLOSS stations
▲ WOCE stations

is, therefore, widespread in southern England, but raised beaches above the high-tide line characterize the shores of much of Scotland.

Where sea levels have risen in the recent past, characteristic coastal landforms occur. These include drowned river valley systems, or ria coastlines, or the fjord coastal landforms of Scandinavia and southern Chile where U-shaped valleys cut by past glaciers line the coast.

In major deltas, particularly those lying above thinner areas of the Earth's crust, sinking of the crust (beneath the accumulated weight of sediments being deposited in the sections below sea level) causes a rise in relative sea level. Where the supply of sediments is not interrupted through dam construction inland, the delta may be in balance, with the newly deposited sediments raising the land surface and compensating for the sinking of the crust beneath.

Human induced changes

Human actions can also cause adjustment of the land surface. In Bangkok, for example, rapid sinking of the land results from the extraction of fresh groundwater for the city's drinking water supplies. Extraction of groundwater results in compaction of the sediments and lowering of the land surface relative to the sea. Extraction of groundwater, oil and gas from coastal and near-shore areas causes problems of rising relative sea level in many coastal areas around the world.

A more recent concern is the potential effect of global warming which may cause the further melting of land-locked ice. In addition to this melting, increased surface temperatures will result in expansion of the surface layers of the ocean which will in turn cause a rise in relative sea level. It has been suggested that global warming may result in a rise in sea level of up to a metre (3 feet) by the year 2100, although the evidence for such a rise is not accepted by all scientists.

▲ **Monitoring the level** of the world's oceans takes place through a series of global networks of tide gauging stations. This map shows the location of WOCE (World Ocean Circulation Experiment) stations and the GLOSS (Global Land-Ocean Survey System) network, which contribute information to the global networks.

The Continental Shelf

▲ The active type of continental margin occurs at a destructive plate boundary where island arcs and ocean trenches occur. Along the west coast of South America the continental plate has been raised to form the Andes Mountains.

▲ The passive type of continental margin is found around the edges of spreading ocean basins. It retains the rift valley structure when the continents split to form the ocean basin. The seaboard of much of the Atlantic Ocean is of this type.

▶ This sonograph shows a system of gulleys at the head of a submarine canyon on the continental slope of the Bay of Biscay. Many such canyons were formed by rivers at times of lower sea level when the continental shelf was dry land.

▶ Currents move material off the continental shelf and into the deep ocean basins. Finer particles are carried in suspension, while coarser materials may be periodically moved by turbidity currents – dense mixtures of sediments and water which move down submarine canyons in the continental shelf and slope. One such turbidity current, started by an earthquake in the North Atlantic in 1929, is estimated to have travelled at a speed of 55 knots for a distance of some 300 km (185 miles).

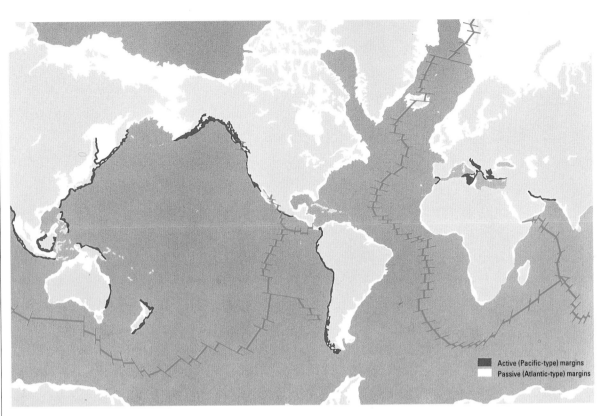

Active (Pacific-type) margins
Passive (Atlantic-type) margins

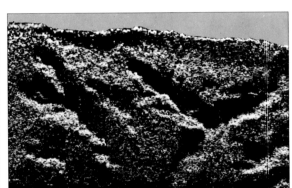

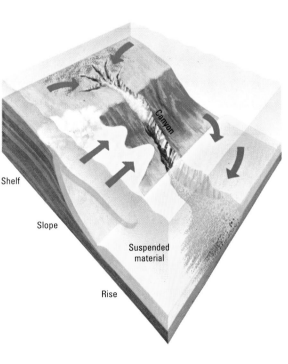

Shelf

Slope

Canyon

Suspended material

Rise

Surrounding the margins of most of the world's continents are terraces with a coastal plain area lying above sea level and a continental shelf area that extends out below the ocean surface. The outer edge of the continental shelf is at a relatively constant depth of 150 metres (500 feet). The constant depth reflects the fact that the continental terraces were formed around 15,000 to 20,000 years ago when the sea level was lower than today. The outer edge is also the real limit of the continental plates, separating them from the oceanic crust of the deep ocean floor which lies at a depth of around 3500 metres (12,000 feet).

Continental margins

Separating the continental shelf and the deep ocean floor is a steep continental slope. Continental slopes generally have a gradient varying between 3 and 20 degrees, depending on the nature of the continental margin. Passive, Atlantic-type ocean margins, generally are marked by a gradual continental rise running up to the foot of the continental slope, and wide continental shelves that are backed by extensive coastal plains. Such ocean margins, formed by the splitting apart of continental blocks, are characterized by down-faulted slopes, with earthquake and volcanic activity infrequent.

In contrast, active, Pacific-type ocean margins are areas of high earthquake and volcanic activity. The continental shelves of such margins tend to be narrow, and they are separated from the ocean floor by deep ocean trenches that mark the point at which one plate is sliding beneath another. The neighbouring land is often backed by high mountain ranges rather than a broad coastal plain. The continental slopes along such ocean margins are formed by compression, with sediments being actively incorporated into them as the oceanic crust slides beneath the continental plate. In some more restricted areas, such as the coast of California, complex continental borderlands with a mixture of plateaux and basins are found. These represent the fragmented pieces of continental plates moved together by different processes.

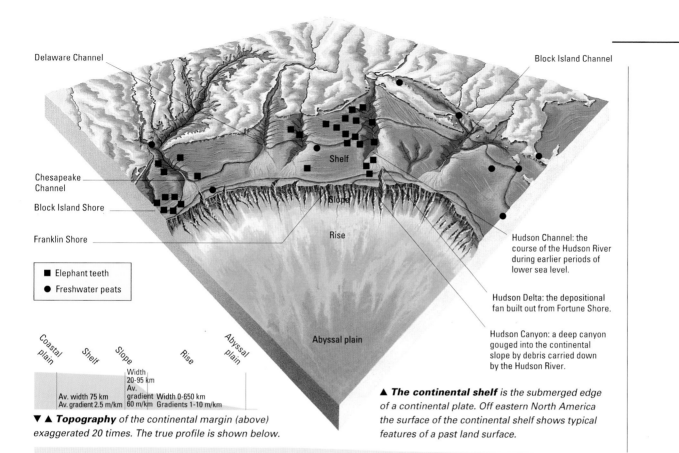

Delaware Channel

Block Island Channel

Shelf

Chesapeake
Channel

Block Island Shore

Franklin Shore

Slope

Rise

Hudson Channel: the
course of the Hudson River
during earlier periods of
lower sea level.

■ Elephant teeth
● Freshwater peats

Hudson Delta: the depositional
fan built out from Fortune Shore.

Hudson Canyon: a deep canyon
gouged into the continental
slope by debris carried down
by the Hudson River.

Abyssal plain

Coastal
plain

Shelf

Slope

Rise

Abyssal
plain

Width
20-95 km
Av.
gradient
60 m/km

Av. width 75 km
Av. gradient 2.5 m/km

Width 0-650 km
Gradients 1-10 m/km

▼ ▲ **Topography** of the continental margin (above)
exaggerated 20 times. The true profile is shown below.

▲ **The continental shelf** is the submerged edge
of a continental plate. Off eastern North America
the surface of the continental shelf shows typical
features of a past land surface.

The changing levels of the sea have left their mark on the
continental shelves. Features such as submerged beaches,
cliffs and river valleys traverse the continental shelf areas,
and sedimentary deposits laid down in previous river valleys and flood plains are widely distributed. These
deposits may form important sources of minerals, such as
sands and gravel, and may include placer deposits (see
p.105) of precious metals or diamonds. The drowned relief
of continental shelves is still being modified by submarine
processes. Waves generated by storms can carry sands and
suspended materials at depth, even on outer shelf edges.

Turbidity currents
Some continental margins are being actively built outwards as sediments are transported across the shelf and
deposited at the edge. Others are being eroded into submarine canyons that are as large as any cut by rivers on land.
On the surface of the continental shelves, river valleys,
formed at times of lower sea level, drain towards the edge
of the continental slope. Such drainage systems continue
to be eroded and may be cut deeper by turbidity currents.

Turbidity currents are dense mixtures of sediment and
water that flow rapidly down the continental slope. They
may be initiated by slumping of loosely consolidated sediments at the head of drainage systems near the continental
slope. The speed of such currents was dramatically illustrated in 1929 when an earthquake shook the Grand Banks
area off the east coast of North America. Submarine cables
were broken by the current and the timing of the successive breaks provided an estimate of the speed of the current. Cables were broken up to 480 kilometres (300 miles)
away from the source of the shock and the maximum
speed of the turbidity current was estimated at 55 knots.

The coarser debris carried to the foot of the continental
slope by turbidity currents may be built up into screes at
the head of the continental rise. Finer particles may be
retained in suspension just above the sea floor. Extensive
fans of material mark the position of canyons leading from
the continental shelf to the deeper ocean floor.

In areas of tidal scour,
the underlying bedrock is
swept clear of sediment.

In areas of strong tidal
current, coarse material is
sorted into strips running
parallel to the flow.

As currents diminish,
sand is deposited in
waves as much as 20 m
(60 ft) high.

Fine sands are deposited in
areas of weak current as
irregular patches.

▲ **Offshore** sediments
are deposited in
different patterns
depending on the

strength of current and
the size of materials, as
illustrated by this series
of sonographs.

The Face of the Abyss

The surface sediments of the deep ocean floor vary both in their composition and their origin. In any one area the layers of sediment reflect the history of that section of the ocean basin. Bottom sediments have two origins: bioclastic materials, which are formed from the dead remains of animals and plants that lived in the surface waters of the ocean; and terrigenous materials, which are ultimately derived from erosion of the land surface.

Land-based sediments

The products of land-based erosion are sorted by the currents, and few materials from such sources end up in the deep ocean basins. Most are deposited on the continental shelves and at the foot of the continental slopes, with larger material being deposited closer inshore and finer materials being carried further away from land. Deep-sea fans of material are located off the continental shelves in front of large rivers. The sizes of many fans indicate that they are millions of years old. Of these the Ganges fan is perhaps the most striking. It extends as a cone beyond the continental shelf for a distance of around 2500 kilometres (1500 miles) and ends at a depth of around 5000 metres (16,500 feet).

The red clays, which are composed of the finest clay particles, are the major type of deep-sea sediment derived from land since the clay particles remain in suspension more easily and are moved farther away from the continental margins. Such clay particles have been found in ocean basins at depths of over 1000 metres (3300 feet).

Bioclastic sediments

Also found at depths of a 1000 metres (3300 feet) or more are the bioclastic oozes, or pelagic deposits, which may be calcareous or silaceous. The former are composed of the chalky remains of foraminifera and pteropods, whereas silaceous sediments are composed of the shells of radiolarians and diatoms, the single-celled phytoplankton characteristic of the surface waters of the ocean. The distribution

of such oozes reflects both the productivity of the surface waters and the chemical and physical characteristics of ocean water masses. Silaceous oozes are found predominantly in tropical and polar regions, while the deepest ocean basins tend to contain only red clay since the calcareous shells of forams tend to dissolve at depth.

The thickness of deep-ocean sediment varies, in part according to the age of the oceanic crust. Newly formed crust, which is close to the mid-ocean ridges, has little or no sedimentary cover and the shape of the newly extruded volcanic lava is easily seen. Moving away from the ridge towards the ocean margin, the depth of sediment increases. At around 3–5 kilometres (2–3 miles) from the ridge the lava is partially hidden, and at distances of 10–13 kilometres (6–8 miles) the ocean floor is a featureless surface, interrupted only by the tops of the largest lava masses. The rate of accumulation of the fine sediments which cover the floor of the deep ocean is extremely slow, ranging from a few centimetres to as little as a fraction of a millimetre per thousand years. Despite such low rates of accumulation, some areas of the ocean basins are covered by hundreds of metres of sediment.

Shifting ocean sands

Deposition through the passive sinking of materials from above is not the only process affecting the distribution and extent of ocean floor sediments. Other processes in the Earth's crust and movements in the ocean water body transport materials across the ocean floor. Earthquakes, volcanic activity and deep-water currents move sediments either gradually or abruptly from one area to another, resulting in more rapid accumulation in some areas and scouring of sediments in others. Where soft, unconsolidated sediments are deposited in areas of tectonic activity, slumping and sliding of sediments is common, but in areas where bottom or strong mid-water currents meet submarine features (such as submerged seamounts or ocean ridges), scouring of the bottom may occur. Acoustic side-

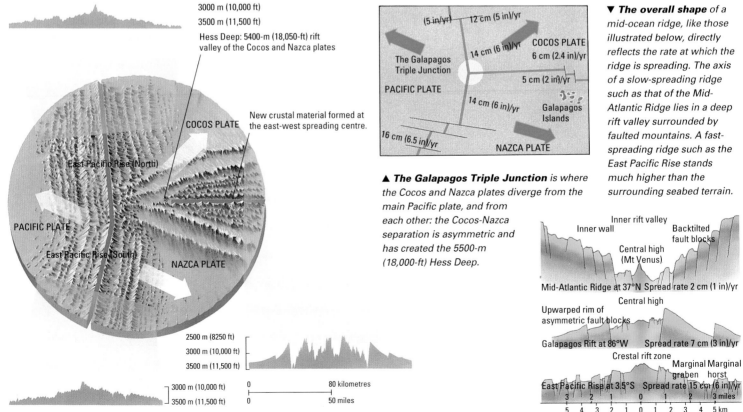

3000 m (10,000 ft)
3500 m (11,500 ft)

Hess Deep: 5400-m (18,050-ft) rift valley of the Cocos and Nazca plates

New crustal material formed at the east-west spreading centre.

COCOS PLATE

East Pacific Rise (North)

PACIFIC PLATE

East Pacific Rise (South)

NAZCA PLATE

2500 m (8250 ft)
3000 m (10,000 ft)
3500 m (11,500 ft)

3000 m (10,000 ft)
3500 m (11,500 ft)

0   80 kilometres
0   50 miles

(5 in)/yr   12 cm (5 in)/yr

The Galapagos Triple Junction

14 cm (6 in)/yr

COCOS PLATE
6 cm (2.4 in)/yr

PACIFIC PLATE

5 cm (2 in)/yr

14 cm (6 in)/yr

Galapagos Islands

16 cm (6.5 in)/yr

NAZCA PLATE

▲ **The Galapagos Triple Junction** is where the Cocos and Nazca plates diverge from the main Pacific plate, and from each other: the Cocos-Nazca separation is asymmetric and has created the 5500-m (18,000-ft) Hess Deep.

▼ **The overall shape** of a mid-ocean ridge, like those illustrated below, directly reflects the rate at which the ridge is spreading. The axis of a slow-spreading ridge such as that of the Mid-Atlantic Ridge lies in a deep rift valley surrounded by faulted mountains. A fast-spreading ridge such as the East Pacific Rise stands much higher than the surrounding seabed terrain.

Inner rift valley
Inner wall
Backtilted fault blocks
Central high (Mt Venus)

Mid-Atlantic Ridge at 37°N Spread rate 2 cm (1 in)/yr

Central high
Upwarped rim of asymmetric fault blocks

Galapagos Rift at 86°W   Spread rate 7 cm (3 in)/yr

Crestal rift zone
Marginal graben Marginal horst

East Pacific Rise at 3.5°S   Spread rate 15 cm (6 in)/yr
3 2 1 0 1 2 3 miles
5 4 3 2 1 0 1 2 3 4 5 km

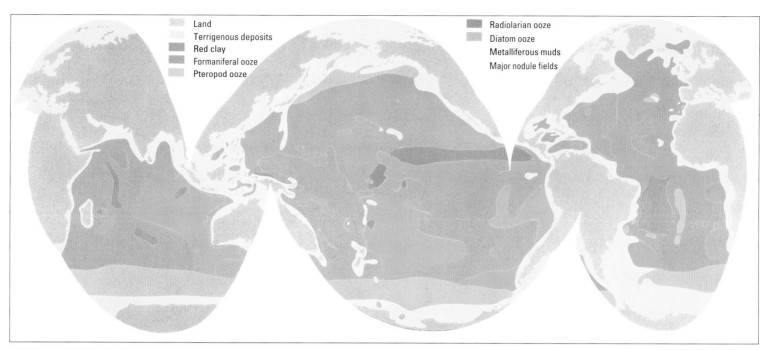

Land
Terrigenous deposits
Red clay
Formaniferal ooze
Pteropod ooze

Radiolarian ooze
Diatom ooze
Metalliferous muds
Major nodule fields

▲ **Pteropod** (Limacina mercinensis), the Eocene, measures 15 mm (0.6 in).

▲ **Radiolarian tests** resemble exquisite glass sculptures.

▲ **Diatoms** are important in the formulation of deep-sea oozes.

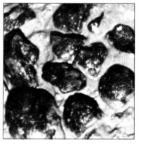

▲ **Fields of manganese** (polymetallic) nodules occur in all ocean basins.

▲ **For many years**, survey vessels have carried out sampling of the deep ocean basins, collecting and analysing the seabed sediments. Although not complete, these surveys have allowed a picture to be built up of the nature and distribution of land-derived sediments, sediments formed from marine organisms' remains and those formed chemically.

scan sonar has revealed that vast transverse and crescent-shaped sand waves move across the surface of the ocean floor in much the same way as similar type of sand features move across continental deserts – such as the Sahara in North Africa.

In addition to the slow rain of detrital material from the surface of the oceans and the movement of finer particles into the ocean basin from the continental margins, other materials are added to the ocean floor. Hydrothermal vents, which are associated with areas of seafloor spreading, may produce metalliferous muds. These muds are formed from the precipitation of metals in solution when the hot molten materials from the underlying crust come into contact with the cooler ocean water. Hydrothermal vents often support a characteristic community of organisms which depend for their existence on bacteria that convert chemical energy into living matter. The bacteria in turn form the basis for the food chain of filter feeding animals that comprise such communities.

Other material found in some areas of the ocean floor includes the remains of volcanic ash, which can fall into the ocean thousands of kilometres from its site of eruption. Additionally, lava from the eruption of submarine volcanoes may also be added directly to the ocean floor. Manganese, or polymetallic, nodules form extensive fields in some areas, particularly in the Pacific. First discovered by the *Challenger* expedition of 1872–6, these nodules formed over millions of years as a result of interacting chemical, physical and biological processes.

▲ **Hydrothermal vents** support unique communities of tube worms, crabs and bivalves. These depend on chemotrophic bacteria which use the sulphur as a source of energy. Some of these bacteria have formed symbiotic associations with the worms and molluscs; others are a food source for the whole community.

Water – the Unique Compound

Despite its commonplace occurrence and familiarity, water is indeed a unique compound – one on which all life on Earth depends. Its physical and chemical properties shape the face of the planet, determine the nature of our climate, and support life on land and in the oceans.

The physical properties of water are unusual. Most substances expand when heated and contract when cooled, and although this is also true of water over part of its temperature range, water does not behave in this way over the full range of temperatures experienced on Earth. As water is cooled it continues to contract until it reaches temperatures of about 4°C (40°F). Below this temperature, however, water freezes and expands. As it freezes, water increases in volume by around 9 per cent. As a consequence, frozen water, ice, is less dense than liquid water and so it floats on the surface of the sea. In this way, ice forms an insulating layer that prevents further cooling of the deeper water layers. If this did not occur, and water continued to contract as it froze, then frozen surface water would sink and the polar seas would eventually become solid ice.

The hydrological cycle

Not only does water occur as a solid and a liquid, but water vapour is a major component of the Earth's atmosphere. When heated, water evaporates to become water vapour, which passes into the atmosphere where it forms clouds. The water vapour eventually condenses, falling as rain or snow, often at considerable distances from the point of evaporation. This transformation of water from liquid to gas and back to liquid, and its passage from oceans and surface waters to atmosphere and land, is known as the hydrological cycle. This cycle is fundamental to terrestrial life, since without a source of freshwater, life on land would not exist.

A second and important characteristic of water is its capacity to store heat. Water has a very high heat capacity, hence it stores heat during the day (or summer) and releases it slowly to the cooler atmosphere at night (or during

▲ **Wind waves** result from energy which is imparted to the ocean surface by the atmosphere. Waves break on the shore when the water on the surface moves forwards more quickly than the water close to the bottom.

The Endless Cycle
The water or hydrological cycle is the process of water exchange between oceans, atmosphere and land. Water in the oceans, evaporates (1), giving water vapour in the atmosphere which may condense and fall back to the ocean (2). Winds carry water vapour over the land (3), and evaporation from rivers (4), lakes (5), the land (6) and plants (7) adds to atmospheric moisture (8). Precipitation is absorbed by the vegetation (9) or flows on the surface in streams (10) and rivers (11). Some penetrates the ground (12) via porous rock (13) and may drain into lakes (14) or rivers; in other areas lake water drains into the groundwater (15). Groundwater may be important for vegetation (16), while in coastal areas it is a component of the drainage to coastal waters (17).

ered compressible, the pressure on the water in the deep ocean basins is such that it is thought that the water column is compressed by around 30 metres (100 feet) under its own weight. If water were incompressible then sea levels would be higher than today, and areas of productive, low-lying land would form part of the sea floor.

The optical properties of seawater are of vital importance to life in the oceans. The primary producers (phytoplankton) – organisms that convert sunlight into chemical energy through the process of photosynthesis – are dependent on sunlight. Ultimately, all other marine organisms are in turn dependent on phytoplankton for their sources of food. Seawater, with its high concentrations of dissolved and suspended matter, absorbs light. Attenuation of sunlight in the oceans occurs to such an extent that below about 200 metres (650 feet) no light penetrates at all. Different wave lengths of light penetrate to different depths and the consequence is that most primary production is confined to a few tens of metres below the surface.

Measuring productivity

The depth to which only 1 per cent of sunlight penetrates is called the euphotic zone. The photic zone, the depth at which photosynthesis occurs, is limited by the penetration of wavelengths of light in the range between 400 and 700 nanometres. In general, red light penetrates least and blue-green light the deepest, although this situation may be reversed where dense phytoplankton blooms occur and high-chlorophyll concentrations absorb the light in the blue-green end of the spectrum. Such differences in light absorption can be used by ocean colour scanners for measuring primary production in the surface of the ocean.

winter). The capacity for the oceans to store heat moderates the climate of neighbouring land areas. Land, with its lower heat capacity, warms up more rapidly than the oceans during the day and cools quicker at night. On local geographic scales, such differences in heating and cooling rates between land and ocean result in local wind patterns that give rise to warm offshore breezes at night, and cool onshore breezes during the day.

Without its extensive surface waters, the Earth would be intolerably hot during the day and frozen at night. The vast extent of the oceans, therefore, acts as an enormous heat storage engine, slowly absorbing and releasing heat. Through the ocean basin circulation systems, heat is absorbed from the Sun's radiation in one area and transported to areas often considerable distances away, before being released to the atmosphere.

Vertical circulation

The mobility of water is clearly an important physical characteristic. It has numerous consequences in terms of the transport of heat and dissolved materials. Changes in density (as a result of expansion and contraction) combined with changes in salinity (due to inputs of freshwater from either land or from the atmosphere) result in the thermohaline circulation (movement driven by density differences) of the oceans. This circulation results, over time, in the movement of both surface waters from the tropics to high latitudes and deep-water flows of cold water from polar regions to lower latitudes. Such current systems play an important role in regulating the climate of the Earth and in determining the distribution of biological productivity in the ocean basins.

Several other physical properties of water have practical and important consequences for organisms both in the sea and on land. Although water in general is not consid-

▲ **Heat from the Sun** is reflected back to the atmosphere (pink arrows), while some is absorbed by the oceans and transported from the Equator towards the poles (purple arrows). Changes in atmospheric composition through burning fossil fuels may affect the climate and hence the circulation of the oceans.

Chemistry of the Sea

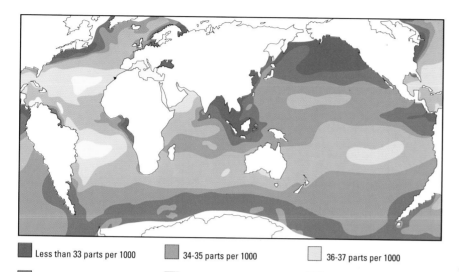

Less than 33 parts per 1000

33-34 parts per 1000

34-35 parts per 1000

35-36 parts per 1000

36-37 parts per 1000

More than 37 parts per 1000

▲ **The map** shows the average surface salinity of the oceans for the month of February. Although salinities change throughout the year, the overall pattern remains similar, with the areas of highest salinity being found in semi-enclosed seas such as the Caribbean and the Red Sea where evaporation of surface water exceeds inputs of freshwater from run-off and rainfall. The areas of lowest salinity are found in coastal seas where large rivers, such as the Yangtse, Mississippi and Amazon, discharge into the ocean. Differences in temperature and salinity drive the circulation of the oceans.

▼ **When light strikes** the surface of the sea, between 3% and 30% is reflected, depending on the angle of the sunlight to the surface. The characteristic blue colour of the ocean results from the absorption and scattering of other wavelengths of light, with the red end of the spectrum being the first to become lost. In areas where suspended sediment is high, scattering of light occurs and the ocean may appear greenish or brown, depending on the nature of the particles present. At depths below 1000 m (3300 ft) virtually no light penetrates at all.

Chemically, seawater is an unusually pure substance, being almost 95 per cent water. The remarkable chemical properties of seawater are, however, a reflection, at least in part, of the substances dissolved in it.

Every year around 330,000 cubic kilometres (80,000 cubic miles) of seawater evaporate from the surface of the oceans, entering the atmosphere as water vapour. Of this total some 100,000 cubic kilometres (24,000 cubic miles) fall as rain, sleet and snow on the surface of the land. The remainder falls directly back into the oceans. Rainfall is made slightly acidic by the carbon dioxide in the atmosphere and this acidic solution attacks the rocks on which it falls. The chemical action, combined with changes in temperature and the physical abrasion of wind and rain, gradually erodes the surface of the land. Erosion products are then carried in solution or in suspension through streams and rivers into the oceans.

Chemical composition

In addition to oxygen and hydrogen, the basic components of water – the two most abundant elements in seawater – are sodium and chlorine, which together produce salt. The concentrations of these two elements are around 1.05 and 1.9 grams per litre respectively. Only ten other elements are present in seawater at concentrations greater than one part per million. The remaining elements are present in very minute quantities.

The total salt concentration of seawater is expressed in parts per thousand, and for open ocean seawater the concentration is around 35 parts per thousand. However, the salinity of the ocean varies regionally according to the local inputs of freshwater as rain and run-off from the neighbouring land, and depending on the rate of evaporation from the ocean surface. In areas of high freshwater input surface salinity may be as low as 32 parts per thousand, whereas landlocked seas may have much higher salinity – between 40 and 41 parts per thousand as in the Red Sea for example.

The ocean displays marked changes in salt concentration from one area to another and with depth. The changes in salinity, and hence density, contribute to oceanic circulation. Therefore, although evaporation of surface water in the tropics may increase the salt concentration, so increasing the density, this may be balanced by thermal expansion and a corresponding decrease in density. As warm surface water, of high salinity, moves away from the tropics and cools, it becomes more dense, eventually sinking to form the dense cold masses of bottom water that flow from high latitudes towards the Equator.

Ocean nutrients

Several nutrients essential for the growth of phytoplankton occur in solution in seawater. These include nitrogen, phosphorus and silica. As these substances are taken from solution in the surface waters of the oceans by the phytoplankton, so the primary production declines. Only when these nutrients are released back into the water can a new cycle of production begin. Human disposal of sewage, and run-off of agricultural fertilizers, may alter the availability of nutrients in coastal waters, leading to bursts of primary productivity in the form of algal blooms.

Once the phytoplankton die they start to sink and decompose. Those that sink below the level of light penetration (the euphotic zone) carry with them the nutrients they acquired during their lifetime. Together with other organic matter, these are decomposed at depth by bacteria and the nutrients are released once more into solution. This process of decomposition results in decline in dis-

50 m (165 ft)

100 m (330 ft)

150 m (495 ft)

200 m (660 ft)

solved oxygen concentration; and where the waters form layers in semi-enclosed bays, the bottom water may become anoxic (without oxygen) and fish die.

Until the nutrients are brought to the surface they cannot be used by photosynthetic organisms, the distribution of which is limited by the depth of light penetration. In coastal areas, mixing of the surface and bottom water may only occur on a seasonal basis under the influence of changes in temperature and wind direction. Some of the most productive areas of the open ocean are those where cold, nutrient-rich water is brought to the surface in the zones of upwelling. In addition to these nutrients, various other elements are essential for plant growth, including copper, iron, zinc and cadmium, the dissolved concentrations of which may be affected by the rates of biological productivity in the area concerned.

Of the other 80 elements known to occur in seawater, a number are concentrated by some marine organisms, in a process known as bioaccumulation. For example, vanadium has a concentration in seawater of less than one millionth of that of sodium, yet some filter-feeding animals have been found to accumulate concentrations 100,000 times greater than that of the surrounding seawater. Other bioaccumulators include oysters, which absorb zinc, lobsters, which concentrate copper, and several shellfish, which concentrate mercury. This capacity to concentrate mercury led to the poisoning of hundreds of people in Minimata, Japan, as a consequence of eating shellfish contaminated with industrial effluents high in mercury.

The carbon cycle

The cycling of carbon between the oceans, the atmosphere and the terrestrial biosphere is an essential process. It controls the environmental conditions on the surface of the planet. Inorganic carbon occurs in the form of carbon dioxide in the atmosphere and is dissolved in the waters of

Trace elements	0.01%
Flouride (F⁻)	0.003%
Strontium (Sr⁺⁺)	0.04%
Boric acid (H_3BO_3)	0.07%
Bromide (Br⁻)	0.19%
Bicarbonate (HCO_3^-)	0.41%
Potassium (K⁺)	1.10%
Calcium (Ca⁺⁺)	1.16%
Magnesium (Mg⁺⁺)	3.69%
Sulfate (SO_4^{--})	7.69%
Sodium (Na⁺)	30.61%
Chloride (Cl⁻)	55.04%

Seawater Elements
Eleven constituents account for over 99 per cent of the salt content of seawater, and most are present in solution as free ions, the relative abundance of which is indicated. Seawater salinity varies geographically and with depth, but the ratio of the constituents remains fairly constant. Nutrients such as nitrogen vary with depth and the low concentration in the surface waters of the oceans limits the production of microscopic plants.

the oceans. During photosynthesis, inorganic carbon is "fixed" by living organisms into organic carbon in the form of carbohydrates. Organic carbon may then be buried in coastal sedimentary deposits or transferred to the deep ocean basins in the rain of plankton and faecal materials that drop from the euphotic zone.

Most organic carbon, however, is recycled through respiration, but some remains unchanged, and over geological time forms deposits of fossil fuels. The burning of fossil fuels is now returning this buried store of carbon to the atmosphere in the form of carbon dioxide which, it is believed, will result in global warming.

◄ *Algal blooms* occur when nutrients are in high concentrations in the surface water. They occur naturally as part of the seasonal cycle in temperate areas, but in coastal waters they may result from sewage inputs and run-off of agricultural fertilizers.

The Frozen Seas

The polar ice sheets, which cover about 12 per cent of the surface of the oceans, have important effects on the world's climate. Ice acts as an insulating layer, preventing further loss of heat from deeper water to the colder atmosphere above. Apart from icebergs, which are calved from the glaciers at the edges of land-based ice sheets, all sea ice is formed as the sea's surface freezes.

Seawater freezes at -1.9°C (29°F), during which the surface of the sea undergoes a series of recognizable changes. At first, small crystals of ice develop which become packed together into either a continuous skin or a series of pancakes with characteristically raised rims. The first crystals to appear are minute spheres which develop into thin disks or platelets known as frazil ice. Slightly later, when the crystals have multiplied, the sea takes on a soupy, matt appearance, known as grease ice. If the sea is calm then the ice crystals freeze together into a continuous semi-transparent skin called nilas. But if the sea is at all rough then this skin soon breaks up into individual plates with raised edges due to their constant collisions with each other. Eventually, either the nilas thickens or the pancakes freeze together to form a flat unbroken floe which continues to increase in thickness, quickly at first, but more slowly as the thickening ice insulates the water beneath.

Salt-free ice

As the crystal lattice forms in the freezing water, salt is excluded and becomes concentrated into liquid brine cells, each less than a tenth of a millimetre (0.004 inches) in diameter. Newly formed ice, therefore, contains less than 10 parts per thousand of salt compared with the 35 parts per thousand found in normal seawater. Eventually, the brine cells join together to form drainage channels through which the brine escapes into the sea.

Layers of ice

As the ice coalesces and consolidates during the polar winter, the sea becomes covered with a layer of first year ice, about 2 metres (7 feet) thick in the Arctic and 3 metres (10 feet) thick in the Antarctic.

During the polar summer the surface ice melts, forming surface pools. If the temperature is sufficiently high, the pools may extend through the thick ice sheet. If the floe survives the warmth of the summer months, it may become multiyear ice – extremely strong, multi-layered ice that has sheets of re-frozen meltwater incorporated into its structure. Such ice is blue in colour and is common in the central Arctic, making the area impassable to even the most powerful ice-breakers.

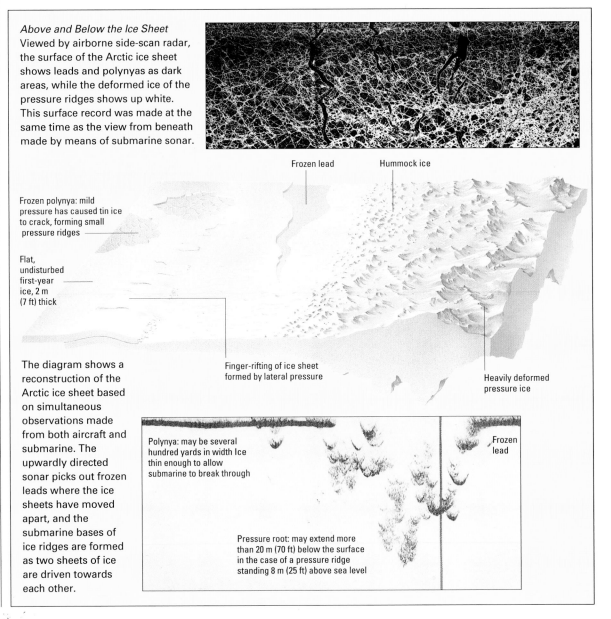

Above and Below the Ice Sheet
Viewed by airborne side-scan radar, the surface of the Arctic ice sheet shows leads and polynyas as dark areas, while the deformed ice of the pressure ridges shows up white. This surface record was made at the same time as the view from beneath made by means of submarine sonar.

Frozen lead

Hummock ice

Frozen polynya: mild pressure has caused tin ice to crack, forming small pressure ridges

Flat, undisturbed first-year ice, 2 m (7 ft) thick

Finger-rifting of ice sheet formed by lateral pressure

Heavily deformed pressure ice

The diagram shows a reconstruction of the Arctic ice sheet based on simultaneous observations made from both aircraft and submarine. The upwardly directed sonar picks out frozen leads where the ice sheets have moved apart, and the submarine bases of ice ridges are formed as two sheets of ice are driven towards each other.

Polynya: may be several hundred yards in width Ice thin enough to allow submarine to break through

Frozen lead

Pressure root: may extend more than 20 m (70 ft) below the surface in the case of a pressure ridge standing 8 m (25 ft) above sea level

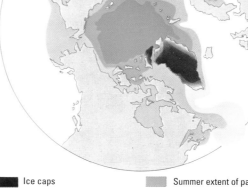

An ice floe does not remain smooth. It is deformed by wind and current action, and may fracture to produce either long, open water leads, or polynya – wide pools of open water which quickly freeze. When winds subsequently converge, the new ice in a lead may become compressed and fractured and pushed up into pressure ridges. Pressure ridges may reach 10 metres (32 feet) in height and lie as deep as 45 metres (150 feet) below the ice sheet.

Wind stress is transmitted by the ice to the underlying water so that ice drift and current flow are identical over the long term. The Arctic has two major drift patterns: the Beaufort Gyre, in which ice floes and bergs may be trapped for up to 20 years, and the transpolar drift stream, which carries ice from Siberia, across the Pole, and down the east coast of Greenland. In the Antarctic, winds blowing from west to east around the globe generate northeast-flowing currents that carry ice into the southern oceans.

Floating giants

Icebergs, such as the one which sank the *Titanic*, represent huge fragments of ice broken from the ends of glaciers or the edges of land-based ice sheets. In the Arctic, some 12,000 icebergs are calved annually from the glaciers. These drift with the current up the west coast of Greenland and return down the western coast of Baffin Bay, eventually passing out into the Atlantic. The survival of such icebergs is highly variable – in 1958 only one reached the Atlantic, but in the following year some 693 were recorded. A typical new berg weighs around 1.5 million tonnes, stands 80 metres (260 feet) out of the water, and extends more than 350 metres (1150 feet) below the surface. By the time it reaches the Atlantic it may have shrunk to a tenth of its previous size.

Antarctic bergs form in the same way but are calved from the floating ice shelves that fringe that continent. They are much larger than Arctic bergs – often more than 80 kilometres (50 miles) long – and they usually retain their flattened tabular form. Thousands may calve each year, drifting north as far as 40°S before melting away.

■ Ice caps			Summer extent of pack ice
Permanent ice shelves			winter extent of pack ice
Permanent pack ice			Limit of drifting bergs

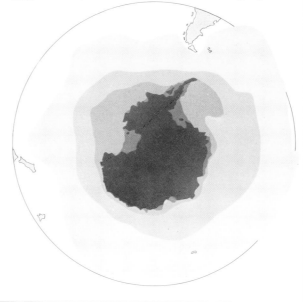

◄ *The pack ice* covering the Arctic Ocean for much of the year is generally thinner and more deformed than that of the Antarctic seas. Ice formed against the Arctic shores, and ice formed by freezing of surface waters, rapidly becomes rafted and broken by the ceaseless movement within the enclosed ocean basin. Antarctic pack ice is characterized by much larger floes, generally undeformed except for crushing of the edges and some rafting and pressure ridging where ice is trapped against the coast.

◄ *Antarctic icebergs* are generally larger than those of the Arctic, and are formed of pieces detached from the floating ice sheets fringing the Antarctic continent, rather than being calved from the land ice sheets.

Tides

▶ **This picture shows** *the tidal bore on the River Severn, England, which results from the rapid movement of water into the funnel-shaped estuary as the tide rises. This bore is 1 m (3 ft) high and travels at 20 km/h (13 mph) for 33 km (21 miles) upriver.*

The ancient Greeks were among the first to note the relationship between the tides and the Moon's monthly orbit around the Earth. However, it was not until Newton presented his gravitational theory, nearly 2000 years later, that the explanation for this relationship was more fully understood.

Everything in the Universe exerts a gravitational force on everything else. In this way, gravity tends to pull the Earth and Moon together. They are kept apart, however, by the centrifugal force of the Moon's orbit. At the centre of the Earth, the gravitational and rotational forces between Earth and Moon are balanced, but at the surface of the Earth they are not. On the side closest to the Moon the gravitational force is strongest, causing the surface of the ocean to bulge towards the Moon. On the opposite side the force is weakest, resulting in a bulge away from the Moon caused by the Earth's rotation. These forces and the response of the ocean would result in twice daily, or semi-diurnal, tides perfectly aligned with the Moon were other factors not involved.

However, since the Earth is spinning on its own axis, the bulge of the tides is displaced and appears slightly ahead of the Moon's actual position. This displacement is a result of the frictional forces between the water mass and the Earth's surface, which slows the oceans' response to the gravitational pull of the Moon. The Sun also exerts a gravitational force on the surface of the ocean, although this is weaker, around two-fifths as strong as that of the Moon. Since the Sun and Moon vary in their relative position to one another and to the surface of the ocean, these two gravitational forces sometimes work in the same direction, and sometimes in opposition to one another.

The lunar cycle

The monthly cycle of tidal amplitude reflects the relative positions of the Sun and Moon. When both the Sun and Moon are pulling together, the tidal range is greatest, resulting in high spring tides. However, when the gravitational forces of the Sun and Moon are at right angles to one another, the characteristic low neap tides result.

The gravitational pull of the Moon when it is above the Equator is not equal over the whole surface of the

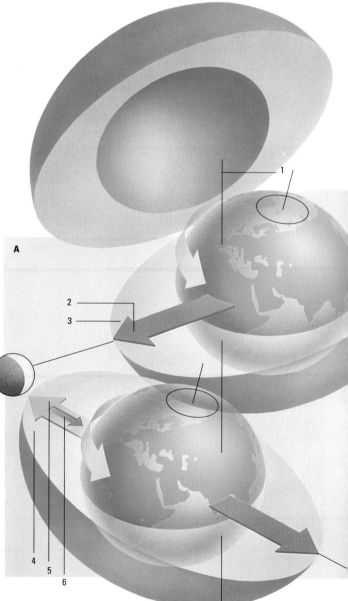

Earth. It is strongest at the Equator and becomes progressively weaker north and south. The further north or south the Moon lies (relative to the Equator), the greater the tendency towards a once daily, or diurnal, tidal pattern.

The Moon's orbital plane changes on a cycle of 18.6 years, reaching a maximum angle of 28.5° to the Earth's Equator. At such times the Moon changes its position from 28.5°N to 28.5°S of the Equator every lunar month. When the angle between the Moon's orbital plane and the equatorial plane are greatest, the two daily tidal heights are different, with one very high tide and one smaller high tide.

Tidal variations

In addition, the oceans' water mass is not free to flow over the entire surface of the globe – being restricted in the different ocean basins, and with internal flow being channelled by the mid-ocean ridges and other topographic features. The oceans are also acted on by the Coriolis forces generated by the Earth's own rotation. The result is that the tidal patterns differ in different oceans, reflecting the interaction of this wide variety of forces.

The tidal forces acting on the ocean water masses produce complex standing and rotating wave systems characteristic of each ocean basin. Nodal points occur at the centres of these wave systems and represent the points of least disturbance of the ocean water mass where changes in the vertical height of the sea surface are generally around zero. Elsewhere the tidal rise and fall of the ocean surface may be between 1 and 3 metres (3 and 10 feet). However, tidal range may be much greater in some semi-enclosed bays, such as the Bay of Fundy in Nova Scotia where the tidal range can alter by as much as 13 metres (42 feet).

Most coastal areas experience semi-diurnal tides and diurnal tides are relatively rare due to the shape of the coastline. Mixed tides are an amalgam of these two tidal types and are characterized by two tides each day, one of which is much stronger than the other. The Atlantic and Indian oceans generally display semi-diurnal tides which are of similar magnitude, while the Pacific Ocean displays a more mixed tidal pattern in which one tide each day is much larger than the other.

▲ **This sea stack** in the semi-enclosed Bay of Fundy in Nova Scotia illustrates the powerful erosive action of tidal forces. The tidal range is amplified by an oscillating wave, giving the area a tidal range of some 13 m (44 ft) – one of the highest in the world.

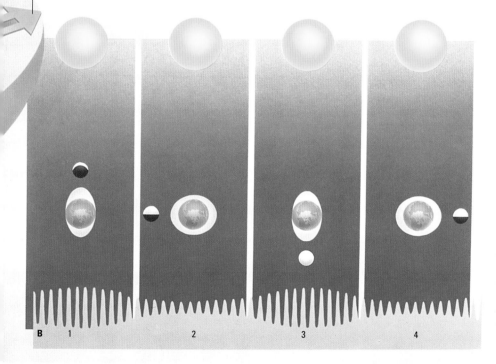

◄ **As the Earth and Moon (A)** rotate about their common centre of gravity (1), the gravitational pull of the Moon (2) makes a tidal bulge (3). The second bulge (4) appears to be caused by an anti-gravitational force (5). However, this second bulge exists because the Moon's pull on the Earth is stronger than its pull on the farthest body of water (6). In effect, the Earth accelerates faster towards the Moon, leaving the body of water behind. Spring tides, (B) the highest tides, occur when the Sun, Moon and Earth are in a straight line (1 and 3). When the Sun and Moon are at right angles to one another (2 and 4), their gravitational forces to some extent cancel each other out, producing neap tides.

Currents and Gyres

The major patterns of water movement in the ocean basins determine the overall direction of current flow. Near a coastline, the pattern of coastal currents is more obviously influenced by tidal forces, the direction of riverine flows and the shape of the coastline itself. Such local patterns of water movement dominate inshore waters in the continental shelves, but the waters of the open ocean move in response to processes occurring over much wider geographic scales.

The great current systems of the open ocean are driven by atmospheric circulation. Once away from the influence of neighbouring land, certain large-scale patterns of water movement dominate the ocean basins. These patterns include the great gyres, or circular movements of the ocean north and south of the Equator, and major surface currents such as the circumpolar current of the Antarctic and the Equatorial currents which cross the Pacific and Atlantic ocean basins.

Atmospheric influences

The wind systems in each hemisphere are dominated by the trade winds (in the vicinity of the Equator), which drive the surface ocean waters to the west and the westerly winds at higher latitudes – which return water to the eastern margins of the ocean basins. The combined effect of west-moving waters at low latitude and east-moving waters at higher latitudes results in the circular current patterns, or gyres. A gyre is a basin-scale, closed current system which consists of a strong western boundary current carrying the water polewards and a less strongly defined return flow to the east.

In the North Atlantic Gyre, the North Equatorial Current flows from Africa towards Latin America, becoming the northerly flowing Gulf Stream, which flows along the North American eastern seaboard before re-crossing the Atlantic as the North Atlantic Current, then flowing south as the Canaries Current which rejoins the North Equatorial Current to complete the cycle.

The speed of the ocean surface currents is generally around 10 kilometres (6 miles) a day, but the western boundary currents, such as the Gulf Stream of the North Atlantic and the Kuroshio Current in the North Pacific, may achieve speeds of up to 95–160 kilometres (60–100

miles) a day. In the Southern Hemisphere the currents are generally weaker than in the north and the more open southern ocean system is dominated by the Antarctic Circumpolar Current.

The Ekman spiral

Although persistent winds drive the ocean current systems, other forces also act on the moving water mass to change its direction and speed of movement. The so-called Coriolis force results from the rotational movement of the Earth which tends to deflect any moving particle, whether water or air, relative to the surface of the Earth. If a water mass moves northward in the Northern Hemisphere the Earth's rotational influence results in it moving clockwise, or to the right; while a southerly moving mass in the Southern Hemisphere moves to the left, or counter-clockwise. This contributes to the flow of water in a gyre.

The spiral movement of water in the gyres results in water being piled up towards the centre. The level of the

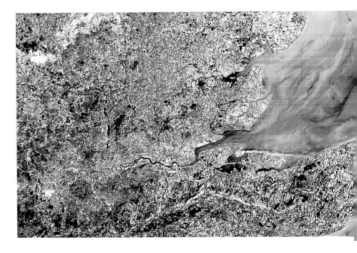

▲ **The sedimentation** in the estuary of the River Thames is clearly defined in this satellite picture of London. It shows that the predominant inshore current around this part of England is northeast. Further out to sea, however, the current flows to the south.

▶ **Long-term records** of weather patterns allow meteorologists to compile maps of the monthly mean conditions of atmospheric pressure and air-mass circulation. The map (right) shows the average situation for July. The map should be compared with the current circulation map (far right). Winds and wind-generated surface currents are almost perfectly matched – their paths show the effect of the Coriolis force. The clockwise rotation of winds around the high-pressure zones of the Northern Hemisphere coincide with the clockwise gyres of the North Atlantic and North Pacific. A similar situation is apparent in the southern oceans.

Atmospheric pressure in millibars

Sea   La

990
1000
1010
1020

water in the Sargasso Sea, for example, is about a metre (3 feet) higher than in adjacent coastal regions. The outward pressure of this dome of water balances the inward pressure created by the rotational forces of the Earth's spin. Changes in atmospheric pressure can also influence the level of the sea and stable areas of high or low atmospheric pressure can cause rotational movements of the surface water around such centres.

When winds blow over the surface of the ocean they impart motion to the surface of the water through friction, thus the surface layer of water tends to move in the same direction as the prevailing wind. As one passes deeper into the water mass, however, the influence of the Coriolis force becomes more apparent and the direction of water movement becomes progressively deflected away from the direction of surface movement. This so-called Ekman spiral involves the upper 90 metres (300 feet) of water in the ocean surface. While the overall direction of movement of the entire water mass is at right angles to the direction of the prevailing wind, the surface current flows at an angle to the wind. At depth the water may be moving in the opposite direction to the surface currents.

Mesoscale eddies and "rings"

As a consequence of the Ekman transport of whole water masses, the trade winds and westerlies actually pile up water in the centre of the gyre and result in the gyres being squeezed towards the western margins of the ocean basins. This contributes to the faster speeds of the Gulf Stream and other western boundary currents which are generally fast flowing and stable; however, as they move to higher latitudes and begin to be deflected away from the coast, they lose their stability and begin to meander. These meanders increase in amplitude and may ultimately be cut off from the main flow as "Gulf Stream Rings".

The existence of mesoscale eddies and rings has only been known for a relatively short period of time and with the advent of satellites the improvements in remote sensing of the sea surface have demonstrated their wide distribution in association with strong ocean currents. The importance of these features in terms of ocean energy transfer and in the patchiness of ocean productivity is only now becoming better understood.

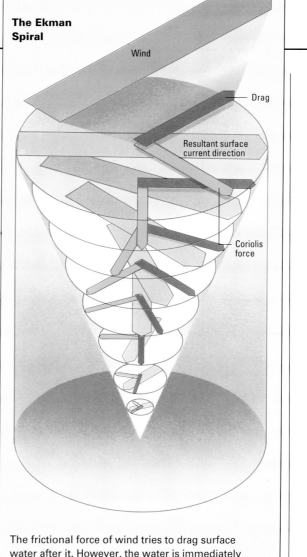

The Ekman Spiral

Wind

Drag

Resultant surface current direction

Coriolis force

The frictional force of wind tries to drag surface water after it. However, the water is immediately deflected by the Coriolis force. The surface layer in turn drags the layer beneath it, which again is deflected. As the movement is transmitted downwards, the deflections form an Ekman spiral, whereby the deepest water may be moving at 180° to the surface flow. These forces result in the surface current flowing at an angle to the wind.

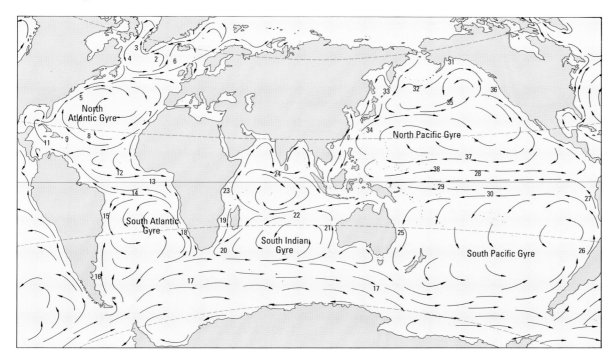

North Atlantic Gyre

South Atlantic Gyre

South Indian Gyre

North Pacific Gyre

South Pacific Gyre

◀ *The 38 major named* currents make up five current gyres.

1 East Greenland Current
2 Irmingher Current
3 West Greenland Current
4 Labrador Current
5 Gulf Stream
6 North Atlantic Current
7 Canaries Current
8 North Equatorial Current
9 Antilles Current
10 Guiana Current
11 Caribbean Current
12 Equatorial Countercurrent
13 Guinea Current
14 South Equatorial Current
15 Brazil Current
16 Falkland Current
17 Antarctic Circumpolar Current
18 Benguela Current
19 Mozambique Current
20 Agulhas Current
21 West Australian Current
22 South Equatorial Current
23 Somali Current
24 Monsoon Drift
25 East Australian Current
26 Humboldt Current
27 Peru Current
28 Equatorial Current
29 S. Equatorial Countercurrent
30 South Equatorial Current
31 Alaska Current
32 Aleutian Current
33 Oyashio Current
34 Kuroshio Current
35 Kuroshio Extension
36 California Current
37 North Equatorial Current
38 N. Equatorial Countercurrent

Wind, Waves and Tsunamis

Waves are a familiar sight. Their awesome power releases a considerable destructive force as they break against a shoreline. Waves are of several different types. They differ in their length – the distance separating wave crests – and their periodicity – the time which separates successive wave crests.

Capillary and wind waves

At one end of the scale are ripples, or capillary waves. They have a periodicity of less than a second and wavelengths of less than 1.74 centimetres (0.69 inches). At the other end of the scale are the tides, with a periodicity of 12 or 24 hours.

Capillary waves are set up by wind blowing gently across the water surface. Such waves are characterized by rounded crests and pointed troughs. Wave shape is determined by surface tension, and gravity only becomes important in wave form when the wavelength exceeds 1.74 centimetres (0.69 inches). As wavelength increases, the waves become more pointed and the troughs more rounded. Waves grow in height as increasing wind pressure acts against the waves' windward side. As the waves grow, their crests become steeper until they reach a point where their crests become unstable and break, producing breaking waves characteristic of strong wind conditions.

Storm waves

Waves move across the surface of the ocean in much the same way as a sailing ship, with the wind behind the wave pushing it across the ocean surface. The energy transferred from the wind to the surface of the water can itself be transferred from one wave to another. During storms, waves receive so much energy that they are whipped into waves of different periods, lengths and directions. The resulting chaotic pattern is termed a "wind sea".

Wave height, the vertical distance between the trough and crest of a wave, is determined by the fetch (distance

▲ This picture shows the destructive force of hurricane Alicia, which devastated the city of Galveston, Texas, in 1983.

Wind velocities measuring over 117 km/h (72 mph) resulted in high waves, causing severe damage to coastal structures.

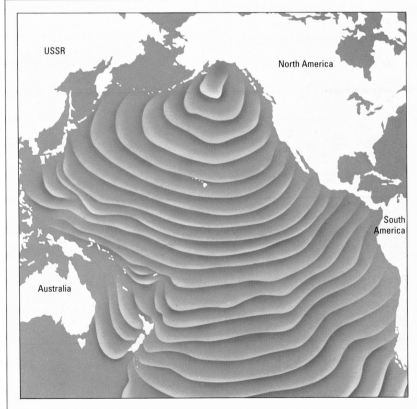

Tsunami
Tsunamis are dramatic waves generated when a submarine earthquake causes a sudden shift in the ocean floor along a fault line. The upheaval creates a bulge that breaks down into a series of waves travelling at speeds of up to 750 km/h (450 mph). The map shows the hourly position of a tsunami which originated just south of Alaska.

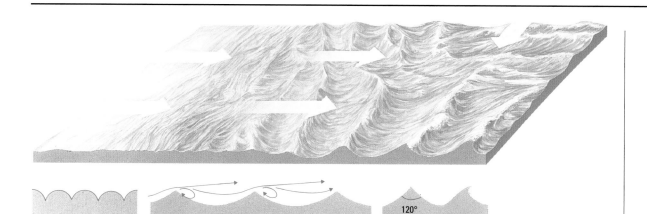

▲ **Wind** blowing across water produces small capillary waves with round crests and pointed troughs. This shape is determined by surface tension.

▲ **Once wavelength** exceeds 1.74 cm (0.68 in), gravity takes over from surface tension as the dominant force on wave form. The crests become more pointed; the troughs rounded. The wind reinforces the wave shape by pressing down on the windward side and eddying over the crest to reduce pressure on the leeward side.

▲ **As the waves** grow their crests steepen until they reach an angle of 120°, at which point they become unstable and break, producing whitecaps.

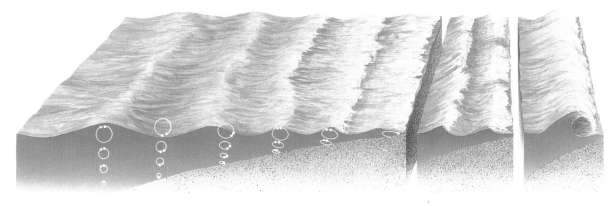

▲ **Waves in the** process of generation are called "forced waves". As they move from the source area, they travel as long-crested swell waves.

▲ **Orbital movement** of water particles at the surface has a diameter equal to the wave height, but reduces to zero at a depth equal to half the wavelength.

▲ **In shoaling water,** of depth less than half the wavelength, the waves heighten and steepen and the wavelength shortens. The particle movements become more elliptical until they can no longer maintain their orbits and the wave breaks.

▲ **Breakers on a** gently sloping shore may either surge up the beach (left) or break gradually as water spills down the wave front. Where the shore steepens abruptly, (right) the wave overturns as a plunging breaker.

over which the wind blows), the duration of the wind and wind speed. In the Pacific Ocean, which has the greatest fetch of any ocean, wind waves reach their greatest height. At the heart of intense storms, waves may reach between 12 and 15 metres (40 and 50 feet), although the largest recorded wave was 35 metres (115 feet) high.

High waves occur most frequently when storm waves drive up against the continental shelves. As waves move away from the storm centre they change shape, becoming less ragged and taking the form of long smooth swells.

Breakers and surf

As waves approach the shore, their height becomes greater and their wavelength shorter, eventually breaking as surf on the shore. The breaking of the wave results from the orbital movement of the water particles inside the wave and the friction caused by the bottom, which slows the movement of the bottom water relative to the surface.

Inside each wave, water moves in an orbital way with particles at the surface moving in a circle of diameter equal to the wave height. At a depth of half the wavelength, no orbital motion is present in the water mass. As

the waves approach the shore and the depth decreases to less than half the wavelength, the particle movements become more elliptical until they can no longer maintain their orbit and the wave breaks.

Tsunami

The so-called tidal waves, or tsunami, are not associated with tides but with submarine earthquakes or volcanic eruptions. A sudden shift in the ocean floor caused by vertical movements along a fault line may create a wave at the surface of the ocean. These waves can travel at up to 720 kilometres (450 miles) per hour. The amplitude of such waves is usually quite low, and much of the wave energy is expended at the continental slope. However, in some areas, where the bottom topography focuses the wave energy, devastating effects on land can occur.

Although tsunami can occur in the Atlantic, they are more frequent in the Pacific Ocean, where they are generated around the Pacific rim, in association with the active subduction of the ocean plate margins. As a consequence systems have been established to provide early warning of the passage of such waves following earthquake activity.

The Air–Ocean Interface

▶ **This vortex** in the clouds is seen in the lee of of Isla de Guadalope, which is 1298 m (4257 ft) high. The upwelling of cold water produced by the California Current results in the cloud formation which is a feature of the air–ocean interaction in this area.

Almost every year the eastern American coastline or Gulf region is hit by a hurricane accompanied by torrential rains and often devastating winds. Such storms are generated far out to sea off the coast of Africa or in the Caribbean region, where the temperature of the ocean surface waters is high. Hurricanes are the most awesome of the many manifestations of air–ocean interaction.

The vast ocean surface, covering some two-thirds of the Earth, acts as a giant thermostat. It absorbs energy in the form of heat during the summer and releases it over winter. The energy released by the oceans heats the atmosphere above, causing turbulent air movements. In turn, the movements of the atmosphere affect the surface of the oceans, establishing wind waves and currents. Such phenomena may be of short or long duration depending upon the strength and persistence of the wind.

The responses of the surface waters of the oceans to movements in the atmosphere are, however, much slower than the response of the atmosphere to the heat released from the oceans. This is due to the fact that the rate of change of water is far less than that of the air above.

Coastal breezes

The surface waters of the oceans are, however, intimately linked with the atmosphere above. This is perhaps nowhere more obvious than in short-term weather events such as the daily pattern of onshore, offshore breezes that occur in many coastal areas all over the world. Because the land is heated more rapidly than the sea during the day, the air above the land rises and is replaced by air flowing from the sea to the land. Cool onshore breezes thus help to moderate the heat of the day. At night, however, the direction of the wind reverses. As the land cools down to a temperature below that of the adjacent sea, gentle offshore breezes spring up.

In addition to the transfer of heat from ocean to atmosphere, water vapour also enters the atmosphere from the surface of the seas. Where wind flows over warm ocean water it picks up moisture and if it subsequently encounters colder ocean water, dense ocean fog forms as water droplets condense in the atmosphere. In the northwest Pacific, for example, during the summer months, southerly winds flow over the warm Kyoshio Current and as they pass over the cold Oyashio Current they give rise to areas of dense ocean fog.

▼ **Unlike a hurricane,** which grows from the surface of the ocean upwards, a water spout grows downwards from its parent thunder cloud. The spin in a water spout is believed to result from strong lateral winds that cause the columns of air to be deflected into a vortex which extends downwards, whipping the surface of the sea into a misty spray.

The fully developed hurricane may extend into the tropopause between 12 and 20 km (9 and 12 miles) above the Earth's surface. Cool, dry air is drawn into the central column of the storm at high level.

Walls of dense cumulus cloud form concentric rings around the eye: they are separated by annular zones of clear air.

Warm, moist air is drawn into the system at sea level, feeding energy into the system as it spirals upwards, finally to be dissipated in the upper air.

The strongly rising airflow creates a region of intense low pressure in the eye, often the cause of severe damage to homes which may explode as the eye passes across them.

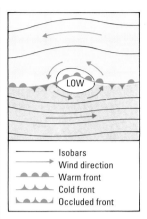

Isobars
Wind direction
Warm front
Cold front
Occluded front

▲ **Cyclones** result in much of the mixed rainy weather of temperate latitudes. They originate as waves in the polar front between polar air and maritime air.

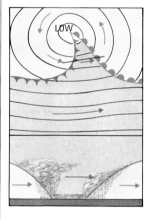

▲ **These weather systems** are closely tied to longer waves in the upper atmosphere which affect the exchange of heat, momentum and moisture between the air masses.

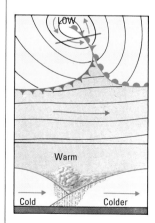

▲ **The depression** tightens as it develops and occludes as the trailing mass of cold air catches and either undercuts or, as above, overrides the warm front.

Cyclonic storms and hurricanes

The formation of storms (and in particular hurricanes) represents another, often violent, interaction between the ocean surface and the air above. In tropical areas, where the surface of the sea exceeds 27°C (80°F), heat and water vapour are discharged to the atmosphere. As the air is warmed it becomes less dense and rises rapidly, forming a spiral pattern. As the warm, moisture-laden air rises and expands, energy is released through the heat of condensation and the speed of the wind increases. The centre, or eye, of such a storm forms a calm region of low pressure, while the dense walls of cumulus-type clouds which encircle the eye make up the regions of highest wind speeds and extremely dense precipitation.

Hurricane Gilbert, which swept through the Gulf of Mexico in 1988, was 1500 kilometres (900 miles) in diameter, and although the storm only travelled at 18–25 kilometres (12–16 miles) per hour, wind speeds near the eye reached over 320 kilometres (200 miles) per hour. Pressure in the eye was a record low at 885 millibars. Wind speeds in such tropical storms are inversely related to the air pressure at the centre – the lower the pressure, the higher the wind speed.

Once a hurricane or typhoon passes over land it begins to lose its power, since the source of energy from the warmer ocean is cut off and the increased friction upsets the pattern of air circulation. Associated with such storms are sea surges and torrential rain. Hurricane Gilbert was accompanied by a wave surge of 6 metres (20 feet) and rainfall of between 250 and 380 millimetres (10 and 15 inches) which fell in only a few hours.

Shaping the coasts

In terms of hours and days, atmospheric movements may result in considerable changes in the surface waters of the oceans. Where strong winds combine with high tides the water level at the coast may be much higher than normal.

This often results in penetration of salt waters into estuaries and coastal flooding.

Most coastal erosion, however, occurs over much longer periods of time as a consequence of storm waves generated during a particular season of the year by strong atmospheric disturbances. In many areas beaches change their shape from season to season. This reflects seasonal differences in the predominant wind direction and consequent changes in the sites of deposition and erosion along the shore. Winds that are at a slight angle to the shoreline result in the lateral movement of the water mass (along the shorelines), which in turn causes drift of sediment materials in the direction of the current.

Studying the air–ocean interface

An understanding of the mechanism by which air and ocean exchange heat and moisture is of fundamental importance to medium- and long-term weather forecasting, which in turn has enormous influence on agriculture and associated industries. Over the last few decades considerable international time, effort and money has been invested in the investigation of weather and the climate system. In the 1970s, the Global Atmospheric Research Programme was established to examine these interactions on an ocean basin scale.

The GATE experiment in 1974 investigated (through a co-ordinated programme of ship-borne, aircraft and satellite observations) air–sea interactions in the tropical regions of the Atlantic. During the 1980s the TOGA (Tropical Ocean Global Atmosphere) programme was established to examine similar air–ocean interactions in the western Pacific. TOGA has, for example, contributed substantially to our understanding of the El Niño phenomenon. During El Niño years a major change occurs in the ocean currents of the southern Pacific with consequences for climate, weather patterns and sea level conditions over wide areas of the tropics.

The Oceans and Climate

The oceans have been described as the flywheel of the climate system. They store energy when it is in abundant supply during the day, or summer, and release it during the night, or winter. Unlike a flywheel, however, the oceans play a much more active role in the global climate system. Through their constant motion, the oceans transport of considerable quantities of energy across the surface of the Earth. The heat storage capacity of the oceans, and the length of time involved in ocean circulation, play a major role in determining regional and global climates.

Temperature variations

When the ocean cools it responds through the process of convection, resulting in heat being brought to the surface. The overall cooling is, therefore, spread over a considerable depth, and the overall fall in temperature of the surface is quite small. As a result the surface of the world's ocean varies in temperature over a much smaller range than does the land surface, from -2° to 30°C (28° to 86°F). At any one location, however, the variation in temperature is even smaller, less than 1°C (2°F) during the course of a day and around 10°C (18°F) over a period of one year. In contrast, the range of temperature over a continental landmass may be as great as 100°C (180°F) from place to place, and as much as 80°C (144°F) in one site over the course of a year.

The thermal inertia of the oceans and their slow response results in delays in the seasonal cycles of ocean areas compared with land. This difference in timing of response gives rise to both short- and long-term changes in atmospheric circulation, of which the great seasonal monsoon cycles are a well-known example.

The ocean heat engine

Since ocean waters move both vertically and horizontally over the Earth's surface, they redistribute considerable quantities of heat around the globe. The movement of warm, tropical waters to both north and south polar regions represents a source of heat transfer that is as large as that of the atmosphere, and which has dramatic consequences for climates on adjacent landmasses. The North Atlantic Gyre represents one such heat engine.

Mild winters

In the tropical Atlantic Ocean, solar heating results in the evaporation of water from the surface and so causes a rise in salinity and hence in density. Some of this water flows north, passing between the coasts of Iceland and Britain. Here it gives up heat to the atmosphere, which, because the winds in this area are predominantly from the west, carries warm air across western Europe. This flow of heat results in the mild winters characteristic of western Europe and distinguish it from other continental areas at similar latitude which are much colder.

In addition, because the water gives up heat to the atmosphere, the surface water temperature drops close to freezing point and the density of the water increases further. In the Greenland Sea, for example, the combination of low temperature and high salinity ensures the surface waters are more dense than the underlying water. As a consequence they sink, occasionally to the bottom, where the cold, high-salinity water mixes with, and slides under, the bottom water, spreading on the ocean floor and flowing south.

This pattern of thermohaline circulation in the North Atlantic, which results in the mild climate of western Europe, may not have occurred at the end of the last ice age. The rapid melting of the Laurentian ice sheet would have released considerable quantities of freshwater into the surface waters of the North Atlantic. Under such conditions the North Atlantic would have been covered by a layer of less saline water which would have frozen in

▼ **The thermohaline**
circulation of the oceans
has been named the Great
Ocean Conveyor belt. In
downwelling areas, heat is
transferred from the oceans
to the air, and the water
sinks to form deep, cold
water. In upwelling areas the
cold water rises and heat is
transferred from air to ocean.

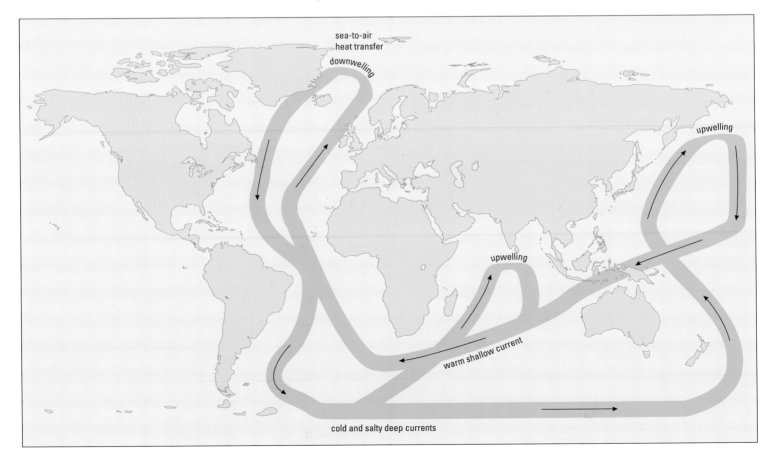

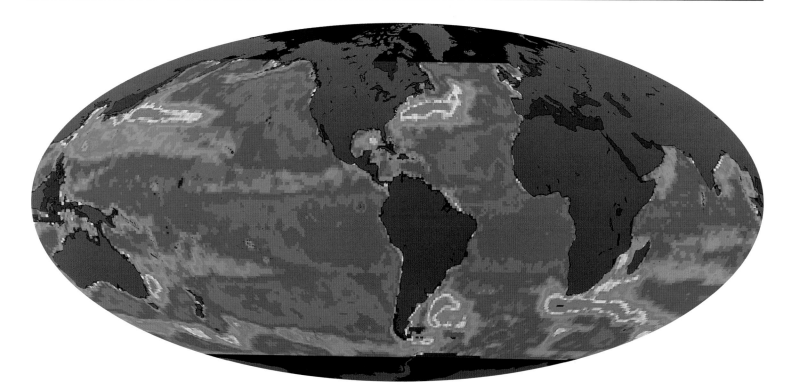

some areas. This frozen water would have prevented the transport of heat from the equatorial Atlantic and therefore resulted in severe winter climates in western Europe.

Large quantities of cold, deep water are also formed in the Southern Hemisphere around the Antarctic landmass. Deep water from the north and south polar regions spreads out through world's oceans, becoming mixed with the warmer water above as fresh supplies of colder water push beneath it. Over the centuries, these cold water masses gradually rise, close enough to the surface to become involved in surface water mixing. Shallow circulation of water, which does not reach very high latitudes, but nevertheless passes polewards, cools and sinks as a consequence of convection, and may circulate at intermediate depths on time scales of a few years or decades.

The circulation of deep-ocean water and the convection currents which cause it to sink in polar regions are not only of great consequence for the global climate system, but they are also vitally important for marine primary productivity and associated food webs. In areas of upwelling, where cold, nutrient-rich water is brought to the surface, high biological production occurs. Short-term changes in ocean circulation patterns can have important consequences for productivity in such areas.

El Niño

An example of this is found off the Peruvian coast, where cold, nutrient-rich, Antarctic water is brought to the surface. The normal winds, which drive the surface water westwards across the Pacific, draw up the colder water from below. During El Niño years, because of weakened trade wind patterns, the warm pool of nutrient-depleted water in the western Pacific flows back towards the Latin American coast. The upwelling is shut off under these conditions and the nutrient-rich, cold water is trapped below the euphotic zone. This has adverse effects on marine productivity

During El Niño years the climate of the neighbouring landmasses is affected, with high rainfall occurring on the Pacific coast of Latin America and major changes in the monsoon circulation of the western Pacific.

▲ *This false-colour map* illustrates the sea surface currents with the strong western boundary currents appearing as red patches. These mark the ends of the ocean gyres and it is their passage that affects the climate of the neighbouring land. Off Japan lies the Kuroshio, whilst the Gulf Stream is easily seen along the eastern seaboard of North America. The Arghulas Current passes south of the Cape of Good Hope, whilst the Brazil Current can be seen off the coast of Latin America and the East Australia Current off the coast of Australia.

◀ *Like the boobies* and cormorants found in the same area of South America, these brown pelicans suffer high mortalities during El Niño years, when the upwelling of cold, nutrient-rich water ceases, and the production of Peruvian anchovetta crashes. El Niño represents a major shift in the ocean-atmosphere interaction drastically affecting climate.

OCEAN EXPLORATION

◄ **The advances in** underwater technology over the years, have gone some way to help us learn about and understand our ocean environment. However, the sheer scale of the world ocean makes exploration a vast task.

First Encounters

▶ **Dugout canoes** such as these in Nigeria were once widespread and probably represent one of the earliest forms of water craft. While simple canoes with a low free board can be used on sheltered inland waters they are not stable enough for use offshore.

For thousands of years people have had a close relationship with the oceans and their resources. Unfortunately, much of the early archaeological evidence of this relationship has been lost following the rise in sea level at the end of the Pleistocene period about 10,000 years ago. In addition, the materials used in the construction of boats, such as wood, skin and fibres, do not survive well in most archaeological sites, and so our knowledge of early boat construction is fragmentary. Early coastal habitation sites are, however, known from accumulated food refuse. In southern Africa and Melanesia such remains indicate occupation of the coast some 40,000 years ago.

Coastal regions offered settlers the combination of fertile agricultural land and abundant protein in the form of shellfish and finfish. Extensive shell middens (areas of accumulated refuse) along old coastlines worldwide indicate that gleaning, or the collecting of food, from the intertidal area has long been important to human societies.

Early craft

The first boats were almost certainly those built for use on rivers, lakes and coastal swamps. In many different areas of the world, canoes of different shape and form, and made from either hollow logs, reeds, skins or bark, were, and still are, used extensively. The bark canoes of the North American Indians and the skin-covered kayaks of the Inuit, for example, were all in wide use at the time of first European contact, while on the Sepik River of New Guinea canoes carved from logs, often with ornately decorated prows, are still a common sight. Skin-covered boats, known as coracles or curraghs, were also in use in France and Ireland until relatively recently.

Such boats, although light and manoeuvrable, lack sufficient stability for use in rough seas. Early Pacific islanders discovered that fitting an out-rigger to a single-hulled canoe greatly increased its stability. This innovation together with the addition of sails rather than relying on paddles or oars, enabled the islanders to venture further offshore in search of migratory fish such as tuna.

The first seafarers

Although we have no information on the earliest ocean-going vessels, we know people successfully crossed from Southeast Asia into Australia and New Guinea at least 40,000 years ago. Archaeological remains from Australia and New Guinea dating from this period indicate that communities had been established by this time at widely different locations. To have achieved this, substantial numbers of people must have crossed from Southeast Asia. Such a crossing would have involved, even during times of lowest sea level, an open stretch of ocean at least 40 kilometres (25 miles) wide.

Seagoing craft were developed independently in many different parts of the world. One of the oldest surviving ships is the Egyptian Cheops ship, a planked craft dating back to 3000 BC. The colonization of the islands of the central Pacific, which occurred around 4000 years ago, depended on the construction of ocean-going craft.

In Europe, maritime trade was quite well established as long as 5000 years ago, with seafarers from the Aegean trading in obsidian around the Mediterranean. The early Greeks and Vikings were also accomplished seamen. The founding of colonies on Greenland and Iceland, and the discovery of America by the Vikings, attest to the versatility and seaworthiness of their longboats.

In the Indian Ocean, the development of trade links stretching from the coast of East Africa to Arabia and the western Indian coastline must have developed quite early, since the Maldives and Sri Lanka are known to have sent emissaries to Rome over 2000 years ago. By the time Europeans penetrated Southeast Asia, the Arabs had established an extensive trading network based on colonies and outposts which stretched from the east coast of Africa to the tip of New Guinea. The very nature of this trading Empire demonstrates not only the existence of ocean-going vessels, but also a knowledge of the monsoon wind patterns of the western Indian Ocean and the development of navigational aids which enabled vessels to cross open stretches of water out of sight of land.

◄ **Arab** **dhows** *were in use over 1300 years ago. They made long trans-oceanic voyages to the East African coast, India, Southeast Asia and China, carrying Arab and Persian traders to Guangzhou (Canton).*

Ocean Pathfinders

▲ *Ferdinand Magellan, a Portuguese explorer, led a circumnavigation of the globe which lasted from 1519–22 and was completed by Del Cano after Magellan's death in the Philippines.*

▲ *Christopher Columbus is credited with the discovery of the New World in 1492, having successfully navigated across to the West Indies and returning to Portugal by running down the latitude before sailing due east to make landfall.*

▲ *Captain Cook's three voyages of exploration between 1768 and 1779 led to the opening of the Pacific for trade and commerce. He was killed by Hawaiian natives on 14 February, 1779.*

Although we are aware of early navigational instruments, such as the stick charts of the Micronesians which were used to great effect some 1000 years ago, the earliest ocean voyagers most likely remained close to land, sailing along a coastline for orientation, or for short distances out at sea to use favourable winds. It is remarkable to note, therefore, that European ocean voyaging is relatively recent, and that the European dominance of the world's oceans is relatively short in terms of human history, no more than 500 years.

This dominance of ocean trade and transport stems at least in part from the extensive investment in voyages of exploration which characterized European history during the 15th and 16th centuries and were initially undertaken by the Portuguese. Sent out by Prince Henry in the 1420s to discover a maritime route to the source of African gold, Portuguese seamen discovered the Azores and Madeira and in 1434 rounded Cape Bajador at 26°N, the previous, southern limit to Atlantic seafaring. To achieve this it would have been necessary to sail south using the trade winds, then due west to encounter winds from the south and then use westerly winds to bring them back to Portugal. The Arabs had long known about and used similar wind patterns in the western Indian Ocean.

Early navigation

The early voyages of exploration were based on the use of compass and sandglass. The compass provided mariners with direction, while the sandglass was the standard marine time-keeper, and was used with log and line to measure distances. These crude but effective instruments, together with written directions, had allowed reliable year-round navigation of the Mediterranean region for some considerable time. The ocean current systems of the Atlantic, however, made calculations using only these navigational aids inaccurate.

The development of nautical astronomy brought about a major breakthrough in ocean navigation. By measuring the altitude of the North Star, and later the Sun at noon, and multiplying by the number of miles in a degree, a skilled pilot could provide a reliable estimate of the distance sailed in a northerly or southerly direction. By 1480, astronomical rules and tables had been developed allowing pilots to calculate their latitude. Using these sorts of instruments and tables, the explorer Bartholomeu Diaz rounded the Cape of Good Hope in 1488, and returned to Lisbon having discovered a sea route to the East. Ten years later, Vasco da Gama, using the anticlockwise wind circulation of the South Atlantic and Indian oceans, reached India in his two specially constructed three-masted, square-rigged ships. Such ships were developed in Europe around 1450 and armed with cannon, gave Europe mastery of the oceans for nearly four centuries.

Age of discovery

At the beginning of the 16th century, the further development of navigational aids enabled mariners to determine their position at sea with increasing accuracy. The sea astrolabe, for example, provided precise measurements of the altitudes of the stars and planets, and when used in conjunction with the plane charts, which provided a latitude scale, pilots were able to sail north or south to the latitude of a known landfall before sailing due east or west to make land. Columbus used this navigational technique for his return after crossing the Atlantic in 1492 to discover the "New World".

During this period of exploration, the Spaniards in Lisbon and Seville continued to chart discoveries and

▲ *This Portolan was prepared by Frederici d'Ancore in 1497 and shows the Mediterranean and Europe in detail. The Portolan chart was used with magnetic compass and sandglass for direction and distance finding. Although pilots could calculate their latitude at sea, longitude could not be measured accurately, hence good maps, sailing directions and skilled pilots were essential for maritime voyages.*

develop navigational techniques, and by 1516 the first world map based on these discoveries was published by Waldseemüller. The vast extent of the Pacific was only discovered during the circumnavigation of 1519–22 led initially by Magellan and completed by Del Cano.

Meridonal tables published in 1599 by the mathematician Edward Wright enabled hydrographers to construct charts mathematically on what became known as the Mercator projection, after the cartographer Gerhardus Mercator. The first sea chart to use these tables was also published in 1599, and theoretically allowed a ship's position to be plotted in terms of latitude and longitude and the distance and direction to landfall to be accurately charted. The problem of accurately determining longitude at sea was not solved, however, until much later and required the development of precise navigational instruments.

From the late 15th century onwards Portuguese pilots determined their latitude by observing the Sun's altitude at noon with an astrolabe, while the cross-staff, developed around 1514, allowed the Sun's altitude to be read directly. Thus, although latitude could be determined with relative accuracy during the 16th century, the determination of longitude was not possible until the 18th century.

Greenwich Observatory

Although Galileo, who invented the pendulum-controlled clock and astronomical telescope, discovered that the satellites of Jupiter could be used for determining longitude with reasonable accuracy, the method proved to be impractical for use aboard a ship. So in 1675 the Royal Observatory was founded at Greenwich, London, with the charge of solving this problem.

In 1760, John Harrison's fourth chronometer and the lunar distance method of calculating longitude provided the solution to the problem of calculating longitude at sea. The publication of the annual nautical almanac from 1767, which gave the lunar distances from the Sun and certain stars for every three hours at Greenwich, combined with John Hadley's invention of the reflecting quadrant in 1731, subsequently improved into the sextant in 1757, enabled lunar observations to be made on board. The sextant works by reflecting an image of the Sun or a star, via a mirror, onto a glass plate through which the horizon is visible. By adjusting a calibrated, sliding arm, the angle between the Sun or star can be altered until the image rests on the horizon, so determining the longitude.

During his three voyages between 1768 and 1780, Captain Cook explored the Pacific Ocean using the lunar distance method for determining longitude and his accurate charts provided the first basis for the economic exploitation of this vast expanse of the world's surface.

▶ **This model** of Captain Cook's vessel the Endeavour shows the cramped quarters endured by seamen on voyages that would usually last for several years. The lack of fresh fruit and vegetables and a diet of salt pork and biscuit on long voyages led to outbreaks of scurvy. The development of the three mast square-rigged vessel in the mid-15th century gave Europe mastery of the oceans for nearly 400 years.

Navigational instruments

Accurate time-keeping and variations in the Earth's magnetic field caused navigators problems, and several attempts were made to improve on the timekeeping of the sand glass. John Harrison's marine timekeepers were essential developments for determining longitude. A copy of his fourth chronometer was used by Captain Cook during his voyages of exploration. Although magnetic variation had been known to compass makers since the 1450s it was thought to be constant wherever it was observed. In 1635, the annual change in the Earth's magnetic field was discovered and in the late 1690s the Admiralty commissioned an astronomer, Edmond Halley, to measure the variations in the North and South Atlantic. He published the first isogonic (lines of equal variation) chart in 1701, later extending it to the Indian Ocean.

Evolution of a Science

Although Aristotle studied the marine life of the Aegean and discussed various theories concerning the salinity of the sea, the early exploration and study of oceans was hampered by problems of navigation and the lack of suitable equipment. It was only during the scientific revolution of the late 17th century that more methodical observations began to be made.

Early development

Sir Isaac Newton used tidal information to illustrate his theory of gravity. The information was compiled on the basis of seafarers' observations, collected according to the instructions contained in a pamphlet published by the Royal Society in 1665 entitled, "Directions for Seamen". The pamphlet provided advice on the systematic collection of information relating to the depth of the sea, its salinity, tides and currents. Around this time, eminent scientists such as Robert Hooks produced ideas for sampling apparatus including depth sounders, water samplers and deep-sea thermometers.

The practical difficulties of marine exploration and the explosion of information from the newly discovered continental areas led to a decline in scientific interest in the sea, and voyages of exploration were only resumed in the 18th century. The solution to the problem of determining a ship's longitude, discovered in the second half of the 18th century, increased considerably the value of information collected by seafarers on ocean conditions. Information collected during the voyages of Captain Cook was extremely significant for developing Earth sciences.

The 19th century

During the early 19th century, several significant, major scientific contributions were made. Among them were Alexander Marcet's study of the salinity of the world ocean, James Renell's charts of Atlantic Ocean currents and Emil von Lenz's studies on variation in temperature and salinity in the deep ocean. The latter observations seemed to confirm the hypothesis that it is density differences in the ocean, rather than surface winds, which are responsible for ocean currents.

In 1855, Matthew Fontaine Maury published the first ocean basin chart. Showing the North Atlantic ocean floor, the chart was reproduced in his book "The Physical Geography of the Sea". It was based on soundings made with lead and line, and although the map contains errors, he showed that some previous, very deep soundings were exaggerated or inaccurate. The science of bathymetry, or the study of the physical form of the ocean floor, has progressed significantly since that time. Accurate bathymetric charts for much of the world's ocean floor are now available. These charts are based on information derived from a wide variety of observational techniques, and our understanding of the physical form of the ocean floor has contributed considerably to the development of the theory of plate tectonics and continental drift.

The *Challenger* Expedition

One of the most important scientific voyages ever made, the *Challenger* expedition of 1872–6 laid the foundations for the modern science of oceanography. Two biologists, W.B. Carpenter and Wyville Thomson persuaded the British Government to equip the expedition to study deep-sea circulation and the distribution of life in the seas. The voyage set a pattern for similar oceanographic cruises during the late 19th and early 20th centuries.

The voyage was the first to discover manganese nodules on the deep ocean floor, which were found at all sites in the ocean basins. The *Challenger* expedition also sampled the benthic fauna and fishes down to depths of 4500 fathoms – 8000 m (27,000 ft) – thus demonstrating the existence of life in the abyssal depths. Hundreds of dredge samples were examined initially on board and the samples were preserved for analysis on return to England.

▲ *The* **Challenger** *expedition* carried a team of scientists on its four year circumnavigation.

◄ **Challenger***'s expedition* report was illustrated with diagrams of organisms such as these radiolarians.

▶ *The* **cramped laboratory** on board HMS Challenger *was used for examining dredge samples, drawing and preserving specimens, and also served as a library.*

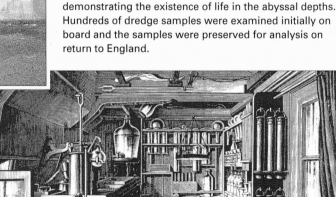

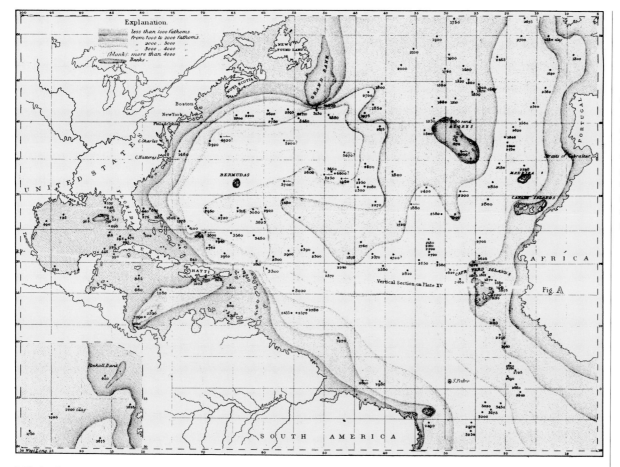

◀ **The first map** to be made of an ocean basin, this charting the bed of the North Atlantic, by Matthew Fontaine Maury, appeared in his book The Physical Geography of the Sea. Maury used soundings made with a lead and line to prepare his map and demonstrated that some previous, very deep soundings were inaccurate or exaggerated, although some errors still remain on this map.

Life in the seas

Until the middle of the 19th century most scientists believed that life could not survive below around 400 fathoms – 700 metres (2300 feet) – but gradually marine biologists extended this limit. In 1869, Wyville Thompson dredged creatures from 2500 fathoms – 4600 metres (15,000 feet) – during the voyage of HMS *Porcupine*. This fascinating discovery led to the voyage of HMS *Challenger* (1872–6), which was considered the first truly oceanographic cruise. Many nations followed this early example and a series of polar expeditions also contributed substantially to the growing body of knowledge concerned with the world's oceans. Studies of deep water circulation by German ships, led to the first accurate model of circulation in the Atlantic.

Although by 1900 the study of the oceans had become recognized as an important area of scientific endeavour, its development was uneven. Many early institutions were little more than seaside stations specializing in biology, although some, such as the institutes established in Paris and Monaco by the Prince of Monaco, were more diverse in research interests. One of the first areas of marine research to attract government interest was fisheries. As early as the 1870s the decline in fish stocks, particularly those of the North Sea, was a cause of concern. By 1925, Britain had established the Discovery Investigations in response to concern about the decline in whale stocks in the Southern Ocean.

During World War II, the science of oceanography received a major boost. Pioneering work on wave forecasting and underwater acoustics was undertaken as, for the first time, war was taken beneath the sea surface in the guise of submarines. Since that time, oceanography has become a sophisticated and in many respects highly technical science, dependent on a number of devices which permit scientists to examine and measure areas, inaccessible directly to human observers.

Improved observations of the sea surface from satellites enable simultaneous estimation of primary production over vast areas and provide extensive data on surface water temperatures, wave and current patterns, and other physical features of the ocean. Just as the scale of observations of the sea surface has expanded, so too has the range and extent of observations underwater. Commercial vessels routinely traversing the ocean basins participate in a co-ordinated "ships of opportunity" programme releasing instruments to measure temperature and salinity profiles at different points along their navigational route. Observations of the marine environment at depth are now possible through the use of remotely controlled, deep-diving equipment. The spectacular rediscovery of the *Titanic* and observations of the diverse marine communities of submarine thermal vents bear witness to our rapidly expanding knowledge of the ocean realm.

Understanding the physical, chemical and biological processes which are occurring in the world's oceans, and determining the oceans' role in the functioning of the global Earth system requires a co-ordinated worldwide approach to observation. One function of the Intergovernmental Oceanographic Commission, the aim of which is to co-ordinate a multi-national oceanographic investigation of the Indian Ocean, is to provide an international mechanism to facilitate the participation of all coastal nations in oceanographic research. Under its auspices nations agree on standard techniques for the collection and handling of data and several World Oceanographic Data Centres are now established to provide operational systems for managing and storing the enormous volumes of data now collected through national, regional and international oceanographic programmes.

Probing the Deep

In 1872, HMS *Challenger* left port on a four-year voyage which marked the birth of modern oceanography. When it left, the ship carried with it numerous new oceanographic instruments, many of which were unproven and which would be considered primitive today. Nevertheless the basic design of today's water samplers, corers, dredgers and current meters are much the same as those developed 100 years ago.

Studying the deep ocean is based on probing and sampling an environment, which until recently could not be observed, even indirectly. Oceanographic sampling instruments fall into two categories: those designed to bring back samples of water, sediment or animals from depths, for subsequent study and analysis, and those designed to record the physical conditions beneath the surface.

Most water sampling devices actually consist of a cylinder, open at both ends, which is lowered to the required depth. A "messenger", a type of weight, is dropped down a guide line to release the two ends which snap shut to close the cylinder. This prevents the water from being contaminated as it is drawn back to the surface. Similar mechanisms are used to open plankton nets in order to sample the planktonic organisms at different depths. Samples may be taken at different depths by a series of nets arranged at points along the same cable, the nets being closed before being hauled on board.

Since one of the requirements of understanding the productivity of the ocean is to be able to relate the numbers of organisms to the volume of water sampled, more sophisticated nets may incorporate some form of flow meter at the mouth to indicate the volume of water which has passed through the net during the sampling period. By counting the number of plankton individuals and relating this to the volume of water, the density of organisms, and hence the overall productivity of particular water bodies, can be estimated.

The most sophisticated plankton samplers are the Hardy continuous plankton recorders, designed by the marine biologist Sir Alistair Hardy. Towed behind moving vessels, the samplers trap the plankton on a continuously moving roll of gauze which passes into a tank of preservative. Subsequent analysis of the samples in relation to the ship's log enable the samples to be located along the path of the ship's voyage and hence a picture of the distribution of different plankton species can be obtained. Ships, towing these samplers through the North Atlantic over many decades, have enabled scientists to draw conclusions about changes in the productivity of the ocean, the onset of spring blooms of plankton and the role of water temperature in the annual cycle of temperate seas.

Samplers and corers

Several types of bottom samplers exist. Those designed to scrape the surface of the ocean floor include a variety of trawls and dredges, which are towed along the bottom to collect either surface samples of sediment and rocks, or benthic animals. This technique is both destructive and frequently results in damaged specimens. In an attempt to overcome some of these problems, more sophisticated grabs have been developed, which are dropped, with the two or more jaws open, from the ship. The jaws close around a sample of the sediment as the grab hits the bottom. Some are operated on cables from the ship, while others can be operated independently. The latter type have weights which cause them to sink and which drop off as the sampler jaws close. Flotation chambers then cause the sampler to rise to the surface.

Most seabed samples taken for geophysical analysis are taken by means of corers. The simplest corers can be dropped over the side of a ship and the core box penetrates the sediment. A device to close the end of the corer and ensure that the sample remains intact is then closed and

◀ **A current meter** mooring being laid. Measurements from moored arrays provide information on sub-surface currents and complement the surface data collected from satellites and drifting buoys over a wider area. Moored arrays also provide time series data enabling scientists to analyse tidal currents and seasonal changes in current strength and direction.

▶ **A carousel water** sampler is lowered into the sea. Such samplers can hold up to 24 bottles, each capable of taking a sample from the sea or ocean at any given time. The samples are taken to measure global distribution of temperature, salinity and chemical traces. The measurements received will provide a vast amount of information on the ocean's water masses.

◄ **ERS-1,** an oceanographic satellite being prepared for launching. Satellite observations provide information on sea surface temperature, surface currents, primary production and a wide variety of other parameters. Satellites are also used as relay stations for a number of instruments such as tide gauges and drifting buoys, which receive information from remote locations and relay it to oceanographic centres.

the device hauled to the surface. Cores of up to 20 metres (65 feet) in length can be recovered by means of piston corers, which incorporate a piston that moves up the tube as the corer enters the sediment, creating a partial vacuum. The development of drilling technology associated with oil exploration has greatly improved sampling techniques, while the *Glomar Challenger*, a research drilling ship, has retrieved cores from 1290 metres (4232 feet).

On-site recorders, designed to measure various physical parameters at different depths, are also varied. They range from the simple reversing thermometers used in conjunction with water sampling devices to record water temperatures, to expendable bathythermographs which measure a range of variables, including temperature, pressure and salinity. Drifting buoys are extensively employed in studies of current systems. Some float on the surface and their signals, which provide information on the strength and direction of currents, are relayed to satellites. Others operate using floats, which maintain them at a precise depth, providing information on underwater currents.

Remote sensing

Although the range of observations which can now be made using satellites has greatly increased, the volume of information available concerning the conditions of the ocean, the scale and extent of on-site observations is still limited. Without a greater investment in ocean observing systems, our ability to understand the functioning of the oceans in the world climate system, and hence our ability to plan for future changes which might result from global climate change, remains severely limited.

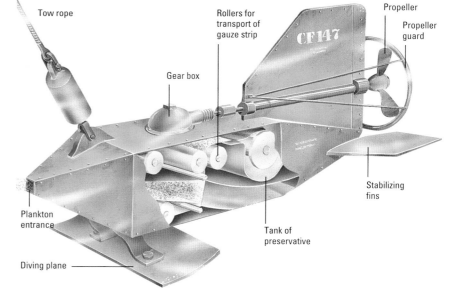

▲ **The Hardy Continuous** Plankton Recorder is towed at a depth of 9 m (30 ft). The flow of water past the propeller drives a series of rollers which wind a continuous strip of gauze.

Plankton enter the front and are trapped on the gauze before passing into a tank of preservative. The gauze is unwound and in conjunction with the ship's log provides a continuous record of the

plankton through which the ship passed. Records from such instruments have provided data on seasonal cycles and longer term changes in plankton in the North Atlantic.

Man Beneath the Waves

▶ **In Japan,** pearl diving was an exclusively female occupation and before the introduction of scuba the length of the dive was limited to lung capacity, which was greater in women than in men.

Oceanography is a relatively young science. Indeed, the comparatively slow development of marine science in comparison with terrestrial and atmospheric sciences is directly related to our inability, until recently, to directly view the deep ocean.

However, in recent years the capacity of underwater photography has greatly increased, with innovations such as electronic flash and television greatly improving our ability to observe the sea bottom. Regrettably, though, conditions in the deep ocean are such that techniques comparable to those used on land, such as aerial photography, are not possible. The images produced from the ocean floor cover only a tiny area, and are often obscured by fine particles stirred up from the bottom. Nevertheless, despite these problems, underwater photography has contributed substantially to our understanding of deep-water animals, and of small, bottom features such as ripple marks and the distribution and density of manganese nodules.

Other techniques for indirect observation of the ocean floor include the use of various echo-sounding devices. The first echo-sounders were developed in 1922. They are based on the principle that the length of time taken for a reflected sound wave to return to the observer relates to the distance travelled. Early sonar was not accurate enough to provide good images. The first oceanographic use of refined sonar equipment occurred in the early 1950s. Modern echo-sounding has now been developed which, together with seismic profiling, provides detailed profiles of the bottom sediments and oceanic crust.

The first divers

Man's physical penetration to great depths in order to observe at first hand is constrained by the fact that we breathe air. Free diving was practised in many parts of the world to collect valuable resources such as pearls in the Persian Gulf or Japan, or for underwater spear fishing, but such dives were inevitably of short duration.

One of the earliest recorded diving operations was that of the Greek Scyllis, who, together with his daughter Cyane, succeeded in cutting the anchor cables of the Persian fleet immediately before the battle of Salamis – a feat which was achieved by swimming underwater, using

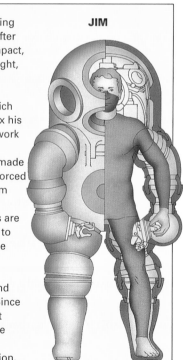

The atmospheric diving suit, known as JIM after its inventor, is a compact, robot-like suit with tight, pressure seals and specially designed articulated joints which allow the diver to flex his arms and legs, and work at depth.

Modern JIMs are made of carbon fibre, reinforced plastic and aluminium alloy. Using such materials, most JIMs are capable of diving up to 450 m (1500 ft). Some JIMs have thrusters allowing the diver to remain in position and work in mid-water. Since the diver is always at atmospheric pressure there is no need for lengthy decompression.

JIM

a snorkel of hollow reed. The first piece of diving equipment to be developed was the diving bell. Initial designs were limited by the amount of air contained in the bell at the time it was submerged. Then in 1690, Edmund Halley designed a diving bell which received its air supply through a leather pipe leading from a weighted, air-filled barrel. As the air supply in one barrel was exhausted, another was lowered down. By the end of the 18th century, however, diving suits had been developed which were fed with air pumped into the suit from the surface.

The invention of the diving helmet led to the development of the Siebe open diving suit in 1819. The suit was used in salvage work on the *Royal George* in the 1830s. Subsequently, in 1837, Siebe produced the first enclosed

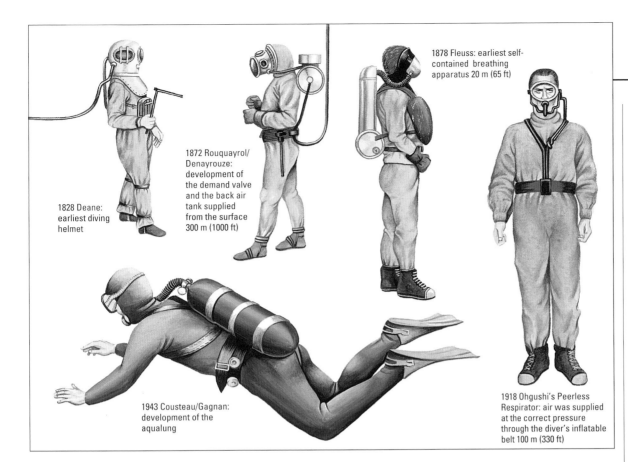

1828 Deane: earliest diving helmet

1872 Rouquayrol/ Denayrouze: development of the demand valve and the back air tank supplied from the surface 300 m (1000 ft)

1878 Fleuss: earliest self-contained breathing apparatus 20 m (65 ft)

1943 Cousteau/Gagnan: development of the aqualung

1918 Ohgushi's Peerless Respirator: air was supplied at the correct pressure through the diver's inflatable belt 100 m (330 ft)

◄ *The development of* diving apparatus has allowed the diver independence from surface-supplied air. H. A. Fleuss developed the first practicable self-contained diving apparatus. It was able to filter out the carbon dioxide from the exhaled air and replace it automatically with the equivalent amount of oxygen.

diving suit incorporating the diving helmet. This remained the standard diving equipment throughout the world for nearly 100 years.

Scuba development

In 1943 Jacques Cousteau and Emile Gagnan carried out the first successful tests of the Self-Contained Underwater Breathing Apparatus, the now well-known Scuba equipment. This represented a major breakthrough in underwater exploration, freeing the diver from the cumbersome diving suit and airlines which restricted the range of movement underwater.

A scuba diver can move freely, carrying an air supply in canisters on a back pack which is breathed through a regulator. Initial trials involved the use of tanks of pure oxygen, but below 8 metres (25 feet), pure oxygen is toxic. Air is normally used to depths of around 60 metres (200 feet), below which nitrogen, which makes up around 80 per cent of air, becomes narcotic. At greater depths, a mixture of oxygen and helium can be used. Since a diver breathes air or a mixture of gases at the same pressure as the surrounding water, nitrogen or helium are absorbed into the bloodstream. If the diver ascends too rapidly, there is not enough time for the gas to diffuse out through the lung surface. Bubbles may form in the blood stream, causing the painful condition known as the bends; in extreme cases, this can lead to unconsciousness or even death. The deeper the dive and the longer the period of time spent at a particular depth, the greater the amount of time needed for decompression. Divers working at great depth are frequently saturated with inert gas prior to the dive and rest in pressurized chambers between spells of work, so that only one lengthy period of decompression is required following completion of the job.

The use of submersible craft is necessary at greater depths than those which can be reached by free divers. Submersible craft also have an added advantage in that the operators do not need to undergo extensive periods of decompression. Nevertheless, skilled divers are still needed, since many of the tasks required in the maintenance of oil rigs and underwater structures could not be achieved without the manual dexterity of a human operative.

▲ *Early diving gear* was heavy and the diver was limited in movement by the need for surface airlines. Modern scuba gear has freed divers from these constraints allowing freedom of movement, essential for working on underwater structures such as oil wells.

Submersibles

▲ **Dr Edmund Halley's** diving bell of 1690 was the first bell in which divers were not restricted to the amount of air contained in the vessel itself. Halley's bell was supplied with air through a leather pipe leading from a weighted, air-filled barrel. As one barrel was exhausted, another was lowered from the surface to replace it.

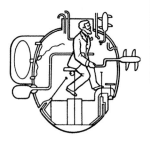

▲ **The walnut-shaped Turtle** was a one-man submersible built in 1776. It was powered by hand-cranked propellers and was used in an attempt to sink a British warship during the War of Independence. The aim was to attach a gunpowder charge to the hull of the ship by means of a screw device. However, the screw could not penetrate the ship's copper hull, and the charge exploded harmlessly.

▶ **The Jason jnr,** a tethered, small, unmanned, remotely operated vehicle is seen here exploring the hull of the Titanic during the Woods Hole Oceanographic Institute's dramatic investigation of the wreck in 1985.

For thousands of years, people have been fascinated with the possibilities of underwater vehicles. Attempts to produce such craft date back to the 4th century BC, when Alexander the Great is believed to have descended beneath the Mediterranean Sea in a vessel resembling an early diving bell.

Early submersibles

Undoubtedly the greatest factor influencing the development of underwater craft has been their military potential. Their first use in such a role was in the 18th century when a small submersible, the *Turtle*, built by David Bushnell in 1776, was used against the British in the American War of Independence. In 1863, the submersible *David* was used successfully in the American Civil War, actually sinking an enemy ship. Thirty years later, Simon Lake constructed the *Argonaut First* – a true submersible in which, unlike the earlier examples, atmospheric pressure could be maintained; the vehicle could move around the seabed on hand-powered wheels. Following these pioneering examples, submarines have been developed by many navies to travel faster, deeper and further afield, thus increasing their military strength.

There are two basic types of military submarine in use today. First, conventional submarines, which use diesel power while travelling on the surface and battery power while submerged, and second, nuclear submarines which are powered by means of a nuclear reactor and are capable of remaining submerged for months at a time.

Exploring the depths

Despite their enormous power, speed and complexity, military submarines are generally unable to descend to depths of more than a few hundred metres. To go deeper, more specialized vehicles are required. The breakthrough for submersible exploration of the deep ocean regions occurred in 1934, when William Beebe and Otis Barton descended to a depth of around 920 metres (3017 feet) in Barton's revolutionary bathysphere. The bathysphere was a heavy steel sphere that was lowered into the ocean on a cable payed out from a surface support ship.

A later version, the bathyscaphe invented by Professor August Piccard, was free of surface cables and largely overcame the limitations of the earlier bathysphere. The bathyscaphe *Trieste* was used by Piccard for a dive of 3170 metres (10,392 feet) off the coast of Italy in 1954, while in 1960 Piccard's son Jacques led a dive in the Marianas Trench in the southwestern Pacific, touching the bottom at a depth of approximately 10,900 metres (35,800 feet), a record yet to be exceeded.

Industrial applications

During the last 30 years or so, the most important developments in underwater craft have come from science or industry. In the 1960s and 1970s, many such vehicles were designed and quite a number were built. However, many of them were scrapped due to a combination of limited capabilities, shortage of funds, and absence of suitable employment. Few have survived to form part of the present generation of highly specialized underwater vehicles.

One of the earliest and most successful of the scientific submersibles is the American *Alvin*. First commissioned in 1964, *Alvin* was capable of diving to almost 3 kilometres (2 miles) and was later upgraded so that it could achieve depths of 4 kilometres (2.5 miles). Alvin submersibles have made thousands of highly successful and productive dives, but none more spectacular than the early in situ observations of hydrothermal vents in the 1970s.

In recent years, *Alvin* has been joined by a relatively small number of scientific submersibles capable of reaching the same depth or deeper. The American *Sea Cliff*, the French *Nautile*, the Russian *Mir I* and *Mir II* and the Japanese *Shinkai 6500* are all able to reach a depth of 6 kilometres (3.7 miles), and thus penetrate all ocean depths except for the deepest ocean trenches.

Remote submersibles

The ability to enter the deep-sea environment is one which is of enormous value to both scientists and engineers. However, manned submersibles have a number of disadvantages, not least the undoubted safety risks, as well as the expense and complexity of the necessary precautions. For these reasons, unmanned, remotely operated vehicles (ROVs) have attracted attention for more than 30 years.

These craft are mostly small, manoeuvrable units capable of transmitting television pictures or other information and of undertaking limited manipulations. They are usually controlled via an umbilical cable from the surface or from a second submersible, either manned or unmanned. Perhaps the best known of these systems is the combination of the towed ROV *Argo* and the tethered small vehicle *Jason Jr*, which was used by Dr Robert Ballard of the Woods Hole Oceanographic Institute for the dramatic investigation of the wreck of the *Titanic* in 1985. But many less well publicised systems are in use routinely; in late 1992, for example, there were 194 heavy-duty ROVs in use in the offshore oil industry worldwide.

By their very nature, most submersibles, manned or unmanned, have limited ranges in both time and space. In order to overcome this restriction, the latest objective in the development of scientific deep submersibles is for untethered vehicles capable of traversing wide ocean areas at all depths. These vehicles would operate independently, following a pre-programmed route, but with the possibility for "artificial intelligence" allowing the programme to be varied according to the conditions. There are technical problems to be overcome, but these are not insurmountable, and several research groups are actively working towards their solution.

◀ ▼ *Modern submersibles* are versatile vehicles used for a variety of scientific and industrial purposes. Most can be fitted with a wide range of equipment, such as video, still cameras and claw-like manipulators. Some types, known as diver lockout submersibles, such as the Johnson Sea Link shown here, are equipped with pressure chambers from which divers can enter and exit, allowing them to perform work and then return to the submersible for gradual decompression.

The hydraulically powered, highly mobile manipulators of today's submersibles are powerful enough to cut through the toughest steel cables, but are also sufficiently sensitive to hold extremely fragile objects. Manipulators are often equipped with powerful lights to help the crew see the work area.

Submersibles are usually fitted with a number of directional propellers called thrusters. These give the craft great manoeuvrability when working underwater.

The control console houses sophisticated computer-aided navigational and recording equipment. With this sort of technology available some submersibles are capable of working either manned or unmanned.

Most submersibles have two buoyancy systems. The larger cylinders are used to raise and lower the craft to and from the main operating area, while the lower, smaller tanks are used to make finer adjustements to the craft's depth.

Maritime Law

Attempts have been made to regulate the use of the seas for over 20 centuries. The Rhodian Sea Law or Rhodian Code, which dates to the 3rd or 2nd century BC, is thought to be the one of the earliest regulating authorities. It was devised to apply to the Mediterranean and its principles were adopted by both Greeks and Romans and observed for 1000 years.

Since the early voyages of exploration during the 15th and 16th centuries, maritime nations have come into conflict over their rights and jurisdiction of the oceans. The early dominance of the Portuguese and Spanish in the exploration of the New World, and their leadership in the European exploration of the Indian and Pacific oceans, led them to divide the world into separate spheres of influence. But the extension of this concept of the world's oceans was soon challenged by the rising naval power of the Dutch and British. In the early 17th century, Hugo Grotius published the treatise *Mare Liberum* which formed the basis for the subsequent adoption by all maritime nations of the concept of the "Freedom of the Seas".

By the 17th century, two areas of jurisdiction were generally recognized. The first, the territorial sea, was perceived as being waters within 5 to 10 kilometres (3 to 6 miles) offshore that were considered to be under the jurisdiction of the coastal state. This was taken as the area necessary for a coastal state to protect itself against attack. It was, however, still open to the right of free or innocent passage by ships of other nations, provided that they did not threaten the security of the coastal state. The second area encompassed what were termed international waters, to which no nation could lay claim, and which were open to individuals of all states for navigation and fishing.

The freedom to exploit marine resources has, with few exceptions, always been taken as a public right. Thus seafarers travelled extensive distances in search of fishing grounds, whales and seals, without thought for ownership of the resources concerned. As demand for fisheries products grew in Europe, particularly during the Industrial Revolution, fish stocks in nearby waters were depleted. To satisfy demand, voyages to the Grand Banks off Canada, to the whaling grounds of the southern oceans, and to Arctic fishing grounds grew in frequency and scale.

The League of Nations
The decline of whales in the 19th century took place at a time when a number of fish stocks were also showing signs of over-exploitation and early attempts were being made to regulate fish catches. These were based on voluntary quota systems and unfortunately failed, because states which did not agree with the quotas simply withdrew from the fisheries commissions or failed to sign the agreements.

In 1930, the League of Nations tried to secure international agreement to the declaration of the 5-kilometre (3-mile) territorial limit. However, discussions broke down when countries failed to agree on the extent of the associated contiguous zone over which they would be granted limited rights.

In 1945, following the development of technology during World War II, and as a result of increasing recognition of the potential of oil and gas reserves in offshore areas, President Truman issued a unilateral proclamation of the United States, declaring exclusive right to exploit its continental shelf. This declaration was soon followed by the declaration of an exclusive 200-mile fishing zone by the Pacific States of Latin America – an attempt to prevent foreign fishing fleets from exploiting the rich anchovetta resources off the Peruvian and Chilean coast.

In 1958, the United Nations convened the first United Nations Conference on the Law of the Sea (UNCLOS I), at which four Conventions were discussed by 86 participating states. The first of these, the Convention on the Territorial Sea and Contiguous Zone, did not set the limits to territorial waters, but it did agree to the principles on which the baselines would be determined. The Conference also established the principle of a 12-mile contiguous zone, within which the coastal state was permitted to enforce customs, sanitary and fiscal regulations. The traditional right of innocent passage was maintained.

The High Seas Convention laid down four basic freedoms of the seas: those pertaining to freedom of navigation, fishing, overflight and the laying of submarine cables. The convention also recognized that any state could register ships with which it could claim a "genuine link", a decision which resulted in the proliferation of flags of convenience.

The Convention on Fishing and the Conservation of Living Resources of the Sea recognized the interests of coastal states in maintaining fish stocks beyond their territorial waters, and obliged other states fishing such stocks to co-operate in observing conservation measures. This convention proved the most contentious and was accepted by only 34 states.

The fourth Convention recognized the rights of coastal states to exploit exclusively the natural resources of the "submarine areas" on the continental shelf to a depth of 200 metres (660 feet) and beyond, where the water "admits of exploitation". The seaward limits to these areas were, therefore, left open and were limited solely by the available technology. The rights of coastal states in this regard were further strengthened by the 1969 pronouncement of the International Court of Justice in the North Sea Case, which recognized that a coastal state has a right to

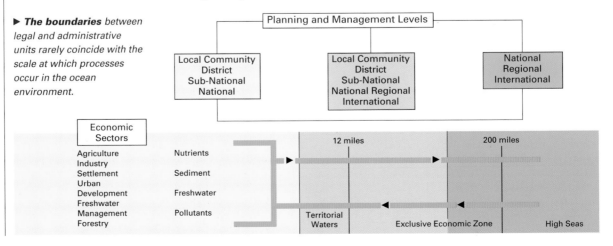

▶ **The boundaries** between legal and administrative units rarely coincide with the scale at which processes occur in the ocean environment.

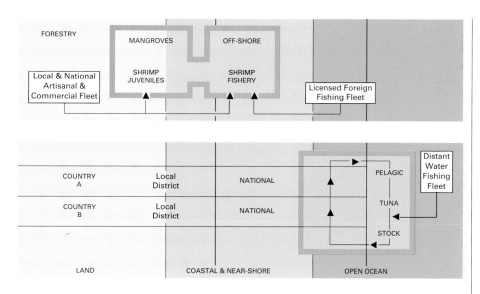

► *In the upper diagram, changes in forestry affect the mangrove habitats of juvenile shrimps thus affecting the trawl fishery. In the lower diagram several neighbouring countries share a transboundary stock of tuna which is also fished by distant fishing fleets on the high seas. In this case management of the tuna stock is impossible unless all countries agree, while in the case of the shrimp, fishermen have no control over land-use changes affecting their resource.*

exploit "the natural prolongation of its landmass under the sea". As a consequence, many states laid claim to the whole of their continental margins.

Throughout the 1950s and 1960s, the exploitation of ocean resources and the use of maritime space continued to grow. Conflicts arose between nations over rights to fish stocks and fishing grounds, and over shared resources; and between individuals and commercial concerns competing for the same space and the same limited resources.

The Third United Nations Conference
During the 1970s, there was a growing belief that the mineral and living resources of the oceans could provide the necessary economic basis for development of the world's growing population. Ambassador Pardo, Malta's distinguished Ambassador to the United Nations, proposed that the resources of the ocean should be held in trust and that no state should extend its unilateral control over ocean space beyond the limits of the territorial seas.

Following the establishment of the Seabed Committee, which was charged with examining the Law of the Sea, the General Assembly of the United Nations called for a moratorium on exploitation of the minerals of the seabed beyond the continental shelf, and unanimously adopted a declaration of principles proclaiming the seabed "the common heritage of mankind", to be exploited only under an agreed international regime. In 1973, the United Nations convened the third Conference on the Law of the Sea (UNCLOS III), the aim of which was to ensure "consideration to ocean space as a whole", and to develop a single consolidated treaty encompassing the intentions of existing international conventions.

The complexity of the task was enormous: 148 participating states, all with their own political and economic priorities, were faced with the job of drafting a mutually acceptable text which would encompass all existing areas of contention and draw some order from the conflicting claims and counterclaims of member states. Three main committees were established by the first session to consider issues arising from the previous Conventions, namely: the preservation of the marine environment; issues of scientific research; and the transfer of technology. Following numerous sessions held over a period of years, a single negotiating text was finally agreed and adopted in 1982 by the United Nations Conference.

The Convention has not yet come into force, since the 60th and final instrument of ratification necessary for its entry into force was only deposited with the United

Nations on 16 November 1993, 11 years after the original document had been signed. The Convention became finally binding on all parties on 16 November 1994, following which various international institutions will have to be created; in particular, the International Seabed Authority and the Tribunal for the Law of the Sea.

The Convention has already provided the standard for state practice and the overwhelming majority of the 130 coastal states have adopted the 12-mile or lesser limit for the territorial sea. Ninety-one states have declared their 200-mile Exclusive Economic Zones (EEZs) within which the states have right to exploit both living and non-living resources but over which they have no territorial or other jurisdictional rights. In addition, a number of states have adopted national legislation following the provisions of this Convention.

Other conventions
The delays associated with the negotiation and final ratification of the UNCLOS have resulted in a proliferation of treaties, protocols and conventions of lesser scope. These were designed to regulate the use of the world's oceans. Among them are some 11 Regional Seas Conventions, negotiated between states with a shared area of ocean space, such as the Mediterranean (the Barcelona Convention) and South Pacific (the Noumea Convention). These conventions encompass multi-lateral actions to monitor and mitigate marine pollution, to conserve resources, to respond to oil pollution emergencies and to co-operate in the use of shared ocean space.

In addition, a number of international conventions have been drawn up by various agencies of the United Nations. They cover such diverse issues as waste dumping (the London Dumping Convention), pollution of the sea from ship-based discharge (the MARPOL Convention), or the regulation of fisheries based on shared, transboundary stocks. These include among others the agreements under the Tuna and Fisheries Commissions established by the FAO (Food and Agricultural Organization).

The need for this increasingly complex system of regulations at the national, regional and international level reflects the increasing pressure which is resulting from the unregulated use of marine resources. Marine resources are limited, and the capacity of the oceans to absorb the ever-increasing levels of pollutants being put into them is being exceeded in some areas. If the oceans are to remain a source of renewable resources for future generations, human use must be controlled.

OCEAN
LIFE

◀ **Life in the oceans**
comes in an incredible array
of shapes and sizes. From
the microscopic plankton,
through the dolphins to the
vast whales that wander the
world ocean.

Plankton – the Basis of Life

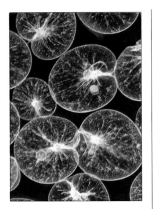

▲ *Dinoflagellates* such as Noctiluca scintillans *are important phytoplankton.*

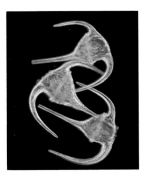

▲ *Ceratium tripos* has *three projections that slow its rate of sinking.*

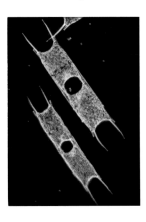

▲ *Diatoms,* such as Bidulphia sinensis, *are also important primary producers.*

Unlike the land, where vegetation dominates and large, complex trees grow to enormous size, the primary producers of the ocean realm are mainly of minute size. Ocean waters support complex communities of plankton, microscopic plants and animals which form the basis for the food webs that support the larger, multi-celled animals. The smallest phytoplankton, picoplankton, are less than 2 microns (a hair's width) thick; the largest, or macroplankton, exceed 2 millimetres ($^1/_{12}$ inch) in length

Only in the shallow coastal waters are larger plants, the seaweeds and seagrasses found. Comparable to the lowly grasses of land, even these do not achieve large size. Since sunlight does not penetrate beyond 200 metres (650 feet) in the oceans, primary production can only occur in surface waters, where it is generally limited by the low availability of nutrients such as nitrogen, phosphorus and in some areas iron.

Where the ocean exceeds the depth of light penetration plants cannot grow fixed to the bottom. For this reason, the vast majority of ocean primary production is based on single-celled, floating plants, the phytoplankton. There are two dominant groups of phytoplankton, the diatoms which are characteristic of colder waters, and the dinoflagellates, which dominate the warmer water areas. Although most phytoplankton are unicellular, some grow in chains, or form spherical colonies of larger size. Large colonial phy-

toplankton are characteristic of areas of upwelling, where nutrients are more abundant than in the open ocean.

Starting the chain

Most of the phytoplankton are eaten by herbivorous planktonic animals, the most abundant of which are copepods. Copepods are small crustaceans whose constantly moving limbs sweep the phytoplankton towards their mouths. The majority of phytoplankton production of the oceans is grazed by these herbivores, in contrast to a terrestrial community where plants store energy in tissues which are not eaten but decompose under the action of bacteria and fungi. The turnover time of the phytoplankton community is quite short because their growth and reproduction are rapid. If the community was not grazed by herbivores then phytoplankton would double in quantity by the process of cell division within one or two days.

▼ *Concentrations* of the global phytoplankton community change seasonally, as is clearly demonstrated by the three-monthly composites. A represents the months of January to March; B shows

months April to June; C, July to September and finally D represents October to December. Red and yellow indicate areas of highest concentration, green and blue areas have less dense populations, while purple

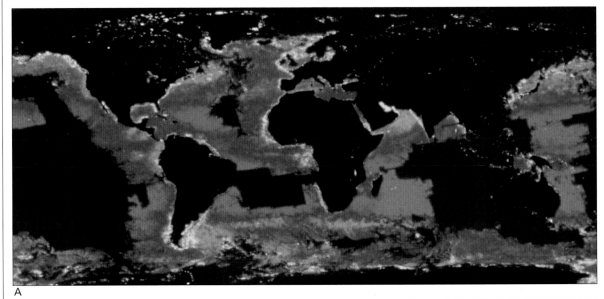

A

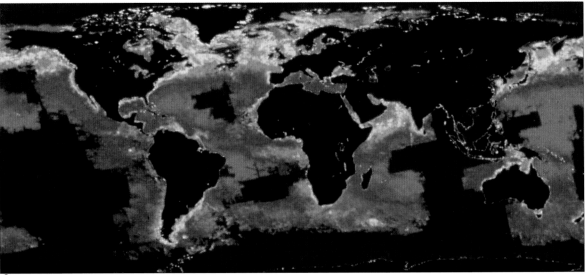

C

As a consequence of the high-grazing rates, the standing stock of phytoplankton in any one area of ocean is normally less than the standing stock of herbivorous zooplankton feeding on them. Although longer lived than the phytoplankton, these zooplankton are also relatively short lived, surviving for a few weeks and breeding throughout the year in low latitudes, but only during the warmer periods of the year at high latitudes. The planktonic herbivores are preyed on by carnivorous zooplankton which are in turn fed on by small fishes.

Feeding

The herbivorous plankton possess straining mechanisms for filtering the phytoplankton out of the water; crustaceans sweep food up using hairy, modified legs, and animals such as salps filter the water through their barrel-shaped bodies. Even larger animals, such as the baleen

whales and basking sharks, rely on plankton for food. In the case of the baleen whales, krill, small herbivorous crustaceans, are simply strained from the water as they pass into the open mouth and through the plates of baleen. The baleen plates have fringes of hair-like structures which trap the krill.

Plankton communities

Although the diversity of organisms found in different planktonic communities is immense, not all species remain in the plankton throughout their lives. Indeed, in shallow water, coastal environments the planktonic community may be dominated by meroplanktonic organisms at certain times of the year. Meroplankton are temporary planktonic residents, and are usually the early stages in the life of sessile organisms that go through a planktonic phase to ensure their dispersal. As a consequence many benthic organisms have quite wide geographic distributions. Their larval stage spent in the plankton ensures they are moved around before settling to the ocean floor and assuming a sedentary adult lifestyle.

Bivalve molluscs, such as mussels or giant clams, may introduce many millions of eggs into the ocean where they hatch into small larvae. The larvae float and feed in the plankton for several days or weeks before maturing and becoming attached as part of the benthic community.

▲ **Copepods,** *such as* Temora longicornis, *are the dominant herbivores of the plankton community.*

▲ **Euphausid crustaceans** *such as* Thysanopoda *feed on the copepods.*

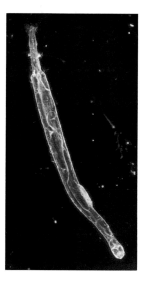

▲ **Arrow worms,** Sagitta *sp. are active carnivores in the plankton community eating a variety of crustacea.*

areas have the lowest concentrations of all. Note the "blooming" of phytoplankton over the entire North Atlantic with the advent of the Northern Hemisphere spring, and seasonal increases in equatorial phytoplankton concentrations in both the Atlantic and Pacific oceans, and off the western coasts of Africa and Peru. See p. 162 for the effect on phytoplankton during the Pacific El Niño years.

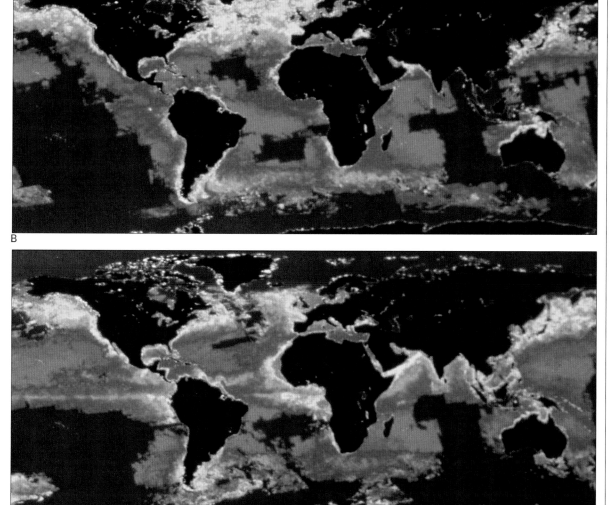

B

D

The Sunlit Surface

J F M A M J J A S O N D
Arctic

J F M A M J J A S O N D
North Atlantic

J F M A M J J A S O N D
North Pacific

J F M A M J J A S O N D
Tropical

▲ *Seasonal cycles* of *plankton production vary according to latitude. The characteristic Arctic cycle has a single peak in the summer – the only period of sufficient light for plant production. In the North Atlantic temperate zone, zooplankton increase must await the spring bloom of phytoplankton, which grazing pressure then quickly reduces. In the North Pacific, however, zooplankton overwinter as a shallow layer of subadults which take immediate advantage of phytoplankton production, and a peak of phytoplankton does not occur. Finally, the tropical cycle shows only minor increases and reductions in plankton throughout the year.*

The term primary productivity refers to the conversion of energy from sunlight into the chemical energy of organic molecules through the processes of photosynthesis. Photosynthesis can, therefore, only take place in the presence of sunlight, and with very few exceptions all marine communities depend on the production of phytoplankton and zooplankton for their sources of food energy.

The intensity of sunlight is not, however, uniform over the surface of the oceans, either in spatial or in temporal terms. At high latitudes, intensity varies according to the season – it is low during the winter periods and high during summer. As a result, primary production at such latitudes is cyclical, and the cycles are reflected in the growth and production of the animal communities that feed on the primary producers, the phytoplankton.

Seasonal patterns of productivity vary according to latitude. In the Arctic, for example, a single peak of primary production occurs in summer during the period of highest light intensity, to be followed by a peak in zooplankton production. In the North Atlantic, the peak of phytoplankton occurs earlier, in spring, and is followed by the zooplankton peak and a smaller second peak of phytoplankton during the autumn. In tropical systems, the peaks of plankton production are far less marked and less predictable.

Given the high intensity of light in the tropical and subtropical zones throughout the year, one might expect that primary productivity in such areas would be greater than at higher latitudes. However, this is not the case. This is because primary production is also limited by the concentration of nutrients, in particular nitrogen and phosphorus (although both silica and iron may be important limiting factors in certain areas). Tropical and subtropical

waters are characterized by low-nutrient concentrations which limits the growth of phytoplankton.

As a consequence, the areas of highest annual production in the world's oceans, with production of more than 200 grams (7 ounces) of carbon per square metre (11 square feet) per year, are found at high latitudes in the temperate zone, but outside the polar regions. In contrast, primary production in the tropical ocean regions is normally less than a quarter of that of high-latitude regions. Since much of the nutrient supply for the world's oceans comes from inputs to the coastal region via land-based run-off, the most productive areas of the world's oceans tend to occur in waters over the continental shelves.

Significant areas of upwelling of cold, nutrient-rich water also occur off the western continental margins under the influence of winds, as well as in mid-ocean areas where the strong and steady trade winds, along with the Coriolis force, interact to draw deeper water to the surface. More localized upwelling may occur in the vicinity of islands, through strong vertical eddies in the wake of the island, or through the shearing force between two opposing currents. At higher latitudes, seasonal vertical mixing may occur as a result of winter storms, resulting in high-nutrient availability during the spring when temperatures rise and sunlight intensity increases.

The species present in the plankton communities which develop in areas of different nutrient concentration are themselves quite different. Highly productive areas tend to support larger sized phytoplankton, and the community of both phytoplankton and zooplankton tends to be composed of fewer species at higher density than in less nutrient-rich areas. In nutrient-deficient areas, individuals

Subarctic assemblage

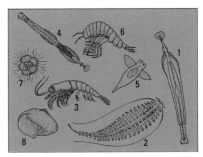

Central assemblage

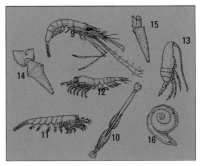

1 *Eukrohnia hamata*	9 *Stylocheiron suhmi*
2 *Tomopteris pacifica*	10 *Sagitta pseudoserratodentata*
3 *Euphausia pacifica*	11 *Euphausis brevis*
4 *Sagitta elegans*	12 *Euphausia mutica*
5 *Clione limacina*	13 *Clausocalanus paululus*
6 *Parathemisto pacifica*	14 *Cavolina inflexa*
7 *Globigerina quinqueloba*	15 *Styliola subula*
8 *Lamacina helicana*	16 *Limacina Lesuerii*

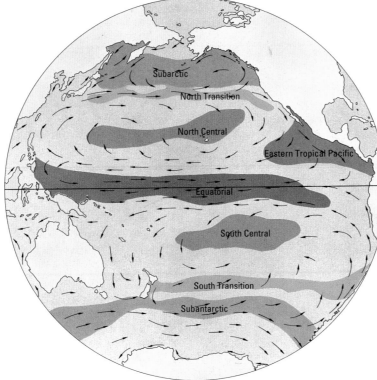

▲ *The Pacific Ocean* has eight major plankton communities, the distribution of which is controlled by the ocean's current systems. The map illustrates the extent of the "core" zones of all these communities and their relationships with the current systems: the outermost contour of each zone extends further and may overlap with other zones. The assemblages show zooplankton species which are characteristic of the subarctic and central zones of the Pacific.

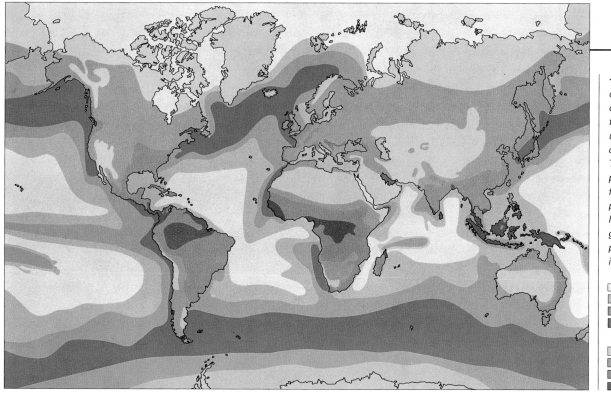

◄ **Primary productivity**
describes the creation of organic matter by plants from inorganic elements, using the Sun's energy captured by photosynthesis. The organic material produced is then available for use by animals. The primary productivity of the land and sea, measured in grams of carbon produced per sq m (11 ft) per year, is illustrated on the map.

- ☐ 0–50 g C /m² (11 ft²)/yr
- ☐ 50–100 g C /m² (11 ft²)/yr
- ☐ 100–200 g C /m² (11 ft²)/yr
- ☐ over 200 g C /m² (11 ft²)/yr

- ☐ 0–100 g C /m² (11 ft²)/yr
- ☐ 100–400 g C /m² (11 ft²)/yr
- ☐ 500–800 g C /m² (11 ft²)/yr
- ☐ over 800 g C /m² (11 ft²)/yr

tend of be of smaller size, the community is more diverse and individual species occur at lower density. As a consequence, in zones of high-nutrient input many phytoplankton are large enough to be consumed directly by small, filter-feeding, herbivorous fishes. At the same time, the proportion of herbivorous zooplankton tends to be higher in areas of high-nutrient availability. Carnivorous species dominate the zooplankton of nutrient-deficient areas.

Another factor affecting the distribution of individual species and communities of plankton is the circulation patterns of the oceans. Eight quite different plankton communities can be recognized in the different water masses of the Pacific Basin, for example. These include subarctic and subantarctic assemblages between 50° and 60°N and 50° to 60°S. The southern assemblage has an extensive distribution, circling the entire Southern Hemisphere, since it is not restricted by landmasses in the same way as the North Pacific assemblage. Since these communities display little variation in their species composition over thousands of kilometres of ocean space, and since the current systems which define them have existed unchanged for at least 26 million years, these communities must represent some of the oldest now existing on our planet.

Pelagic species and ocean productivity

While the smaller organisms of the surface oceans are dependent on the movement of the water masses for their transport, and indeed are confined to individual current systems, larger, pelagic organisms are not. Fish, such as tuna, for example, undertake lengthy, trans-Pacific and trans-Atlantic journeys each year following the seasonally changing patterns of productivity in the surface waters. In addition, species such as the bigeye tuna spawn in areas where the eggs and larval fishes will be carried with the ocean currents and associated plankton community towards areas of high productivity. Whales also migrate over entire ocean basins following the changes in production, and the breeding of many seabirds is timed to correspond with the peak in production of the small fishes on which they prey.

On a smaller spatial scale, many coral reef fishes spawn in areas where the local current patterns will ensure that the larvae remain close to the reef systems. These local current systems also dictate the movement of plank-

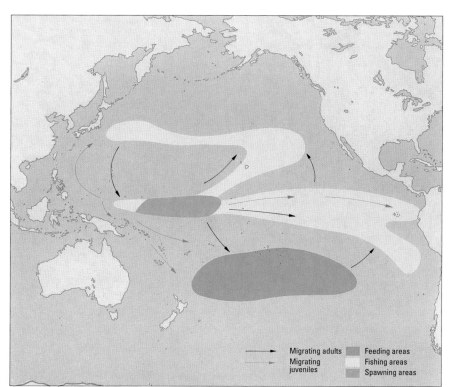

Migrating adults → Feeding areas
Migrating juveniles → Fishing areas
Spawning areas

tonic organisms and consequently, the diurnal migrations of fish which feed on them, but which rely on the reef system as a hiding place at night. Local movements of smaller tuna species around islands in the Pacific, for example, are known to follow local current systems. These local currents determine the distribution, on a daily basis, of the small, plankton-feeding fishes on which the tuna themselves feed.

Despite their uniform appearance when viewed from the air, the oceans vary considerably. Some areas, such as the Gulf Stream, have strong currents; others, like the Sargasso Sea, have low rates of water movement. Nutrients are unevenly distributed, and sunlight varies seasonally and with latitude. The plant and animal communities which inhabit the sunlit surface are, therefore, adapted to this wide range of different environmental conditions.

▲ **Many other creatures** *inhabit the Pacific Ocean besides the zooplankton communities. The fast-swimming tunas are a special group, since because of their mobility they are not obliged to remain associated with one particular planktonic community.*

The Web of Life

▲ **The Antarctic krill,** Euphausia superba, *is a carnivorous member of the plankton community. It is fed upon directly by the baleen whales that may take as much as two tonnes of krill a day. The reduction in whale populations is presumed to have led to an increase in krill populations which are now harvested commercially to produce animal feed.*

▶ **The feeding habits** of *the herring change during the different stages of its life. As the animal grows it is capable of feeding on larger-sized prey and the range of different species taken increases. The adult herring feeds on a wide variety of species including animals such as arrow worms, sand eel larvae and amphipod crustaceans, which are themselves predators competing with the herring for smaller species of prey.*

In its simplest form, a food chain can be seen as a series of links. Each species in the chain depends on the species below as a source of food energy. At the bottom of the chain lie the primary producers, or autotrophs. They convert sunlight into complex, organic chemicals which they use for growth and reproduction. The growth of individuals and the increase in their numbers provides the food source for herbivores, which are in turn eaten by carnivores in the level above.

Since all life processes involve the expenditure of energy through the process of respiration, not all the energy produced by autotrophs is available for consumption by the herbivores. Some energy is lost through heat, and some phytoplankton die before they are consumed. Similarly, actively moving herbivores and carnivores use energy merely in locomotion. In addition, animals do not totally digest all the food they take in, and so some energy is lost through faecal material. Dead animals and plants, and faecal matter provide the energy for decomposing bacteria and for sedentary, filter-feeding animals on the ocean floor. These animals feed on the rain of organic material dropping down from above.

Marine food webs

In terrestrial communities, simple pyramids of numbers, or biomass (the weight of living tissue), demonstrate this decline in energy as one passes up a food chain. The numbers, or biomass, of primary producers exceeds that of the herbivores, which in turn is greater than that of the carnivores. Marine communities are, however, quite different. They display inverted pyramids of numbers and biomass, in which the numbers of primary producers are less than those of the herbivores, which are, in turn, less numerous than the carnivores. If, however, the production over an entire annual cycle is considered, then the production of autotrophs exceeds that of the herbivores. This reflects the rapid rates of production of individual phytoplankton, which reproduce in a few hours or days, compared with the zooplankton, which may take weeks or months to reproduce, or the smaller predatory fishes, which take a year or more to produce new individuals.

In reality, however, complex marine communities rarely consist of a simple chain of single species, with each feeding on the species in the level below. Such communities would be extremely unstable and liable to fluctuate considerably in numbers, if not collapse entirely if a single link in the chain should disappear. In practice, therefore, food webs, rather than food chains exist, with a complex inter-linkage between different members of the same community. A single species such as the flounder, for example, may feed on several different species of annelid worms and molluscs. It may compete for some of these resources with predatory starfish, eels and other fishes. The herbivorous species at the bottom of the food chain may themselves be competing for the detrital material on which they feed. To avoid competition, species at the same level have devised different strategies to divide the available resources between them. The filter-feeding

▲ **Feather stars** *are just one of many species of bottom-dwelling animal that feed on the detritus which sinks from the surface, or capture their food from the* water column. The feather star catches its food by extending its arms and trapping suspended particles by means of a sticky mucus.

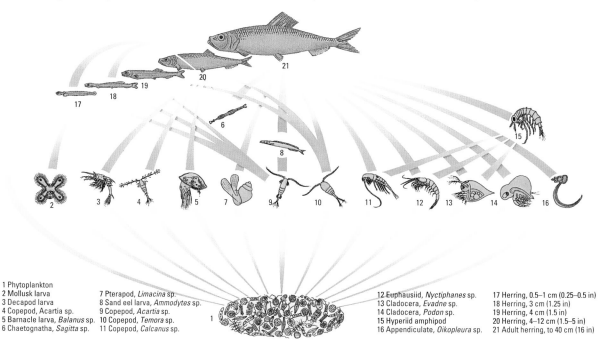

1 Phytoplankton
2 Mollusk larva
3 Decapod larva
4 Copepod, Acartia sp.
5 Barnacle larva, Balanus sp.
6 Chaetognatha, Sagitta sp.
7 Pteropod, Limacina sp.
8 Sand eel larva, Ammodytes sp.
9 Copepod, Acartia sp.
10 Copepod, Temora sp.
11 Copepod, Calcanus sp.
12 Euphausiid, Nyctiphanes sp.
13 Cladocera, Evadne sp.
14 Cladocera, Podon sp.
15 Hyperiid amphipod
16 Appendiculate, Oikopleura sp.
17 Herring, 0.5–1 cm (0.25–0.5 in)
18 Herring, 3 cm (1.25 in)
19 Herring, 4 cm (1.5 in)
20 Herring, 4–12 cm (1.5–5 in)
21 Adult herring, to 40 cm (16 in)

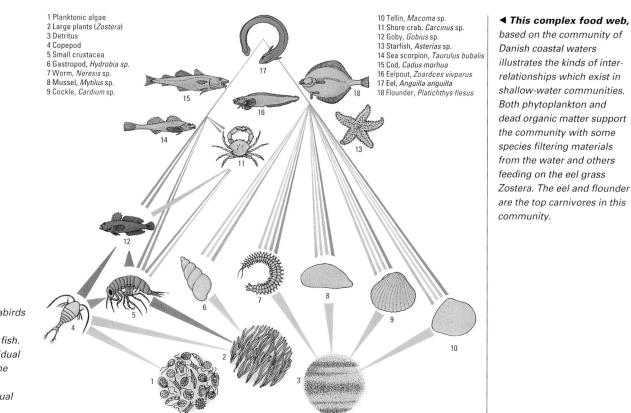

1 Planktonic algae
2 Large plants (*Zostera*)
3 Detritus
4 Copepod
5 Small crustacea
6 Gastropod, *Hydrobia* sp.
7 Worm, *Neresis* sp.
8 Mussel, *Mytilus* sp.
9 Cockle, *Cardium* sp.

10 Tellin, *Macoma* sp.
11 Shore crab, *Carcinus* sp.
12 Goby, *Gobius* sp.
13 Starfish, *Asterias* sp.
14 Sea scorpion, *Taurulus bubalis*
15 Cod, *Cadus morhua*
16 Eelpout, *Zoardces vivparus*
17 Eel, *Anguilla anguilla*
18 Flounder, *Platichthys flesus*

◄ **This complex food web,** based on the community of Danish coastal waters illustrates the kinds of inter-relationships which exist in shallow-water communities. Both phytoplankton and dead organic matter support the community with some species filtering materials from the water and others feeding on the eel grass *Zostera*. The eel and flounder are the top carnivores in this community.

▼ **The presence** of seabirds usually indicates the presence of a school of fish. The diving of one individual bird attracts others to the area and increases the chances of each individual finding food.

bivalves, for example, which are dependent on the suspended detrital material in the water, will select different particle sizes, or will position themselves in different locations, to make use of different sections of this resource.

The position of a particular species within a food web is not necessarily constant, and may change with age. Herrings, for example, feed on a wide range of different species at various stages of their life cycle. In general, the range of species eaten increases with age, as does the size of the prey.

Upwelling food chains
Areas of upwelling appear to have relatively simple food webs, with a more limited range of species in each level. The upwelling off Peru is a good example. Here, the anchovetta, from the larval stage to adulthood, feed on the various plankton in the region. Once fully grown, the anchovies, in turn, are prey to large numbers of boobies, cormorants and brown pelicans. However, this apparently simple, three-tiered food chain involves a wide range of other species as well. The anchovetta are prey to bonito and hake, as well as being an important food source for marine mammals such as seals and dolphins. The relationships between the different species in the Peruvian system are, therefore, more complex than a simple linear chain, and fluctuations in the abundance of one species are reflected in changes elsewhere in the web. The dramatic decline in productivity following El Niño years not only causes reduction in the numbers of anchovies, and hence bird species, but also affects human use of the resource.

Over-fishing represents one of the greatest threats to the stability of marine ecosystems. Many fishing techniques not only damage and destroy the environment, but the links in the complex web of feeding relationships are disrupted by the selective removal of certain species. Regrettably, our understanding of marine food chains is not great enough to enable us to predict the impact of over-fishing on other species in marine communities.

Feeding

A major feature of the marine environment is the small size of the primary producers. Many marine organisms have consequently developed a mode of feeding that is unique to the aquatic environment: filter feeding.

Filtering food

Filter feeding is the process of feeding on suspended particles, or straining food organisms from the water, which are generally considerably smaller than the animals that feed on them. One of the most extreme contrasts in size between an animal and its food is that between the blue

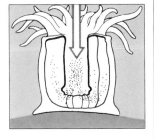

▲ ► **Condylactis passiflora**, *a sea anemone, catches its prey with the stinging tentacles that surround the mouth. The tentacles ensnare the prey and pass it into the sea anemone's digestive cavity.*

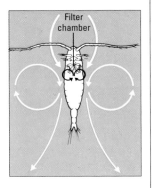

▲ ► **Calamus finmarchicus**, *a copepod extracts tiny food particles from the water by means of a fringe of hair on its limbs.*

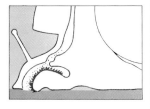

▲ *Herbivorous sea snails scrape their algal food from the surface using a tough, file-like radula moved by a muscular tongue. The smilarly equipped carnivorous whelk, right, uses its foot to hold its prey and bores through the shell.*

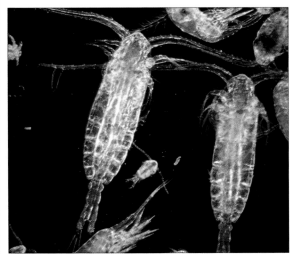

whale, which can reach lengths of over 30 metres (100 feet), and its food, the krill, a crustacean which is no more than 2.5 centimetres (1 inch) long. The whale filters these small animals from the water as it swims. This mechanism is also used by many planktivorous fish, which filter small organisms from the water current which passes over the gills. They do this by means of gill rakers – projections on the inside of the gill bars which trap the plankton inside the mouth and prevent it from being swept out with the current of water.

On a smaller scale, the herbivorous copepods filter feed by means of the rowing action of their limbs, which are fringed with hairs; the hairs strain the minute phytoplankton from the water. The related barnacles, which remain attached to rocks, sweep their limbs in and out to draw suspended particles into the mouth. Aggassiz, the French zoologist, aptly described a barnacle as "nothing more than a little shrimp-like animal standing on its head in a limestone house and kicking food into its mouth". The copepods are themselves fed on by Euphausid shrimps, which comb them from the water in the same way that copepods collect their phytoplankton food.

Sedentary filter feeders

Many sedentary and sessile organisms are also filter feeders, and the structures which they use to collect food from the water are as diverse and curious as the animal groups themselves. Some sedentary filter feeders, such as the barnacles and many tube dwelling worms, actively sweep the catching mechanism through the water to collect their food. Others use a more passive mode of food collection. Feather stars, for example, produce mucus strands or threads which float above the collecting apparatus and trap suspended particles. The mucus strands are then drawn in and rolled down towards the mouth by delicate tube feet, which line the feather-like arms.

Other animals such as the sponges, salps, many bivalve molluscs and tunicates, or sea squirts, draw a current of water into their body by means of cilia, which beat inwards, drawing water into an internal chamber. Here however, the resemblance ends, since in the case of the sponges, individual cells lining the body cavity ingest each particle, whereas in sea squirts the particles are gathered into a mucus strand which is taken into the digestive tract, and the products of digestion are absorbed through the gut wall. Sponges also provide a home for a variety of other filter-feeding organisms, such as small brittle stars and worms, which live inside the chambers of the sponge and feed on the supply of food drawn in by the host.

The volumes of water filtered by such animals are in relative terms extremely large, and their importance in cleaning the water is therefore considerable. Equally important, from the animal's perspective, is the need to keep out large and inedible particles which might otherwise clog up the filtering system. Many filter feeders have, therefore, developed primary filtering mechanisms which exclude large particles and permit only appropriate-sized materials to enter. Perhaps the most sophisticated primary filtering device is displayed by the salps, which filter the water through a net that only permits particles 1 micron in diameter to enter. This enables these animals to feed on picoplankton that are too small to be taken by most other filter-feeding animals.

Feeding with tentacles

A variation on the filter feeding theme is seen in many tentaculate animals, such as the bryozoa or sea mosses and the corals. Here, tentacles are used in feeding. In the case

▲ *Baleen whales* have lost their teeth and evolved plates of horny material which hang from the upper jaw. These baleen plates have fringes of fine, hair-like material which serves to trap the krill on which they feed.

of corals these are used to capture larger planktonic organisms, and the tentacles are armed with stinging cells, the nematocysts. These cells are used to immobilize and trap the prey before they are transferred to the mouth by the movements of the tentacles. The tentacles of the lophophorate animals, such as the sea mosses, function in a different way. Rather than trapping the prey, they serve to increase the surface area on which cilia can beat, moving water and particles down towards the mouth.

Detritus feeders

Bottom-dwelling animals are adapted so that they are able to catch the rain of detritus from above, as well as to feed on material which passes through the outstretched net of filter feeders and actually reaches the sediment. Once the faecal rain which descends from the sunlit surface of the ocean reaches the bottom, it provides the energy source for a complex community of detrital-feeding organisms, of which the bacteria are an important link. The surface sediments of any area of the ocean floor contain large numbers of bacteria which serve as a food source for small interstitial animals of many phyla. These animals are all adapted to moving between the sand grains, using the surface film of water which separates each grain. Effectively, the sediment represents a mixed organic and inorganic soup which is consumed directly by many burrowing organisms that digest the organic materials and pass out a clean stream of inorganic particles.

Bottom-dwelling sea cucumbers are more selective. They search the surface for suitable, larger-sized fragments which are picked up by the tentacles and eaten individually. Many crabs, including single large-clawed fiddler crabs, pick up individual sediment particles and brush them clean with the setae, hair-like structures on the mouthparts. These retain the edible materials and discard the sediment particles.

Small brittle stars often lodge two of their five arms underneath a rock, or in a suitable crevice, and sweep the area in front of their retreat with the remaining three arms. Modified tube feet on the undersurface of the arms produce a sticky mucus which is used to collect food particles; these particles are then passed back along the underside of the arms to the central mouth.

Grazers

Grazing, as a mechanism for feeding in oceans, is restricted to shallow coastal waters, where single-celled algae grow over the surface of the bottom, and larger seaweeds and seagrasses provide a source of larger plant food.

The algal film which covers all surfaces in shallow waters provides a rich source of food for many species which have developed different mechanisms for scraping this plant production off the surface of the rocks.

The radula of molluscs represents one such adaptation. Small teeth are arranged in rows over the surface of a muscular tongue, forming a file-like structure. This is rasped over hard surfaces, removing algae and other debris. Because of the constant wear on the file-like surface, these animals have evolved a mechanism for continuous replacement of the radula. Sea urchins, meanwhile, have developed a system of muscles and supporting ossicles, the Aristotle's lantern, in which are set five teeth. The teeth can be rotated in and out and used, like the radula, to scrape unicellular algae from rock surfaces.

Grazing animals are more limited in numbers and diversity than filter feeders. However, among them are included a variety of small herbivorous fish which crop the algae that form a short turf in many tropical reef systems. Such herbivorous fish are important in maintaining diversity in the coral reef ecosystem because their action reduces the growth of the algae and prevents them from smothering the delicate, juvenile corals.

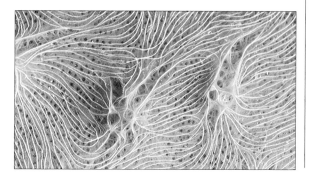

◀ *The entire body* of a sponge is used to filter food. Water is continuously drawn into the body by means of cilia, the microscopic food particles are extracted and the water passed out through larger openings.

Predators and Prey

The predators of the oceans differ considerably from those found on land. A wide range of sessile (fixed) animals abound in the shallow waters of the continental shelves. These provide the opportunity for predation which resembles more the grazing of herbivores than the active chase and capture associated with predator-prey interactions on land.

Browsing on sedentary, or sessile, animals is widespread in marine benthic communities, reflecting the abundance of attached animals, particularly in shallow water environments. Corals are "grazed" by crown-of-thorns starfish, *Acanthaster plancii*, which evert their stomachs over the surface of the coral colony, digesting out the polyps from the hard, calcareous skeleton. In contrast, the strong jaws of the parrot fish are adapted for crunching off coral branches, which are then ingested and ground up by modified plate-like teeth at the back of the jaws. Many smaller animals, including some molluscs, browse on the surface of coral colonies by delicately extracting individual polyps from inside their protective skeletons. A wide variety of nudibranch molluscs are specialized feeders on sea pens, sea whips and sea fans. Some sea slugs even store the stinging cells collected from their coral prey and use them for protection against other predators.

Because sessile animals cannot escape from their predators by moving away, many have developed methods of deterring attack. These include the development by soft-bodied animals of strong, calcareous skeletons into which they can retreat when danger threatens. However, despite their strong protective shells, bivalve molluscs are still vulnerable to predation. Many gastropod molluscs have modified their radula to form a drill, which is used to pierce a neat circular hole through the mollusc's shell. Once through the shell, the gastropod secretes digestive enzymes into the mussel, clam or oyster, and sucks out the resulting protein soup.

Starfishes such as *Asterias* have developed another mode of attack. Grasping the two valves of the bivalve shell with opposing arms, they pull the two halves slightly apart and evert the stomach into the shell. Then, like the oyster drill, they secrete enzymes into the shell, digesting the mussel externally and absorbing the digested food through the stomach wall. Free-swimming scallops, which move by rapidly closing the two halves of the shell and expelling a jet of water from the mantle cavity, react violently to the chemicals produced by *Asterias*, moving rapidly away when a starfish approaches.

Active hunters

More traditional predator-prey interactions are displayed by the vertebrates. Fish predators display classic strategies which involve chasing active prey species. Thresher sharks will hunt in packs, using the long whip-like tail to herd schools of mackerel and herring into tight shoals, before moving quickly in to feed. Many small angel fish in coral reefs avoid predation by remaining close to a branched *Acropora*, into which they dart rapidly when threatened. The brightly coloured anemone fish adopt a similar strategy, retreating to the protective cover of the anemone's stinging tentacles when threatened.

Active predators often hunt by sight, and the coloration of potential prey is designed to reduce their visibility under different light intensities. Many planktonic organisms near the surface are transparent, making them more difficult to detect. Those at depth are often red, but appear black, since red light fails to penetrate as deep as other wavelengths of light. In the lighter surface waters, many fish species are countershaded, with darker blue or grey dorsal surfaces and pale or silvery bellies. This reduces their visibility when viewed from below against

▼ **Many surface schooling** fish, such as these mackerel, are counter shaded with darker dorsal surfaces and lighter, silvery undersides. This pattern of counter shading provides protective camouflage from above and below. The silver lower surface blends in with the sunlight from above, making it harder for predatory fish swimming below to spot them, while the darker dorsal surface blends with the deep blue of the ocean water providing protection from seabirds.

the sunlit surface and when viewed by seabirds from above against the darker ocean.

The cephalopods, the squid, cuttlefish and octopus are all predatory animals, and while octopuses often sit and lurk in dens waiting for unsuspecting snails and fish to approach, the squid and cuttlefish are active predators of crustacea. Some species of squid will direct a gentle flow of water onto the sand surface to expose hidden shrimps, which are then seized with an elongated pair of tentacles. Cuttlefish, squid and octopus, when threatened by larger predators, will avoid capture by expelling a large cloud of ink, darting rapidly away and immediately changing their colour pattern. These various responses confuse the predator and enable the cephalopod to escape.

Patient predators
Sit-and-wait predators are also numerous in the seas and oceans, particularly at depth, where the numbers of potential prey are few. Such predators are often camouflaged themselves, relying on the inability of the prey to detect their presence to bring them within reach of capture. The deep-sea angler fish encourages smaller fish to swim within striking distance by means of a lure, a small structure on the angler's head which is either luminescent, or may resemble a small worm. Prey approaching in search of a meal find themselves under attack.

As a defence against predation, many species of marine organisms have developed toxic or noxious chemicals and adopted brightly coloured patterns to warn potential predators of their characteristics. Many of the venomous sea snakes are striped black or dark blue and white, although this does not prevent them from being a major item in the diet of tiger sharks. Sea snakes such as *Laticauda colubrina* use their highly toxic venom to subdue the eels on which they prey. Females are much larger than the males and feed on large congrid eels in deeper water, while the smaller males capture moray eels.

The Angler Fish (*Lophius piscatorius*)
The sit-and-wait predatory angler fish is perfectly camouflaged against the bottom, where it lurks in wait for unsuspecting prey. When a fish is attracted within reach by its luminescent lures, the angler quickly opens its mouth sucking in its unfortunate victim. The angler's teeth are well developed, ensuring prey cannot escape.

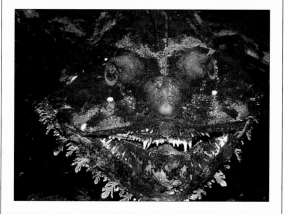

▲ The great white shark is among the most fierce of active predators. Sharks have an acute sense of smell and are attracted to wounded or dying fish. As sharks cannot chew, they tear chunks from larger prey by grasping a mouthful and then twisting in the water to gouge out a piece of flesh. Their razor sharp teeth are set in rows parallel to the jaw and are continuously replaced from behind as the older teeth at the front are either lost or damaged.

Reproduction

Reproductive strategies in marine animals range from the "broadcast" spawners, which produce millions of small eggs, to the more conservative species, which produce only a few large eggs.

The broadcast spawners rely on the survival of a few individuals from the many millions produced. Most are lost either through predation, both at the egg and larval stages, or through being carried by currents into unsuitable areas. Differences in larval survival from year to year can vary enormously and, therefore, dramatically affect the adult population of these animals.

Species that produce fewer, larger eggs, generally rely on the fact that their young, when hatched, are larger and can therefore avoid the intense predation which occurs in plankton communities. The adults of such species may also protect the eggs, as in the case of anemone fish and many molluscs. These species also tend to show less wide fluctuations in numbers since the mortality rate of their juveniles is comparatively lower.

Mass fertilization

Many sessile animals are broadcast spawners. Fertilization of the eggs occurs as a result of chemical stimuli produced by the first ripe individual to release its eggs. In giant clams, for example, each individual is hermaphrodite (having both male and female reproductive cells). On spawning, the ripe individual first produces sperm, which are released in milky clouds into the water. Chemicals in the sperm then stimulate neighbouring clams to begin spawning. Following the production of sperm, eggs are produced. The timing is such that the eggs are released just as neighbouring individuals start producing sperm, thus maximizing the chances of cross-fertilization. A single ripe giant clam may produce many millions of eggs which, if successfully fertilized, will hatch into swimming, planktonic larvae.

Sessile animals use the planktonic larval phase of their life cycle for dispersal. The larvae float passively in the plankton community, where they form the food for larval fish and other predators, such as arrow worms. Since they are carried passively by the currents, and are capable of only limited movements in the water column, they depend on the currents to bring them into a suitable area for settlement when the larval stage is over. To improve the survival rate, many species spawn simultaneously at neap tides, periods in the lunar cycle when tidal currents are at their lowest. The length of larval life, therefore, represents, for many species, a compromise between the need to remain in close proximity to the areas of suitable habitat, which requires a shorter larval life, and the need to disperse over longer distances and colonize new habitats, which demands a longer larval, dispersal phase. In species with longer larval life stages, the larvae often feed, grow and undergo a series of changes while they are still members of the plankton community. In contrast, species with a short larval life tend not to feed and to be merely transitory members of the plankton community, using the larval stages merely as a mechanism for dispersal.

Because they share the common problem of having to remain afloat in the plankton during the dispersal phase, the larvae of different groups of marine organisms may share similar features. They often have, for example, bands of cilia, which allow limited movement (and which are also important in feeding), or they may have a relatively large surface to volume ratio which slows their rate of sinking. In addition, these larvae often have eyespots to allow them to orient towards the sunlit surface.

On settling, the larvae of sessile species must choose a suitable site on which to anchor themselves. For this reason, they have chemical and tactile sensors which enable them to select suitable locations. Some have special glandular areas which secrete sticky substances for initial

Invertebrate Reproduction
Ocean invertebrates reproduce in a variety of ways. Simple animals reproduce by budding, whereby a new individual grows from the body of the original. More complex forms of reproduction involve fertilization of the eggs. Some species, such as crabs, pass through planktonic larval stages, important for dispersal as the larva are moved by surface currents from one area to another.

Seaslug, Doto fragilis, *and egg spirals*

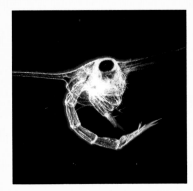
Larva of corysteid crab

Budding of soft coral, Alcyoonium sp.

Ribbon-like bands of seaslug eggs

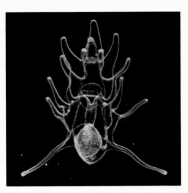
Larva of starfish, Asterias rubens

▲ *Turtles* mate at sea, before the female moves inshore to find a suitable beach. Here she buries the eggs in the sand, out of the rich of high tide.

▼ *Hermaphroditic* giant clams reproduce by releasing thousands of sperm and then eggs into the water. The sperm and eggs fuse to develop tiny larvae.

attachment. Site selection is critical to the survival of the adult. If the larvae settle in an unsuitable area, growth may be retarded or the animal may die.

Finding a mate

Reproduction in mobile animals of the seas and oceans involves, first of all, finding a mate. This can present marine organisms with formidable problems, particularly in the deeper ocean, where individuals are often widely dispersed. Marine animals live in a three dimensional world of water, often under conditions of little or no light, and where eyes are of little use in finding mates. Some animals, therefore, have evolved chemical means of finding and attracting members of the opposite sex. Others, such as the angler fish, have solved this problem in a different way. The males are small, and are permanently attached to the females.

Several species of coral reef fishes, such as the groupers, form spawning aggregations, returning each year to particular spawning sites where the animals, which are normally widely dispersed over the reef system, produce their eggs and sperm simultaneously. In many island environments, these spawning sites are located in areas where the local current patterns will carry the eggs and larval fishes in a circle so that by the time they are ready to become reef dwellers they have been brought back to the vicinity of the reef environment.

In tropical coastal environments, mangroves and sea-grass beds form important nursery areas for many species of fish and shrimps; the complex fronds and root systems provide a multitude of hiding places. Many species of smaller, tuna-like fishes, which spend their adult lives offshore, will come inshore to spawn in close proximity to mangrove systems. Here the larval fishes hatch before moving gradually offshore. Penaeid shrimps, which inhabit soft-bottom areas of deeper water, also use mangrove areas as nurseries, where the juveniles feed and develop during the early stages of life.

Invertebrate Locomotion

▶ **The squid**, **Pyroteuthis**, has a muscular cavity (1) that completely encloses its gills (2). To propel itself the squid expands the muscular cavity, sucking in water through a wide slit at the front of the body (3). Water cannot flow up the funnel (4), because of a one-way valve. To move the squid contracts the muscular cavity, forcing water out of the funnel. The overlapping edges of the split (5) ensure the water is forced out in as narrow a stream as possible, so creating better thrust.

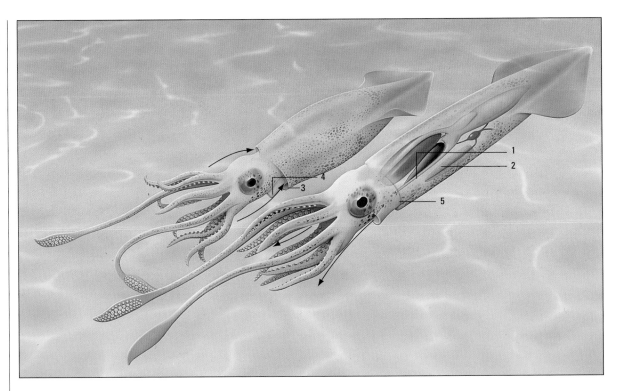

Marine animals appear in all sorts of shapes and sizes. The wide diversity of body forms reflects the many varied lifestyles, and in particular the many modes of locomotion. Pelagic species are often passive, being carried by the surface water currents or, in the case of animals with floats such as *Physalia*, the Portuguese man-o'-war, drifting with the wind. Many invertebrate groups in the marine environment are sessile, or sedentary, moving only sufficiently to retreat within a burrow or protective shell, although such movements may be rapid, particularly when the animal is threatened by a predator. Many species are also soft bodied, and the problems of locomotion without a supporting skeleton have been solved by marine animals in a diversity of ways.

In benthic animals, some form of creeping characterizes many surface animals, of which the gastropods, or snail-like molluscs, are familiar examples. These animals have a large muscular foot, and move by means of waves of contraction, which pass along the undersurface of the foot, combined with cilia which operate in a film of mucus secreted by the foot itself. Movement by means of cilia, small hair-like structures, is common in many groups, including the flatworms; these can also swim by means of flapping movements of the body, and move more rapidly by muscular contractions of the entire body. The most sophisticated locomotory system is perhaps that of the echinoderms, the starfish, urchins and their relatives. These animals possess small suckered tube feet which can be extended forwards, attached to the substrate ahead of the animal, and then shortened to pull the body along. The brittle stars have developed jointed arms which can be flexed from side to side in snake-like movements. These arms are covered with short spines, and by curling them around projections on the bottom, the animals can lever themselves over the surface of the substrate.

Locomotion beneath the surface of the sediment involves, for larger animals, some form of burrowing. Many soft-bodied worms burrow by extension and contraction of the segments of the body which operate using a hydrostatic skeleton. The individual segments of the worm's body contain fluid which is incompressible, so that when the circular muscles of the body wall contract the longitudinal muscles relax and the segment is extended, since the overall volume cannot change. This forces the forward end of the body into the sediments. Then contraction of the longitudinal muscles causes relaxation of the circular muscle and the segment becomes short and fat, anchoring the animal in its burrow. Sequential contractions passing along the length of the animal move it through the sediment. Locomotion in this way requires a great deal of energy and hence many burrowing animals live in permanently constructed burrows.

Sedentary animals
Sessile animals, which live attached to the surface of the substrate, are incapable of moving from one place to another, but sedentary animals, such as feather stars which normally remain attached in one position, may change position if the food supplies are insufficient or if the wave action is too strong. To do this, they detach their basal cirri and sweep the arms up and down in the water, swimming in a rather clumsy manner for short distances before relaxing, sinking to the bottom and re-attaching with the jointed cirri. Some sea anemones will also move by thrashing the body column from side to side. Some slowly creeping sea slugs such as the Spanish dancer, when threatened, open the mantle and move the flaps to swim away. This also displays a vivid pattern of coloured eyespots which startles the predator, allowing the animal to make good its escape.

In comparison with vertebrates, the movement of soft-bodied animals is slow, but the diversity of forms of locomotion is considerable, and the structural adaptations of the animals, numerous. A major advance on the soft-bodied animals is seen in the crustacea, which, like their terrestrial relatives the insects, have jointed skeletons that can be moved by means of opposing pairs of muscles. These animals have developed a wide range of walking and swimming styles based on the use of jointed legs and swimming paddles. These styles range from the stately walk of the sea spiders to the scuttling of crabs and the frenetic swimming of small copepods, which kick their jointed legs to move upwards in the water column.

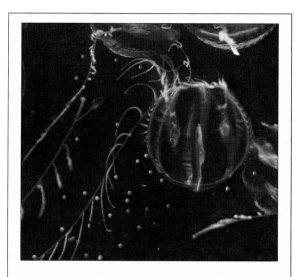

Invertebrate Swimming

Comb jellies swim by using comb-like plates called ctene (1). Each of the eight ctene is covered with rows of rhythmically beating cilia which propel the animal. The ctene beat away from the mouth (2), thus the animal swims mouth first. Simple muscles provide some control over the tentacles (3) which are armed with sticky cells to catch prey.

▼ **Molluscs** such as this cone shell, Conus aulicus, move by means of waves of muscle contraction which pass along the foot.

Locomotion is aided by hair-like cilia and mucus, which is secreted by special glands found on the undersurface of the foot.

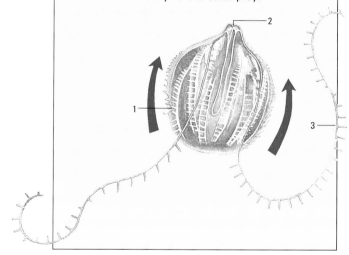

◄ **The jellyfish,** Cassiopeia, is commonly found in the warm waters of Florida and the West Indies. To swim it contracts the flattened round bell, which may grow to more than a metre (3 ft) in diameter, forcing water out from underneath. The resulting movement is a jerky upward form of locomotion.

Active Swimming

▲ **The hatchet fish,** Argyropelecus*, inhabits the deep sea where food is scarce. Although small in size, 9 cm (3.5 in), its mouth can gape widely, enabling it to seize large prey which are held firmly by the backwardly pointing teeth.*

▲ **The tuna** *is a muscular, long-distance swimmer whose capacity for speed is indicated by its torpedo shape, and lunate tail and pectoral fins. Tuna are capable of swimming up to 240 km (150 miles) in a day.*

▲ **The flattened body** *and camouflaged pattern of the plaice,* Pleuronectes platessa, *are adaptations for life on the seabed. Larval plaice are shaped like normal fish. However, as the fish develops, the eyes move to the upper surface as the body becomes flattened.*

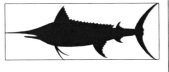

▲ **The marlin** *are aggressive, fast-swimming predators which can achieve a length of over 4 m (14 ft). Like the tuna they are streamlined with crescent-shape tails for high-speed swimming. The spear-like bill is used to maim prey as the marlin shoot through a shoal of slower fish.*

The predominant mode of locomotion among marine vertebrates is swimming, and in particular powered swimming, which depends upon a source of forward thrust, the ability to steer and to change course, and the ability to stay afloat. Aquatic vertebrates have achieved large size, in part due to the buoyancy or support provided by the water in which they live. Nevertheless, animals are denser than water, and thus have a tendency to sink unless they counteract this by swimming upwards. Sharks, rays and other elasmobranch fishes reduce their density by storing oils in the liver. However, they will still sink unless they swim continuously.

A major advance shown by the teleost fishes is the development of the swim bladder – an air- or gas-filled sac lying below the vertebral column which can be used to achieve neutral buoyancy. Gas can be secreted into or reabsorbed out of the gas bladder according to the depth of the fish, and a teleost fish with neutral buoyancy neither sinks nor rises to the surface. This allows the animal to conserve energy, as it can maintain its position merely by small adjustments of the fins which counteract the movements of the surrounding water.

To achieve efficient forward movement, an actively swimming animal requires a source of power located behind the centre of mass of the animal, hence most active swimmers have developed some form of tail. This varies from the horizontal flukes of the whale to the homocercal tail of the teleost fish, which has lobes of equal size above and below the mid-line of the body. In contrast, the tail of sharks is heterocercal – the dorsal lobe is larger in size than the ventral lobe. Such a tail not only provides the power for forward movement, but also provides lift to the rear of the animal. This lift counteracts the tendency of the animal to sink. However, if it were not itself counteracted, this lift would result in the animal swimming tail upwards. Thus the paired pectoral fins at the front of the shark are held at an angle to the water, providing lift to the front end of the animal which balances the lift provided by the tail.

Since the teleost fish achieve neutral buoyancy by means of the swim bladder, the need for upward movement is reduced, and both lobes of the tail are equal in area. The fossil reptile-like Ichthyosaurs had tails which were the reverse shape to those of sharks, the ventral lobe was greater in area than the dorsal one. In these animals, which were air breathing, the problem was not one of sinking, rather of floating to the surface. Their tail shape was an adaptation to provide a downward thrust, enabling the animal to move more easily below the water surface.

Movement and stability

The paired fins of fish are used to steer the animal and to prevent pitching, while the median fins, the dorsal and anal or ventral fins, prevent rocking from side to side. By the combined use of these fins, stable forward movement is achieved. The tail or caudal fin is moved from side to side by the contraction of the segmentally arranged muscle blocks, or myotomes, on each side of the animal. As blocks on one side contract, the tail is moved towards that side; when the opposing blocks contract, the tail moves the other way. Recovery after each contraction is aided by the vertebral column which has a tendency to resume a straight, linear shape. In whales the movement of the tail fluke is achieved by long muscle blocks which lie alongside the vertebrae and work in an up-and-down movement rather than from side to side. Movements of the tail provide rapid forward movement, and one problem facing high-speed swimmers is one of braking. Advanced teleost fish have shifted the position of the rear pair of pelvic fins, such that they lie immediately below the pectoral fins, which have themselves been shifted to a higher position on the body of these animals. By extending all four fins simultaneously, four-fin braking is achieved, enabling the animal to stop abruptly.

Not all active swimmers rely on tails for forward movement. The cephalopods have developed jet propulsion for short, rapid movements and lateral fin folds for

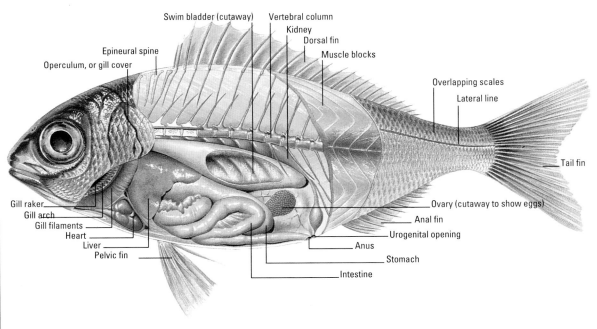

Swim bladder (cutaway) · Vertebral column · Kidney · Dorsal fin · Muscle blocks · Epineural spine · Operculum, or gill cover · Overlapping scales · Lateral line · Tail fin · Gill raker · Gill arch · Gill filaments · Heart · Liver · Pelvic fin · Intestine · Anus · Urogenital opening · Stomach · Anal fin · Ovary (cutaway to show eggs)

▲ **True bony fishes** *are diverse in shape and type of locomotion. They all possess a tail with equal upper and lower lobes providing no up or down thrust. These fish achieve neutral buoyancy by adjusting their density using the swim bladder. The fish can expand or contract the bladder by secreting gas into or absorbing gas out of it. so adjusting the volume and external pressure and counteracting the tendency to sink or float to the surface.*

Evolution of Swimming

◄ *The Ichthyosaurs (A)*
were a group of marine
reptiles resembling modern
porpoises in generally body
form although their tails
were vertical rather than
horizontal and the enlarged
ventral lobe provided a
downward thrust to aid in
diving. Modern whales (B),
like the Ichthyosaurs are air
breathing, in contrast
however, the flukes of a
whale's tail are arranged
horizontally and the tail is
moved by up-and-down
flexure of the backbone
rather than by side-to-side
movements. Modern sharks
(C) like Ichthyosaurs and
whales are streamlined,
however, the upper lobe of
a shark's tail is larger than
the lower lobe providing an
upward thrust and
countering the tendency of
the animal to sink in the
water. The plesiosaurs (D),
another extinct group of
reptiles, although adapted
for active swimming, moved
by means of large paddle-
shaped limbs like modern
sea turtles rather than by
means of a tail fin.

more sustained swimming. In these animals, water held in the mantle cavity is expelled rapidly through a specially modified siphon, shooting the animal backwards in the water. The torpedo-shaped squid have triangular fins at the rear of the body which are moved like the "wings" of eagle rays to provide power for forward movement.

In contrast, the more rounded cuttlefish have a continuous fin fold around the edge of the body. Waves of muscular contraction pass along this to provide forward movement. This form of swimming is paralleled in some flattened, bottom-dwelling fishes such as plaice and flounder which have modified dorsal and anal fins and in which the tail fin hardly functions in locomotion. Many smaller teleost fishes which hide in reefs, or among seaweed, swim using the paired fins in a rowing movement, rather than using the tail, while the elongate eels swim by means of lateral flexure of the entire body. This anguiliform mode of locomotion is also used by sea snakes, which also

have a laterally compressed paddle-like tail. Modification of the fore-limbs to form strong paddles providing forward thrust is seen in extinct animals, such as the Plesiosaurs, and in modern sea turtles. In turtles and in penguins the fore limb is paddle shaped and powerful thrusts of these limbs provide the forward movement, while the hind limbs serve only for steering. The streamlined shape of the turtle's carapace, and indeed of the body of all active swimmers, provides a smooth surface over which water flows in smooth layers, preventing turbulent currents which would interfere with efficient swimming. The high speed swimmers such as albacore tuna have a smooth, torpedo-shaped body and the large, lunate caudal fin enables these animals to achieve high speeds. Unlike most fishes, the red muscle of tunas contains haemoglobin-like chemicals which store oxygen, and a network of blood vessels retains heat in the muscle, enabling these animals to swim at high speeds for extended periods of time.

Return to the Sea

Since all life evolved in the sea, although under conditions quite different from those of the present marine environment, the invasion of land by animals led to the evolution of numerous special adaptations: for supporting the body weight in air; for locomotion using limbs, not fins; for breathing air, rather than absorbing dissolved oxygen; for producing eggs or young resistant to desiccation; for surviving in air, where the conservation of body moisture becomes a priority; and for coping with the wide range of diurnal and seasonal temperature extremes experienced on land. Having successfully adapted to such conditions, it is hardly surprising that the number of animal groups which have made the transition from the land back into the marine environment is limited.

Nevertheless, a number of living and extinct vertebrate groups have successfully made this transition, including representatives of the mammals and the reptiles. Among the birds, only the penguins might be considered marine, having lost the ability to fly and modified their forelimbs for underwater swimming. In general, animals which have made the transition back to the sea only relatively recently show few major structural differences from their land-based relatives. The marine otters, for example, spend their lives in kelp beds, feeding on molluscs which they open by smashing them against an anvil stone held on the chest. Even though these animals sleep in their kelp bed habitats, they show little difference from their amphibious cousins the freshwater otters.

The salt water crocodile, *Crocodylus porosus*, a native of Southeast Asia and New Guinea, inhabits estuarine areas as an adult; however, during its subadult life many individuals migrate along the coast in search of territories not occupied by other individuals. During such dispersal, individuals may cover long distances, and one individual is known to have arrived in Fiji during the last century, presumably from the Solomon Islands, where the nearest resident population occurs. These animals swim using powerful side to side strokes of the greatly enlarged tail.

Among the living marine reptiles are two groups: the sea snakes, abundant in the Indo-west Pacific, and the sea turtles, which are mainly tropical and subtropical in distribution, although some species may be found in cooler temperate waters. The sea snakes are more adapted to the marine environment than the iguana and crocodile, since all species have a flattened, paddle-shaped tail. The majority give birth to a few live young, which are immediately able to swim and feed for themselves, and most species never come on land. The laticuadine sea snakes, in contrast, come on land regularly, to lay their eggs and digest the eels on which they prey. These animals are host to ticks and lung mites, in contrast to the more marine species, which are often host to colonies of bryozoa and stalked barnacles which grow on their surface. The marine species are predators of fish or fish eggs, and all species are highly venomous, being related to the terrestrial kraits and cobras. The turtles, although they display considerable adaptation for locomotion in the marine environment, are more tied to land and must return there to lay their eggs in nests excavated at the head of sandy beaches.

Mammalian hair and insulation

The polar bear, like the otters, has retained its mammalian characteristic of hair. However, hair is an inefficient insulator in water, since it becomes wet and water penetrates to the surface of the skin. The insulation properties of hair depend on the layer of air trapped between the individual hair fibres; once wet, a terrestrial mammal loses heat rapidly. Although most mammals swim well, they must leave the water and groom and dry the fur in order to maintain its insulation properties. In contrast, truly marine mammals such as the seals, whales and dolphins rely on the subcutaneous fat or blubber for insulation. Although the fur seals have retained their fur, they also have a thick layer of fat for insulation, and most seals and sea cows have lost their dense coat, retaining only a sparse covering of body hair and sensory bristles around the mouth. Seals

▶ *The polar bear* of the Arctic swims using its powerful limbs but spends much of its time on land or on the ice. Although the long fur is an efficient insulator on land it provides little protection against the cold when the animal is in the water. Since the polar bear shows few adaptations to a marine existence it is likely that it has adopted its semi-amphibious mode of life only relatively recently.

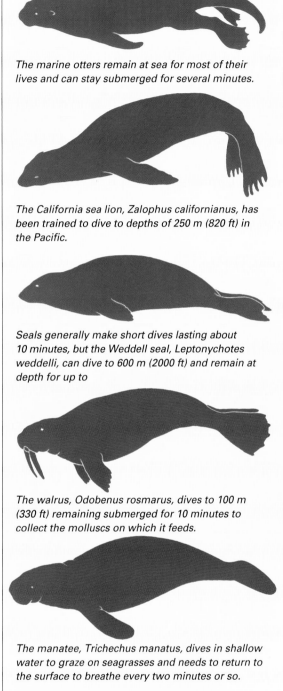

Underwater Expertise

Marine mammals dive and swim easily, however, their capacity for extended dives varies considerably.

The marine otters remain at sea for most of their lives and can stay submerged for several minutes.

The California sea lion, Zalophus californianus, has been trained to dive to depths of 250 m (820 ft) in the Pacific.

Seals generally make short dives lasting about 10 minutes, but the Weddell seal, Leptonychotes weddelli, can dive to 600 m (2000 ft) and remain at depth for up to

The walrus, Odobenus rosmarus, dives to 100 m (330 ft) remaining submerged for 10 minutes to collect the molluscs on which it feeds.

The manatee, Trichechus manatus, dives in shallow water to graze on seagrasses and needs to return to the surface to breathe every two minutes or so.

▲ **Despite its name** the northern fur seal, Callorhinus ursinus, here photographed on St George Island in the Pribilof Archipelago, Alaska derives most of its insulation from the thick layer of blubber beneath the skin.

▼ **The sea snake,** Laticauda colubrina, has a flattened paddle-shaped tail to aid in swimming. Unlike most sea snakes, which never come on land and give birth to live young at sea, this species prefers to stay near-shore.

are of two groups, the eared and earless seals, while the Arctic walrus is considered to be separate from both groups. In the eared seals, the hind limbs are held together when swimming and moved in an up-and-down manner like the flukes of whales. On land, the limbs can be rotated forwards and the animals move by means of a clumsy but nevertheless effective gallop. In contrast, the hind limbs of the earless or true seals cannot be rotated forwards; in the water these are used in a side to side manner like the tails of fishes. On land these animals must drag themselves along using the short forelimbs. Seals come on land to breed and have their young, which, in the case of earless seals, are born in such an advanced stage of development that in some species they can swim within a few hours of birth. In contrast, the young eared seals are born at a much earlier stage of development and must spend a long time

on land before taking to the water. These differences between the two groups suggest that the earless or true seals returned to the sea before the eared seals.

The sirenians or sea cows, which include the living dugong and manatees, are an unusual group of aquatic mammals which inhabit estuarine and shallow coastal seas of the tropics. Although distantly related to elephants, these animals are highly adapted for an aquatic existence. The skin is generally hairless, the hind limbs have been lost, and the end of the body has developed horizontal tail flukes like those of whales and dolphins. These animals never come onshore, and are vegetarian, grazing on seagrass meadows and living in social groups or herds. Unlike the whales and dolphins, however, they are unable to dive for extended periods and must return to the surface every five minutes or so to breathe.

Whales and Dolphins

Among the living groups of animals which have returned to the sea from a terrestrial mode of life, the whales and dolphins are the most specialized. Their physical adaptations to the marine environment are now so complete that they are incapable of surviving out of water, and their degree of adaptation suggests that the group must have returned to the sea around 65 million years ago at the beginning of the Tertiary period. Nevertheless, they are undoubted mammals, being warm blooded, with a four-chambered heart, air breathing and giving birth to live young which are suckled by the mother and tended and protected by all members of the pod, or family unit.

The high degree of specialization of whales, which includes the loss of hair and hind limbs, the development of powerful horizontal tail flukes moved up and down by flexure of the vertebral column, places the whales and dolphins in a distinct order of mammals: the Cetacea. The living representatives of this group can be divided into the Odontoceti, or toothed whales, which generally include the smaller species; and the Mysticeti, or baleen whales, which have lost the teeth and replaced them with horny fringed plates which are suspended from the upper jaw.

When feeding, the baleen whales swim through the plankton with the mouth open, then close the jaws and raise the tongue and floor of the mouth, forcing water out through the sides of the jaws. The krill on which they feed are trapped by the fringes of the baleen plates. A single mouthful may contain no more than a few pounds of food. A large whale feeding in this way may harvest two tonnes of krill a day.

Diversity of Size

The blue whale (above), Balaenoptera musculus, *is not only the largest whale, but also the largest animal to have ever existed on Earth. Individuals may exceed 30 m (100 ft) in length and weigh 150 tonnes. In summer blue whales may eat 3 tonnes of krill a day.*

The sperm whale (below), Physeter catodon, *is the largest of the toothed whales, reaching 18 m (60 ft) in length. They dive to considerable depth in search of their giant squid prey, and have been found tangled in submarine cables at depths of 1100 m (3600 ft).*

The bottle-nosed whale, Hyperodon ampullatus, *is found in the Atlantic, migrating southwards in the winter. It reaches a length of around 10 m (33 ft) and possesses only two teeth. The whale becomes lighter in colour as it ages.*

The pilot whale, Globicephala melaena, *is found in the Atlantic, Pacific and Indian oceans, and is around 7 m (22 ft) long. It moves in pods or schools and feeds on both squid and fish. It has a very distinctive rounded forehead and a deep dorsal fin.*

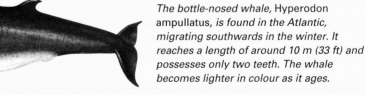

The common dolphin, Delphinus delphis, *grows to around 2 m (8 ft) and has a distinctive beak which carries an astounding 200 teeth. Distributed throughout warm and temperate waters, many have been trained successfully.*

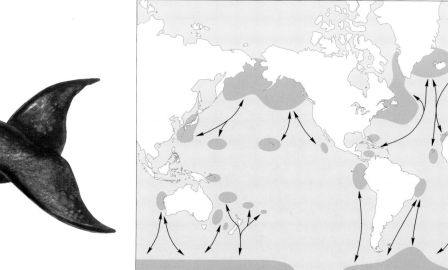

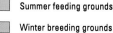

Summer feeding grounds

Winter breeding grounds

▲ **The migration** of *humpback whales is generally north–south throughout all the world's oceans. During summer months, the humpbacks remain in high-latitude, plankton-rich feeding grounds, heading to warmer-water breeding areas during winter months.*

In contrast, the toothed whales often have large numbers of rather simple, conical teeth forming an extremely efficient fish-catching device, and they are active predators of fish and squid. Some have reduced the numbers of teeth to a single pair, while others may have as many as 200. The largest of the toothed whales is the sperm whale, which may reach 20 metres (60 feet) in length and which dives to considerable depths to capture the giant squid on which it feeds. Smaller whales and dolphins may use a sonar system to detect their prey and for navigation in low light intensities. High-frequency sound, in the form of clicks, is beamed out through the melon, a raised lump-like structure of the forehead, and reflected sound is detected through the enlarged lower jaw.

The complex songs of humpback whales are believed to be part of a signalling system which enables individuals to detect each other at considerable distances and to recognize members of the same family unit. Whales and dolphins are highly social animals occurring in family units, with members of the group sharing responsibility for protecting the young which are born singly, tail first. Gestation is long, around 11 months for baleen whales and more than a year for toothed whales. The young are born at an advanced stage of development since they must swim immediately to the surface to take their first breath. Early growth is rapid; the fin whale calf, for example, which is around 7 metres (20 feet) long at birth, measures around 14 metres (40 feet) by the time it is six months old.

The diving ability of whales is remarkable considering that they are air breathing. Before a deep dive, a whale empties the lungs completely, thus eliminating any risk of the bends. To reduce oxygen consumption during the dive, only the heart and brain receive a continuous supply of oxygenated blood. The powerful swimming muscles of these animals are rich in oxygen, stored in myoglobin – a red pigment similar to the haemoglobin of the blood, providing oxygen to the muscle tissue during the dive.

Recovery from exploitation

Whales are found in all oceans of the world, and many larger species undertake regular migrations from breeding to feeding grounds, following the seasonal cycles of productivity in the ocean basins. Smaller species, such as the Ganges River dolphin, may even inhabit large freshwater river systems in the tropics, and in many such areas these animals are an important subsistence food for local communities. The level of harvest in such areas is low and the impact on the populations is generally not great. In contrast, the history of Antarctic commercial whaling is one of unmitigated greed and over-exploitation which has resulted in the severe depletion of many of the larger species. The blue whale, once believed to number in excess of 20,000 individuals, now numbers only around 500. In recognition of these problems, a moratorium on commercial whaling has been generally agreed under the auspices of the International Whaling Commission. The slow rate of growth to sexual maturity and reproduction of these large mammals means that it will be many decades before we know whether or not the species are recovering. It has been estimated that under proper management these resources could have produced a sustainable yield of two million tonnes of protein annually.

Birdlife of the Oceans

Hunting Techniques of the Seabirds

Aerial piracy
Skua

Aerial pursuit
Skua

Surface plunging
Tern

Surface plunging
Brown pelican

Surface plunging
Gannet

Surface plunging
Tropic bird

Dipping
Frigate bird

Dipping
Frigate bird

Dipping
Gull

Skimming
Skimmer

Surface filtering
Cape pigeon

Scavenging
Gull

Surface seizing
Albatross

Surface seizing
Phalarope

Pattering
Storm petrel

Hydroplaning
Prion

Pursuit plunging
Shearwater

Pursuit diving
Diving petrel

Pursuit diving
Cormorant

Pursuit diving
Scoter

Pursuit diving
Auk

Pursuit diving
Penguin

Probably the most truly oceanic of the 285 species of seabirds are the albatrosses, some species of which may spend up to nine months at sea and which cannot walk on land. Their narrow wings are of considerable length in proportion to their width and are too inflexible to maintain flapping flight, making these birds extremely efficient gliders. By turning upwind they obtain maximum lift, rising effortlessly into the air or turning downwind to achieve a powerful dive. The stronger the wind, the more efficient their flight, and during conditions of calm weather they may have difficulty in taking off. The petrels, a group of widespread smaller species, have a different form of flight but are no less accomplished, skimming above the surface of wave crests or turning into the comparative calm of troughs to snatch food from the surface.

In addition to the true seabirds, numerous ducks, geese, divers and wading birds inhabit the coastal areas of the world, feeding extensively in estuarine areas where long-billed forms probe the soft sediments in search of worms and molluscs, diving ducks catch small animals in shallow waters, and geese graze on saltmarsh vegetation. Some species, such as the oyster catcher, have developed techniques for feeding on the rich variety of invertebrates on rocky shorelines using the chisel-shaped bill to knock bivalve molluscs from the rock and crack them open.

Detecting food from the air requires high visual acuity. Given the counter shading of many smaller fish, the occasional flash of silver as one fish in the school turns may be the only clue available to a surface flying bird of the availability of food. Penguins appear capable of using a form of sonar, listening to the echoes of the noise made by bubbles collapsing in their wake to detect their prey at depths of several hundred metres. Penguins have become adapted to an aquatic existence so completely that they have lost the power of flight and the wings are modified to form solid paddles for rapid underwater swimming in pursuit of fish.

Auks are also capable of using the wings in swimming beneath the surface. The wings are held half open, with the primary wing feathers closed, to provide the main source of underwater propulsion. Webbed feet provide steering ability rather than propulsion. This group includes the guillemots, razorbills, auklets and little auks. Several extinct species had adapted to this underwater mode of existence such that they had lost the power of flight. Regrettably, flightless species such as the great auk were extensively hunted for food by sealers and the great auk became extinct in 1844. These flightless auks may be considered the Northern Hemisphere equivalent of the Southern Hemisphere penguins.

Diving for fish

Diving from the air to catch fish is a mode of feeding which some species, such as the gannets and boobies, have developed to a considerable degree. These birds may plunge from heights of 30 metres (100 feet) or more, so the skull is strengthened to withstand the impact, the body is protected by a layer of fat, and modified pneumatic air sacs provide a cushion. Many other species plunge, rather than dive. Flying close over the surface, they capture fish just below the surface rather than at depth, while the skim-

◄ *The main hunting techniques used by seabirds vary according to species and also the conditions of the sea. Terns make shallow dives; gannets and boobies plunge below the surface* *from 30 m (100 ft); while auks and penguins pursue fish below the surface. The cormorant uses its feet for swimming, while penguins and auks are much more likely to use their wings.*

mers actually feed at the surface. In these birds the lower jaw is greatly enlarged in comparison with the upper and is held just below the surface as the bird flies along, being used as a scoop to capture small fish and invertebrates.

For most areas of the world the density of available food resources is comparatively low, and the energy required to collect them is high. As a consequence, many seabirds rear only one or at the most two young at a time. Most species are colonial nesters with species such as gannets and fulmars nesting on cliffs, gulls and terns on the ground in sandy areas, and species such as the Manx shearwater in burrows underground. Colonial nesting appears to provide a stimulus for synchrony of breeding and hence improved success. Winter flocks may also serve to improve survival since among the flocks a single individual diving for food stimulates the others to follow.

An important adaptation of seabirds is their ability to drink seawater. Seabirds have therefore developed specialised salt excreting glands to cope with high salt intakes. These glands lie in a bony socket above the eye and the duct empties through the nostrils. Some species possess a fold of skin with which the nostrils can be closed to prevent seawater from entering when the animals dive. Like most birds, seabirds possess an oil gland at the base of the tail from which oil is squeezed and spread along the feather during preening. The oil prevents the feathers from becoming waterlogged; diving birds will preen several times a day to maintain the waterproofing of their feathers. Some species such as cormorants and shags do not produce sufficient oil to ensure efficient waterproofing and must spread their wings to dry the feathers after diving. Frigate birds, meanwhile, simply avoid getting wet, snatching their fish prey from the water surface.

Seabird populations are affected by man's activities, with scavenging birds being positively affected by the increased volumes of offal jettisoned at sea by factory trawlers. Others are adversely affected by reduction in the

size of the fish populations on which they feed. On a localized, but nevertheless large scale, populations of seabirds have been dramatically reduced by large-scale oil spills such as those of the Exxon Valdez and the slicks resulting from the Gulf War. In addition, seabird populations which depend on stocks of a single species in zones of upwelling, such as the cormorants, boobies and brown pelicans which feed almost exclusively on the Peruvian anchovetta, have been dramatically reduced by natural environmental fluctuations. Following the 1972–3 El Niño event their numbers had declined to some 6 million from an estimated previous level of 30 million in 1950. A further decline followed the 1982–3 event which reduced the population to an estimated 300,000.

▲ **The gannet,** *Sula bassana, is a colonial nesting species. It prefers cliffs and rocky outcrops such as Bass Rock which lies off the Scottish coast from which the species derives its latin name "bassana".*

◄ **The puffins,** Fratercula arctica *spend winter at sea, returning each year to nest on land. A favoured food for the young birds is sand eels which the adults, being adept swimmers, catch underwater.*

Migration

Migration is the term applied to a regular journey made by a particular species of animal, either on an annual or on a lifetime basis. Migratory movements of marine animals are made in response to breeding patterns or to seasonal changes in the availability of food. The long distance migrations of baleen whales across whole ocean basins are made in response to seasonal changes in the availability of their planktonic food, the animals moving into higher latitudes during the summer and into lower latitudes or the opposite hemisphere during the winter. Other migration patterns may involve moving from breeding to feeding grounds. In some cases, the individual animal makes the journey only once, in others the journey may be made year after year.

Among marine mammals and birds, individuals aggregate at suitable locations to court, give birth and care for the young. Such breeding aggregations also occur among many species of fish and may involve movements over relatively short distances, as in the case of reef fish, or over long distances in the case of tuna, herrings and many pelagic species. Some small, coastal wading birds undertake spectacular annual migrations. Knots, for instance, breed in Arctic and subarctic regions before migrating 16,000 kilometres (10,000 miles) or more to wintering

areas at the southern end of Africa, South America and Australia. Such migrations follow the same, generally coastal routes, and the same staging areas are used in each successive year, reflecting the specific requirements of these species in terms of food resources and the location of suitable feeding grounds in coastal areas.

For most marine animals which are members of the pelagic community, migratory patterns are an integral part of their lifecycle. Spawning in many open-ocean fish species takes place where currents will carry the larvae in a planktonic community that provide the young animals with food. Thus tuna migrate annually across the Pacific and Atlantic basins to spawn in specific areas before continuing on to adult feeding grounds. The juvenile tuna float with the plankton to nursery areas and ultimately join the adults in the feeding grounds. In the case of turtles, migration is again associated with reproduction, individual female turtles returning to nest on the beaches on which they were hatched. Suitable nesting beaches seem to be the main determining factor in the egg-laying migrations of these animals, as is the case for many seabirds, which return to nest on the same cliff year after year. Some breeding migrations, such as that of the grey whale, are not determined by the requirements of the young, which

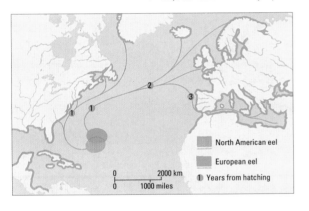

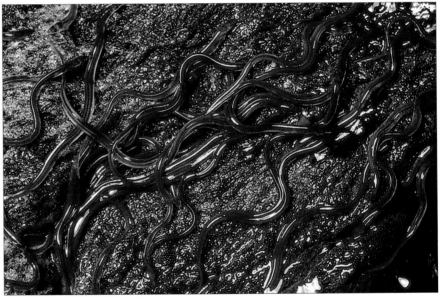

▲ **Both the North** American and European freshwater eels spawn in the Sargasso Sea. The larval eels are carried by the Gulf Stream towards North America and Europe, where

they enter streams and rivers and spend their life feeding and growing. On reaching maturity some years later, the eels return to the Sargasso Sea to spawn and die.

North American eel

European eel

① Years from hatching

0   2000 km
0   1000 miles

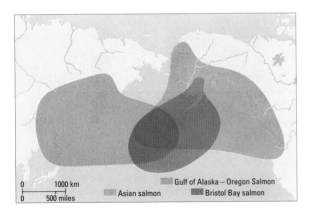

▲ **Unlike the eels,** salmon spend most of their life at sea, returning to freshwater on reaching sexual maturity to spawn some distance inland. Atlantic salmon may

make two or three spawning runs in a lifetime, while Pacific salmon, which are divided into three distinct stocks, spawn only once, and die soon after.

0   1000 km
0   500 miles

Gulf of Alaska – Oregon Salmon
Asian salmon
Bristol Bay salmon

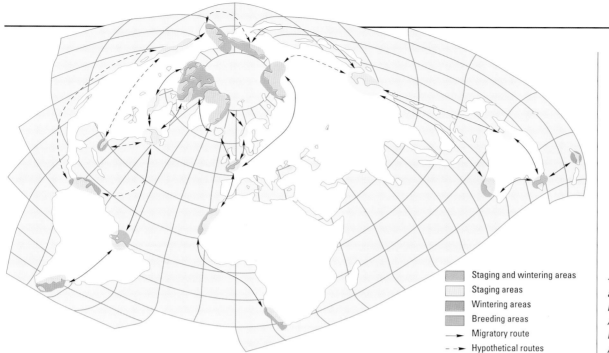

Staging and wintering areas
Staging areas
Wintering areas
Breeding areas
→ Migratory route
- -→ Hypothetical routes

◄ **Knots**, *undertake long annual migrations from their breeding grounds in the Arctic to the Southern Hemisphere during the Arctic winter.*

remain in the company of the parents throughout life. This species makes an annual migration of some 8000 kilometres (5000 miles) from the Bering Sea in the North Pacific to the warmer waters off Baja California to breed.

A large number of different fishes migrate between freshwater and salt water, of which the various species of salmon in the Northern Hemisphere are well-known examples. Spawning takes place in freshwater, usually some distance inland where the rivers are shallow and well oxygenated. Eggs are laid in gravel banks and the larval fish spend a long time in the freshwater reaches of the river, gradually moving downstream during their development to spend their sub-adult life at sea. On reaching sexual maturity they return to the rivers from which they were spawned and the cycle repeats itself. In the case of the sockeye salmon of North America, the adults die immediately after spawning.

In the case of the North Atlantic salmon, however, individuals survive, returning to the sea after spawning; they may make two or three migrations in a lifetime. It has been shown that salmon recognize the river in which they were spawned by subtle chemical cues which distinguish water from one river to the next. The evolutionary advantage of identifying the same river is obvious: only successful spawning results in individuals completing the cycle. When adults enter an unsuitable river and spawn, few or no individuals will survive to return and breed. Restocking rivers which have been restored following degradation from pollution requires that the juvenile migratory species are exposed to water from that river so that they can identify it and return there to spawn.

Atlantic crossing

In some species, the process is reversed; thus freshwater eels from North America and Europe move to the Sargasso Sea to spawn. The larval eels are then carried by currents northwards along the North American seaboard and across the Atlantic to western Europe. Here the young elvers enter rivers, moving upstream to spend their life feeding and growing in the freshwater environment before reaching sexual maturity and undertaking the lengthy migration back across the Atlantic to the spawning grounds of the Sargasso Sea.

Although few invertebrates undertake true migrations, since most larvae drift passively with the ocean currents,

settling out in any suitable area when their larval life is completed, there are some spectacular exceptions. Many tropical land crabs which live throughout the year inland, away from the marine environment, migrate annually in enormous numbers to shed their eggs into the sea, while tropical spiny lobsters migrate inshore to spawn, often forming lines or columns of marching individuals across the seafloor. Once inshore, the eggs are released and the larval lobsters are generally carried passively offshore by the prevailing currents before settling on the bottom. Similar inshore migrations are undertaken by penaeid prawns, the larvae of which use mangrove habitats as nursery areas. The juveniles migrate offshore to deeper water to complete their growth before returning inshore on reaching sexual maturity.

▲ **These female** *Olive Ridley turtles in Costa Rica, Central America, are returning to the sea at dawn following egg laying. Marine turtles mate offshore, after which the females come ashore on sandy beaches to dig a nest, into which they lay their eggs. The eggs incubate beneath the sand until they hatch, whereupon the young turtles make a perilous journey back to the sea.*

Vertical Distribution

In the pelagic environment, vertical gradients, including those of light, temperature and salinity, determine the distribution of organisms. The greatest changes are seen at the boundary between the mixed surface waters and the deeper water masses. In some shallow-water areas a thermocline, or boundary, develops between the warmer surface and the cold water below. Across such thermoclines temperatures may drop rapidly. The epipelagic region, or zone of light penetration, stretches from the surface to a depth of around 200 metres (650 feet) and is followed by the mesopelagic zone between 200 and 2000 metres (650 and 6500 feet). Below this the bathypelagic zone extends from 2000–6000 metres (6500–21,000 feet) in depth, while the deepest ocean trenches contain an abyssopelagic zone greater than 6000 metres (20,000 feet) deep.

Each of these zones contains a somewhat distinctive community of species adapted to the different conditions of salinity, temperature and light intensity. The vertical differences are, however, much greater in the benthic community than in the pelagic community, where many deep-water species range over considerable depths.

Environmental conditions are more stable at greater depth in the ocean than they are at the surface; temperatures vary little over great distances compared with the surface where vertical mixing of the water column occurs. In some instances species of animal found at or near the surface in polar seas also occur in tropical ocean areas at depth, where water temperatures are cool.

At depth, the strength of water currents is low and from the perspective of animals living there, can be considered negligible. Although nutrients tend to be at higher concentrations at greater depth, this has little impact on the organisms living there, since they are only used directly by plants in the epipelagic zone. Except for isolated communities based on the chemical energy supplied by thermal vents, the entire community below the epipelagic zone depends on the rain of faecal materials and dead organisms which sinks down from the epipelagic zone.

The epipelagic zone contains the most abundant life, which depends, either directly or indirectly, on the primary production of the plankton community. Passing down through this zone, the proportion of carnivores increases in

▶ **An imaginary profile** of the typical coastal and oceanic zones is shown, with a selection of the life forms that might occur in the water off the Pacific coast of Central America. The animals illustrated are not drawn to scale as the range of sizes is too great. Plant and animal plankton, the basis of life in the ocean, occur in great quantities: their presence has been indicated, therefore, and examples of the major types have been illustrated. The density of life in general is very high in the upper sunlit zone so, in order to accommodate a reasonable selection of the vast numbers of inhabitants, the depth of the body of the diagram has been distorted.

the pelagic community as the density of primary producers declines. In addition, eye size increases as the light intensity declines.

With the decrease in light intensity, colours change, from blue at the very surface through transparent in the upper layers to red and black at depths of 100 metres (330 feet) or so. The surface waters down to depths of around 150 metres (500 feet) are dominated by swift predatory fishes such as tunas and swordfishes, together with swarms of smaller species such as the lantern fishes.

The abyss

In the aphotic or abyssal depths the only light available is that produced through bioluminescence – light signals produced by animals such as angler fish either to lure prey within reach or by the lantern fishes, which use bioluminescence to signal between members of the same species. Many of the bioluminescent organs of deep-sea animals rely on the presence of symbiotic bacteria contained in these special structures to produce the flashes or continuous beams of light.

Sunlit Zone 200 m (650 ft)

Twilight Zone 1000 m (3000 ft)

Dark Zone 6000 m (19,500 ft)

Trench Zone 10,000 m (33,000 ft)

Flying Fish

Plant Plankton

Animal Plankton

Albacore

Dolphin

Porpoise

Blue Whale

Whitetip Shark

Bonito

Swordfish

Hatchetfish

Sperm Whale

Bristlemouth

Deep-sea Squid

Gulper Eel

Anglerfish

Brotulid

Lamp Shell

Venus's Flower Basket

Isopod

Short-armed Starfish

Sea Cucumber

Tripod Fish

A World of Darkness

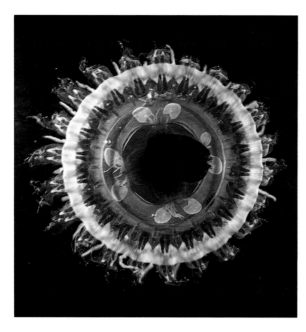

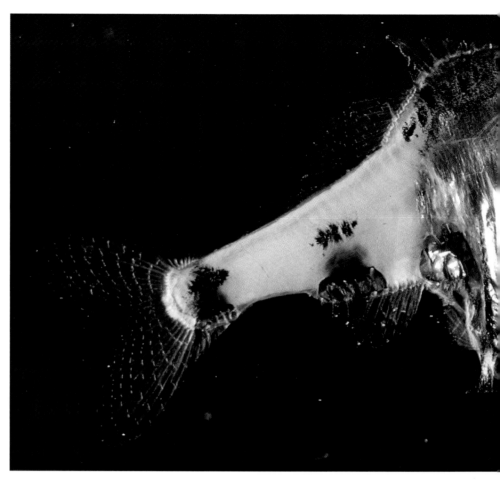

▲ **Atolla**, *a beautifully coloured deep-sea jellyfish, is widely distributed throughout the abyssal zone of the world's oceans. Most species of deep-water jellyfish are reddish brown in colour as this does not reflect the predominantly blue ambient light.*

▶ **The hatchet fish**, Argyropelecus *grows to about 10 cm (4 in). It has huge jaws and is capable of swallowing quite large prey, despite its small size and flattened body. It must rely on occasional meals due to the very low density of prey at depth.*

B elow the euphotic zone of light penetration lies the aphotic zone of total darkness, which stretches almost from pole to pole, extending to the ocean floor. This zone includes almost 90 per cent of the total water column in most areas beyond the continental shelves, and as much as 95 per cent in the area of the deepest ocean trenches. In the absence of light, the characteristic spots, bands and colour patterns of the shallow-water species cannot be seen, and most species are brown, black or violet and generally quite dull. Below 900 metres (3000 feet), most fish have small degenerate eyes or none at all.

Within this region, which comprises the vast bulk of the ocean volume, conditions vary little over great distances. Temperature varies little with depth; day and night and seasonal changes are not apparent; water circulation and current speeds are trivial; and the abyssal plain is covered with a uniform layer of fine ooze, derived from the skeletons of planktonic organisms which have settled slowly to the ocean floor from the euphotic zone.

Only where mid-ocean ridges or volcanic cones occur is any significant topographic feature present, and in the regions of the mid-ocean ridges thermal vents are found, areas where superheated water and lava emerge. These vents are the home to curious communities of organisms, including metre-long tube worms, filter feeding bivalves, and crustacea which depend on the presence of chemotrophic bacteria. These bacteria use sulphur from the emissions of the vents as a source of energy, thus providing a food source to support the entire community. Some of the bacteria form symbiotic relationships with other organisms, such as the tube worms and molluscs, supplying a large proportion of the energy requirements of these animals. Other bacteria are free living, being eaten by filter-feeding organisms within the community.

Given the uniform conditions found at depths in the ocean, it is not surprising that the fauna shows little change over great distances. Of the 11 species of the upper continental slope community found at a depth of 450 metres (1500 feet) off the West coast of Africa, no fewer than eight are important members of the same community off western France. The environments in these two areas differ by less than 1° in average temperature, although they are separated by 40 degrees of latitude. At the foot of the continental slope on the abyssal plain, the characteristic community of animals differs considerably between ocean basins; less than a quarter of species from these areas are common to all oceans. At greater depths, however, a larger proportion of the benthic community is distributed throughout all ocean basins. Pelagic animals, though, appear to be more widespread than benthic species, suggesting greater exchange between these communities during the evolutionary history of the ocean basins.

The decline in diversity

Below the lighted zone, food supplies generally decrease with depth, and there is a corresponding decrease in the abundance of animal life. A sharp division occurs close to the top of the mesopelagic zone, where the density of animal life is much the same as at abyssal depths. The decline in diversity occurs across the boundary between the epipelagic and mesopelagic zones. Below the epipelagic zone the density of individual species is generally low, although the diversity of species remains high. In this region meals are infrequent and animals found in the mid-waters and abyssal depths of the oceans have reduced skeletons, muscles and other body tissues, reflecting the low availability of food. Between 1000 and 2000 metres (3200 and 6400 feet), many fish lack swim bladders,

▲ Orange or bright red bodies are characteristic of deep-water crustacea such as this euphausid shrimp Benteuphausia *sp*. Because these creatures live at depths below the penetration of red wavelengths of light, where only blue light is present, they actually appear black – an effective form of camouflage.

▼ Many deep-sea fishes possess large heads and mouths, and long thin bodies capable of great expansion. The viper fish, Chauliodus *sp*., has huge recurved teeth and widely gaping jaws which enable it to catch prey larger than itself. Its long thin body is capable of considerable extension to accommodate large prey.

maintaining neutral buoyancy by other means such as the retention of ammonia. Many deep-sea fishes possess large heads and mouths, and long thin bodies with an extraordinary capacity to expand. Deep-sea angler fish, such as *Linophryne polypogon*, rarely grow more than 8 centimetre (3 inches) in length, but have extremely elastic stomachs. The stomach of one individual was found to contain a deep-sea eel, two bristle mouths, five shrimps and a hatchet fish. While some angler fishes cruise through the water at depth in search of prey, others remain stationary to conserve energy, and wait for animals to come to them.

Low metabolism, long life

The great swallower, *Chiasmodon niger*, can consume prey almost the same size as itself, while in the gulper eels and ratfishes the head is huge and the remainder of the body is reduced to a long, almost whip-like, tail. The thin body requires less energy to move through the water. Although sparse, the abyssal fauna is highly diverse, the pelagic community of abyssal fishes numbering some 2000 species in contrast to the 200 or so inhabiting the upper reaches of the ocean.

The benthic community displays progressive changes in composition with depth to a much greater degree than does the pelagic community. On the continental margins, the diversity of benthic fishes is much higher than in the abyssal depths. Surveys during the *Challenger* expeditions yielded: 150 individual fish of 47 species, from 40 sites, at depths between 180 and 900 metres (600 and 3000 feet), compared with: 24 individuals of 6 species, from 25 sites, at depths of 4500 metres (15,000 feet). Filter feeding animals decline in abundance with depth, while animals which ingest sediments continue to occur even at the greatest ocean depths. Burrowing animals predominate on mud bottoms, but on the softer, unconsolidated sediments of the deep ocean floor a variety of species occur, including such bizarre forms as the tripod fish, with their enlarged spines that support the body above the surface of the sediment, while sea pens and glass sponges are supported upright in the sediments, filtering particles from the water column. Sea cucumbers also move across this sediment, leaving trails in the soft ooze of the seafloor.

As a result of the low temperature of deep waters, between 2°C and 5°C (35°F and 41°F), the species have low metabolic rates. Individuals grow slowly and live for a long time. Bivalve molluscs at 3000 metres (10,000 feet) off the east coast of North America are up to 250 years old, although only 2.5 centimetres (1 inch) in length.

Life between the Tides

▲ *The mussel,* Mytilus edulis, *lives in dense colonies. They attach themselves to the rocks on which they live by means of strong byssus threads.*

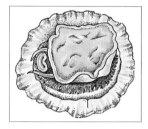

▲ *The common limpet,* Patella vulgata, *has a broad base and clamps on to the rock surface using its muscular adhesive foot, sealing the shell to the rock.*

▲ *The barnacle,* Balanus balanoides, *cements itself to rocks. When submerged, the barnacle opens up its six overlapping plates and kicks food into its mouth with its long feathery limbs.*

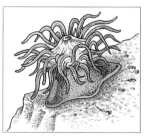

▲ *The headelt anemone,* Actinia equina, *has an adhesive base with which it sticks to the rocks. It withdraws its tentacles into the body cavity when the tide is low.*

The intertidal zone represents one of the harshest environments for life on the surface of the planet. Animals and plants living here must be able to withstand the marine conditions of inundation by salt water which alternates on a daily cycle with exposure to terrestrial conditions.

Intertidal animals and plants therefore display numerous adaptations to resist desiccation during low tide, including remaining tightly clamped to the surface of rocks, as in the case of limpets; closing the shells with a tight fitting operculum, as in the case of snails; or retreating to tide pools, crevices, burrows and beneath seaweeds to avoid the drying action of the wind and the direct heat of the Sun. As the tide returns, the organisms must be capable of withstanding the battering of the waves. Seaweeds have developed strong holdfasts to remain attached, while animals such as mussels have developed byssus threads. Others, such as oysters and barnacles, cement themselves to the surface of the rocks. At high tide, predators such as fish move in to feed on the soft bodied animals which become active under water. When the tide is out the shore is a feeding ground for birds.

Life on and under the sediments

The nature of the bottom sediments determines which species can inhabit a particular stretch of coast. Animals adapted to clinging on to wave-battered rocky shores are quite different from those found burrowing in sand or muddy sediments. On soft shores the fauna is divided into two groups: the epifauna, which are active on the surface, and the infauna, which spend most, if not all of their life underground. The burrowing mode of life requires not only the construction of a suitable burrow, but also some mechanism for collecting food. Many burrowing animals extend a crown of tentacles above the surface to feed on particles in the water column, while others plough through the sediment, consuming it like earthworms. The smallest animals living in sediment, the interstitial animals, move through the sediment, using the film of water which surrounds individual particles. Although consisting of minute, microscopic organisms, this community is very diverse, including ciliated protozoa, rotifers, gastrotrichs, tardigrades, nematodes and many others which feed on the bacteria and organic detritus contained in the sediments.

Since environmental conditions across a shoreline follow a gradient, from the purely terrestrial conditions above the zone of influence of salt spray, to the purely marine conditions found below the level of the lowest low tide, most shoreline organisms occur in recognizable zones. One obvious feature of this zonation is seen in the distribution of plants, with yellow, white and grey lichens occurring in the splash zone and green seaweeds in the upper inter-tidal. Below the zone of green seaweeds are found the brown species, including the characteristic bladder wracks (*Fucus* spp.), plants with gas-filled bladders which float when the tide is in and keep the fronds apart.

Below the brown seaweed zone occur the red seaweeds in the lowest sections of the intertidal. In the subtidal are found the large kelps of genera such as *Laminaria* which grow to large size at rates of up to 1 metre (3 feet) a day. Each seaweed zone forms a separate micro-habitat for different species of animals, with the brown fucoids of the mid-shore serving as nurseries and feeding areas for the abundant, brightly coloured *Littorina,* and the holdfasts of the kelp serving as refuges for numerous small worms, molluscs, crustaceans and brittle stars.

Animal species at the top of the beach display the greatest resistance to desiccation, with species such as

▲ *Rocky shores provide a stable basis for attachment, while the cracks and* crevices provide moist hiding places for soft-bodied animals at low tide.

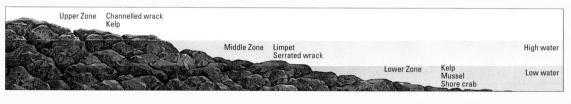

```
Upper Zone    Channelled wrack
              Kelp
                                    Middle Zone  Limpet                                    High water
                                                 Serrated wrack
                                                              Lower Zone  Kelp             Low water
                                                                          Mussel
                                                                          Shore crab
```

The Rocky Shore

Rocky shores support a wide variety of animal and plant species which are zoned in relation to the tidal height. Green seaweeds are found on the upper shore, brown and red lower down, with the large kelps below low tide. They provide both food and shelter for a wide range of animals which hide beneath them to avoid the Sun at low tide, coming out to graze when the tide is in. The sides and roofs of dark caves, often found on rocky shorelines, provide a safe environment for ceatures such as sea anemones. Rock pools harbour a large variety of animals, some of which cannot survive the temporary dryness of the beach at low tide.

The Sea slater, Ligia oceanica, *is related to the terrestrial woodlice and is found living in crevices.*

The limpet, Patella vulgata, *grazes on the algal film of rock surfaces at high tide.*

Upper Zone   Sandhopper

Middle Zone   Mole crab   High water
                Unarmored worm
                Tellin

                                    Lower Zone   Sea cucumber   Low water
                                                  Blue crab

The Sandy Shore

Because the surface is constantly moving under the action of the tides and currents, a sandy shoreline provides no place for attachment of surface-growing seaweeds and no inviting crevices, but it can hold water between its minute particles. Beneath the surface, the environment is unaffected by the weather. At low tide the beach appears dead and inhospitable. However, a wide variety of animals inhabit the space between the sediment particles, while others burrow in the sediments, extending food-gathering structures above the surface to filter food from the water at high tide or digesting organic materials from the surface of sediment particles.

Large numbers of sandhoppers, Orchestia *spp., scavenge among the debris found at the high-tide mark.*

The mole crab, Emerita talpoida, *is highly modified for a burrowing existence on sandy beaches.*

▲ *The lugworm,* Arenicola marina, *is one of the most common sand worms. It lives in a curved burrow and feeds on sand and mud, digesting out the organic food materials.*

Upper Zone   Algae
               Eelgrass

                    Middle Zone   Fiddler crab   High water

                                                                       Low water
                                  Lower Zone   Soft-shelled clam
                                                 Otter shell
                                                 Rag worm

The Muddy Shore

Muddy shores are found in gently sloping areas where the fine sediment is allowed to accumulate. These areas have low water movements and are rich in organic matter, often supporting dense growths of seagrasses, such as Zostera and algae, that provide a plentiful food supply for grazing animals. Because the particles of muddy shores are so fine there is little space between them and so oxygen is at low concentration Burrowing animals, such as some clams, may extend siphons above the surface down which a current of oxygenated water is drawn. Other species, by rythmically moving the body in the burrow, draw down a current of water.

The common otter shell, Lutraria lutraria, *has a siphon which extends above the sediment into the water.*

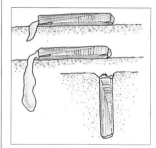

The fiddler crab, Uca *sp., is a familiar inhabitant of tropical and subtropical muddy shores.*

▲ *The common cockle,* Cardium edule, *burrows using its muscular foot, coming to the surface to filter feed on plankton using its two siphons when the tide is in.*

▲ *The razor clam,* Ensis solen, *burrows rapidly to escape danger; the enlarged foot occupies more than half the shell when retracted. In burrowing, the foot extends downwards and retractor muscles pull the animal down towards the foot.*

Littorina saxatalis able to survive with only occasional inundation by the highest tides. These animals are replaced by a succession of closely related species in a sequence down a rocky shore. Competition for space between sessile species may also play an important part in determining the zonation of species, and the distribution of the two barnacles, *Balanus balanoides* and *Chthamalus stellatus*, illustrates this. The latter species is generally found in the top third of rocky shores, while *Balanus* occurs lower down. Both species have planktonic larvae which settle across a much wider stretch of the shore. *Balanus*, however, is less resistant to desiccation and individuals of this species settling at the head of the beach die from desiccation. At the bottom of the intertidal *Balanus* is preyed upon by a predatory gastropod, *Thais*, which is itself limited to the lower third of the intertidal. Larval *Balanus* settling too low on the beach are consumed by *Thais*, which limits its distribution down the shore.

Chthamalus, however, is more resistant to desiccation than is *Balanus*. Larvae which settle high on the beach outside the range of *Balanus* grow and have high survival; those which settle lower on the shore within the range of

Balanus must compete with that species for space. This competition results in higher mortalities of *Chthamalus*, since the more upright shell of this species is undercut by the flatter profile *Balanus*. As the animals grow and the shells touch, that of *Balanus* grows under the other, levering it away from the surface.

Predators and diversity

Predation is also important in maintaining the diversity of species in shoreline communities, and it has been shown that when a major predatory starfish, *Pisaster*, was removed from a stretch of rocky shore, the diversity of the community declined. Predation had previously kept the density of individual species at low levels so that competition for space and other resources did not occur. Once the predators were removed, the populations expanded and those which were more efficient came to dominate the less diverse community.

The complex interactions of different species in the intertidal community therefore reflect the interplay between the different organisms and the physical features of this harsh environment.

The Living Reef

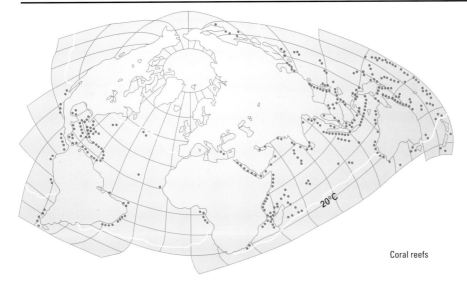

Coral reefs

▲ **Coral reefs** have been estimated to support approximately a third of all the world's fish species, and possibly as many as 500,000 different animal species altogether. Coral grows in warm waters over 20°C (68°F), thriving at about 24°C (75°F). Although the optimum depth for growth is a few metres below the surface, where oxygen and sunlight are abundant, coral can grow at up to around 40 m (132 ft).

▲ **A major predator of** coral is the crown of thorns starfish which may grow to 40 cm (16 in) across. A single large individual is capable of consuming up to 5 sq m (54 sq ft) of coral a year. Population explosions of this coral predator have occurred on a number of reefs in the Pacific and Indian oceans in recent years, devastating large areas of living coral.

Coral reefs have been referred to as the rainforests of the sea; and like rainforests, they are extremely diverse, being home to an enormous variety of animals and plants. Although they are generally found in areas of low marine productivity, reefs themselves are highly productive. The high productivity is due to the reefs' efficient cycling and re-use of nutrients, which are in short supply in the surrounding oceanic waters.

The primary production of coral reefs is between 30 and 250 times as great as that of the open ocean and coral reefs may produce 1500–5000 grams of carbon per square metre (44–146 ounces per square yard) per year.

How reefs are formed

The living reef is composed of a thin veneer of living coral colonies growing on the surface of older, dead coral skeletons. Coral colonies which die or are broken off during storms break down to form sand, which fills the spaces between the frame-building corals. If the land on which the reef is growing sinks, or sea level rises, as is happening at the moment, the reef continues its upward growth. Eventually, with the passage of time, the living reef community may be growing on many hundreds of metres of solid coral rock. The living reef at Enewetak atoll in the Marshall Islands, for example, grows on a base of 1370 metres (4500 feet) of coral limestone. This limestone has been built up on top of a volcanic cone that rises some 5000 metres (16,400 feet) above the ocean floor.

Reef-building corals do not grow well below 20–30 metres (70–100 feet), since they contain microscopic algae, or zooxanthellae, which require sunlight for photosynthesis. The remarkable association between the coral animal and the zooxanthellae benefits both partners, with the algae deriving nutrients and carbon dioxide from the coral while speeding up the rate of skeleton formation of the coral. Other algae, both unicellular and macroscopic, are important to the productivity of reefs, while encrusting, calcareous forms cement loose material together, thus stabilizing the surface of the reef for the colonization of larval corals.

Solitary corals are widely distributed, but do not form reefs, which are best developed in the tropics and subtropics where water temperatures range between 20°C and 30°C (68°F and 86°F). Although reefs grow well at 18°C (64°F) in the Florida Keys, and above 33°C (91°F) in the Northern Great Barrier Reef of Australia and the Persian Gulf, most reefs grow in areas with water temperatures of around 24°C (75°F).

▲ **The Pacific island** of Bora Bora is an extinct volcano which is gradually subsiding. In time, the inner island will disappear and a true atoll will be formed.

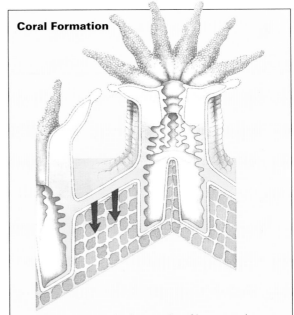

Coral Formation

Coral colonies consist of large numbers of polyps which secrete a skeleton of calcium carbonate. Individual polyps secrete their own theca or cup, into which they can withdraw. Each individual is connected to its neighbour by tissue.

sources for other organisms, the diversity of which is also high. Coral reefs are believed to support a third of all living fish species, for example. While some animals, such as the crown of thorns starfish (*Acanthaster plancii*) and the parrotfishes (*Scarus* spp.) feed on corals directly, many other animals use the reef as a place of attachment. Sea fans, feather stars and sponges grow attached to the surface and filter feed on the plankton and suspended matter contained in the surrounding water. Up to 80 per cent of the plankton in the water passing over a reef may be removed by corals and other filter-feeding organisms.

In sandy areas sea cucumbers feed on surface detritus or eat the sand directly, digesting the bacteria and microorganisms which themselves live on dead organic matter. Grazing animals such as sea urchins and many molluscs feed on the film of algae which grows on all dead surfaces of a reef, while small fishes crop the fine algal turf and are themselves eaten by predators such as moray eels. The moray eel is not immune from predation, being eaten by the venomous sea snake, *Laticauda colubrina*. To avoid predation, many reef animals are brightly coloured, advertising the fact that they are distasteful or poisonous. The bright colours of many other active reef animals serve, however, as signals, enabling individuals to recognize members of their own species of the same or opposite sex.

Present threats

Unfortunately, the diverse and beautiful world of the coral reef is as threatened as the tropical rainforests. It has been estimated that as much as 10 per cent of the world's coral reefs have been degraded beyond recovery by human activity. A further 30 per cent is likely to decline seriously within the next 20 years and it has been suggested that more than two thirds of all reefs may collapse ecologically within the next 80 to 100 years.

Fringing reef

Barrier reef

Coral atoll

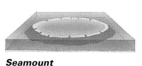

Seamount

▲ *An atoll* starts to form as soon as a volcano rises above the sea. Corals colonize and form a fringing reef around the island. The reef grows upwards as the island sinks, forming a barrier reef enclosing a shallow lagoon. Eventually, the island disappears leaving a ring-shaped reef or atoll. If subsidence occurs faster than coral growth, then the reef dies leaving a seamount below the surface of the sea.

The coral animal

Corals are actually colonies of tiny individual animals or polyps which secrete a skeleton of calcium carbonate. The colonies grow by adding new individuals and the growth may take quite different forms in different species. Corals range from the compact brain corals found in areas of high wave energy, through heavy branching and plate corals in deeper water of the reef edge, to smaller, finely branched forms found in more sheltered water behind the reef crest. The most active coral growth is generally on the outer edge of the reef. Here water movement is greatest, carrying with it plankton on which the coral feed.

In addition to reproducing by simple budding, so increasing the size of the colony, corals also reproduce sexually, producing eggs and larval forms. The larvae are dispersed by currents before settling to form new colonies.

Coral reef diversity

While it is true that coral reefs are among the most diverse marine habitats in the world, their diversity is much less than that of rainforests, which have over 250 species of tree per hectare. A single large reef system may support around 200 coral species and only about 1000 species of reef-building corals have been described worldwide.

The centre of coral diversity is the Southeast Asian region, with some 700 species being found in the Indo-West Pacific, compared with only around 35 in the Atlantic. Over 400 species of hard corals are believed to occur in the Philippines, and as one moves away from this centre of diversity, the numbers of species decline.

The wide diversity in growth forms of different corals provides a multitude of micro-habitats, refuges and food

▲ *Coral reefs* are home to an amazing variety of spectacularly coloured fish.

The coral provides the fish with numerous hiding places and abundant food supplies.

▲ *The aptly named* Parrot fish (Scarus sp.) feeds on coral. It uses its beak-shaped jaws to bite off pieces of coral colonies which are crushed by plate-like teeth at the back of the mouth. The coral polyps are digested and the skeleton excreted as fine sand particles.

OCEAN
RESOURCES

◀ *The world's seas* and
oceans are teeming with fish.
For thousands of years, we
have been exploiting this
rich resource with little
impact on populations.
However, more recently, our
demand and our ability to
find, catch and process
finfish and shellfish has
brought about a situation
of over-exploitation.

Artisanal Fishing

Fish, both shellfish and finfish, are a vitally important source of food for many societies living around the world's coastal regions. In 1985, the FAO (Food and Agricultural Organization of the United Nations) estimated that for 60 per cent of the population in tropical developing countries, between 40 and 100 per cent of their animal protein came from fish.

The diversity of marine animals consumed in different parts of the world reflects, at least in part, the distribution of varying species. In temperate regions most fisheries rely on large stocks of single species which are fished commercially using modern technology. At lower latitudes and in inshore areas the wider diversity of fish is reflected in the range of techniques used to catch the different species. As much as 75 per cent of the world's harvest of fish comes from within 9 kilometres (6 miles) of the shore. The artisanal fisherman, therefore, is a major producer of edible marine products around the world.

The diversity of techniques employed to harvest fish is enormous. Spears, and bows and arrows may be used to catch single fish in shallow water, while hook-and-line techniques are used on set lines, with rods or trolled behind a boat. Other methods of catching fish involve the use of nets. These range from hand-held, frame nets operated by a single fisherman in shallow water, to large nets that require several men to set and haul.

Gorges, hooks, lures and nooses

The simplest fish-catching device attached to a line is the gorge. Made from wood or bone, the gorge is pointed at both ends and when swallowed by a fish it lodges in the gullet, allowing the catch to be hauled into the boat or landed.

Hooks represent a more sophisticated catching device. They come in a variety of shapes and sizes, depending on their place of origin. The shank, for example, may be straight or curved, and the point barbed or plain. Many hooks are small, having been ground from a single mollusc shell. Others, such as *Ruvettus* hooks, are large structures made from shaped wood and shell.

Rock-wall traps were built on gently sloping shores between high tide and low water to strand fish above the receding tide. This simple trap has been used since prehistoric times in the Pacific.

Fish species that swim parallel to the coast may be deflected into nets by strategically placed fences. The seaward end of the fence leads to a heart-shaped trap, from which the fish cannot escape.

Barbed hooks and gorges made of bone or wood are used to catch fish which strike a baited line. The simple gorge is hidden in the food, and rotates and lodges in the fish's gullet.

Small shoals swimming close inshore may be encircled with a beach seine paid out from a small boat. Fish are frightened into the central bag by shouting and splashing.

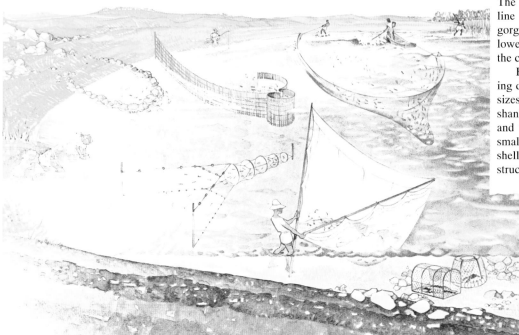

In shallow muddy water, or where the seabed is too rough to use nets, fish are often caught using plunge baskets. The fishermen work slowly forward, trapping the fish and crustaceans under hand-held pots.

Fyke nets consist of a series of conical nets leading one into the next and ending in a cylindrical chamber. Wings of netting are spread out from the entrance into the current to increase the catch of fish.

In shallow water, where the seabed is smooth, a net may be skimmed across the bottom. The rounded ends of the frame ensure that the net glides easily over the bottom as it is pushed through the water.

Lobster pots are generally constructed from heavy netting or basketwork on a stout wooden frame. The pots are usually baited with fish offal and the darker they are, the better, as lobsters like to hide in the dark.

A springy branch, bent and held by a quick-release device, may be used to activate traps and automatic fishing lines. The triggering mechanism of this Congolese trap is dislodged by the fish tugging at the bait.

The lift net is used to catch small surface-swimming fish, often for use as bait. The lift net is lowered into the water, and when the fish swim over it, the net is raised before the shoal have the chance to swim on.

Dredges take shellfish from muddy areas. When towed behind a boat, the teeth at the front prise loose clams and other molluscs.

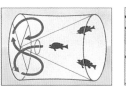

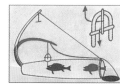

Trolling, or dragging, a lure behind a boat catches fish that mistake the lure for potential prey. Active predatory fish are caught in this way, and traditional lures of shell, bone and feathers were widespread among the Polynesian fishing communities of the Pacific. Today these have been replaced by plastic lures and steel hooks.

In the Pacific, sharks are caught in fibrous nooses suspended from wooden floats. The shark is attracted to shells in the noose, and swims into it. The noose lodges against the pectoral fins and when the shark dives, the float forces it to the surface. After a while, the shark tires and can be lifted into the canoe.

Traps and nets

An ancient element of the fishing technology of Southeast Asia, spreading as far as New Guinea and into the foothills of the Himalayas, is the thorn-lined trap. Spiny branches are tied together in a cone, with the spines pointing towards the tip of the cone where the bait is secured. When the fish enters, the spines lodge behind the gill covers, preventing the fish from wriggling out backwards. Larger traps include fish fences constructed in mangrove and estuarine areas, weirs in river channels and stone traps on reef flat areas. Such traps rely either on catching fish as they move into and out of feeding areas with the tide, or on the behaviour of those fish which tend to swim close to fences, and thus into traps rather than crossing open water.

Traditionally, nets, both large and small, were made of vegetable fibres. Recently, however, nylon has replaced the original materials. Small, hand-held nets are used to dip fish from the water, and larger nets are set as temporary traps, catching fish by entangling them. The circular cast net is a widespread element of fishing technology. Its outer fringe is weighted and the net is cast by a single fisherman. The weighted rim sinks faster than the centre, trapping fish inside.

▲ *This photograph* shows artisanal fishermen setting traps for shrimp and small fish in a mangrove creek in Southeast Asia.

◀ *These fishermen* are using a circular cast net from an outrigger canoe to catch schools of small fish for use as bait.

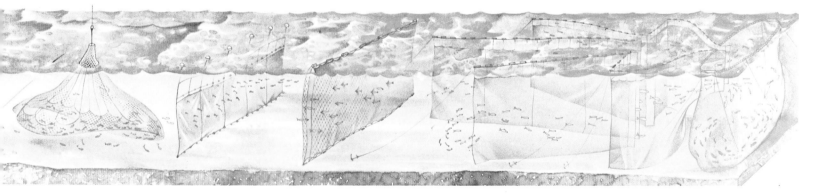

Throwing a cast net 9 m (30 ft) across, so that it opens out to its maximum extent as it

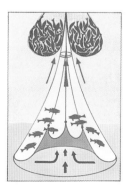

hits the water, requires a high degree of skill. The weighted outer edge pulls the net down over the fish, entangling them and pinning them to the bottom. The net lines are then hauled in, slowly closing the net around the catch. More sophisticated cast nets are equipped with a number of pockets around the edge, which are similarly closed as the lines are slowly pulled tight.

Gill nets restrict the catch of fish to one size, allowing the harvesting of fish in a particular area to be controlled. The nets are made of fine cord which is invisible to the fish, which then become trapped by the gills.

The trammel net consists of two coarse outer nets on either side of a fine entangling net. Bulky and conspicuous, it would normally be avoided by the fish, which are driven blindly into the net by people splashing in the water.

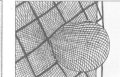

Fish traps that do not depend on a mechanical triggering system include some form of non-return device, designed to prevent the fish from escaping back into open water. In the case of the pound net, used in Japan to catch sardine, the fish are led up a ramp of netting, which takes them well above their normal level. The fish then emerge into a deep bag-like compartment in which

they are free to swim back down to their usual depth. But because the fish do not normally leave their accustomed level unless forced to, they remain at this level – completely unaware of the escape route just above them.

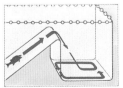

Commercial Fishing

While artisanal fishermen generally operate singly or in small groups on small vessels, commercial and industrial-scale fishing is based on much larger vessels, operating at considerable distances from home ports. These long-distance fishing fleets not only catch fish, but also process the catch on board. The operation of such fishing ventures involves extensive investments in technology and energy-intensive techniques.

Commercial fisheries account for about 90 per cent of the total world fish catch, of which 75 per cent is for human consumption. The remaining 25 per cent is used in the production of fish meal for animal feed. In contrast to the catch of the artisanal fisherman, which is largely consumed fresh or iced, the fish catch of large, commerical fishing fleets is usually blast frozen or canned rather than marketed fresh.

The main target species of the industrial fisheries are pelagic species, such as the various tunas, sardines and anchovies. However, large-scale trawl fisheries do exploit the demersal (bottom-dwelling) fish stocks, including cod and plaice, which are found in the continental shelf areas.

Development of commercial fisheries

Large-scale commercial fishing began as early as the 15th and 16th centuries, when inshore fishermen of Northern Europe were drawn to the rich offshore fishing grounds, and the countries surrounding the North Sea often came into conflict over the exploitation of herring. By the 17th century, fishermen from Devon had already crossed the Atlantic to fish for cod off the coast of Newfoundland.

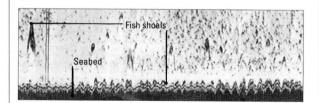

▼ **Trawlers,** which target demersal fish, are faced with declining resources in many areas. The density of fishing vessels is too high for existing fish stocks and the trawls destroy the bottom. The benthic sediments are disturbed and communities of non-target species affected.

▲ **Fish are detected** by sonar equipment on board fishing vessels. A detector, located in the ship's hull, emits pulses of sound and measures the time taken for them to be echoed back by the seabed; when a shoal of fish swims within range, the sound waves are echoed back in a shorter time. The shoal appears on the screen of a recorder as a hyperbola, rising above the seabed.

The introduction of steam and diesel engines fuelled a rapid expansion of long-distance fishing. Not only could such vessels travel further and faster, but by carrying ice as a preservative they could extend the time that fish could be held without salting. This led to a rapid increase in fish catch in the North Atlantic, and fishing vessels were soon penetrating the Arctic circle in search of cod and haddock.

Following World War II, innovations in refrigeration technology enabled vessels to remain at sea until their holds were filled, so increasing the length of voyages. Improved catching technology and the increasing size of vessels also resulted in the exploitation of new stocks, and commercial fishing extended outwards from the North Atlantic to other areas.

Commercial vessels

Three major types of new vessel were developed for these fisheries. The giant purse seiners evolved from small Norwegian mackerel and herring ring netters and were developed for fishing schooling, pelagic fishes including sardines, anchovies and tuna. The large-scale American vessels of this type are capable of encircling an entire tuna school and holding approximately 2000 tonnes of frozen fish.

Oceanic long liners were developed to fish the deeper-swimming tunas. Using, as their name suggests, long lines of up to 30 kilometres (20 miles) in length fitted with multiple hooks, these vessels range the entire extent of the

◄ *The* **Sozidanie** *is one of seven 5000 dead weight tonne vessels used as a factory freezer ship for long-distance fishing trips that last for many weeks. The catch, which comprises a variety of fish species, is processed at sea and the discharge of fish offal from the processing plant will often attract large flocks of scavenging seabirds. While some species of bird benefit from modern commercial fishing, other species must compete with man for their food – in particular the small pelagic species, such as the Peruvian anchovy, which are increasingly fished to produce fishmeal and oil.*

▼ *Purse seine nets (below) are used to encircle entire schools of surface-swimming pelagic fish such as tuna. The net is laid in a wide circle around the school and the ends are drawn together, hence the term purse. The weighted bottom of the net is then drawn tight and the entire net is drawn in towards the vessel. The otter trawl net (bottom) is hauled along the seabed to capture fish which live and breed on or near the sea bottom. To cut down water resistance, the net is normally of wide mesh near the mouth, becoming much finer towards the cod end of the net.*

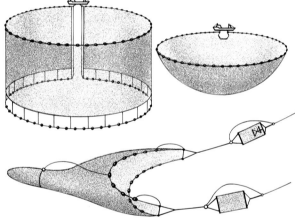

tropical oceans in search of tuna, marlin and billfish. A recent innovation is the drift nets operated by some fleets in the Pacific. The nets may be tens of kilometres in length, and are based on the small-scale North Sea drift nets that were set for herring. They catch a number of non-target fish, marine mammals and turtles, and are banned under a regional convention in the South Pacific.

Factory trawlers are large-scale vessels developed to exploit demersal fish. They are used extensively on the continental shelf areas by long-distance fishing fleets from the former Soviet Union and by many Asian states. Large trawl fleets operate in many offshore tropical areas where the target is frequently penaeid prawns and tropical spiny lobsters, the stocks of which have been dramatically over-harvested in many areas.

The by-catch, or non-edible species, in many of the commercial fisheries operations is a major source of concern to environmental groups and development economists in many tropical countries. In particular, large-scale trawl operations for penaeid shrimps and tropical lobsters not only damage the benthic community, but result in a large catch of fish which are normally thrown back in order to retain freezer space for the more valuable prawns. Such by-catch is normally dead when returned, and could alternatively be processed for animal feed or in some instances sold in those developing countries where the prawn trawling occurs and where protein shortages are frequent.

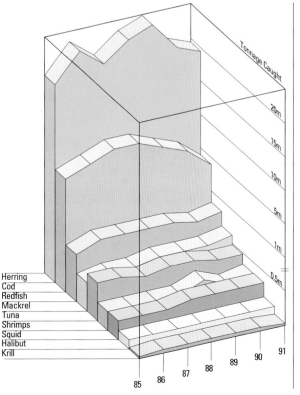

◄ *There has been a trend towards increased fish catches over the last few decades, although the signs are now that catches have stabilized around the estimated maximum yield of 100 million metric tonnes a year. Without improvements in the efficiency of use and regulation of capture fisheries, marine fish stocks are likely to decline as world population and hence demand for sea food grows. Virtually all known marine fish stocks are currently now being exploited, many of them at unsustainable levels.*

Declining Resources

Although living marine resources currently provide between 5 and 10 per cent of total world food production, they supply between 10 and 20 per cent of the world's animal protein.

In a report published in 1967, the Food and Agriculture Organization of the United Nations stated, "at the present rate of development few substantial stocks of fish accessible to today's types of gear will remain in another 20 years". We now have ample evidence that this statement was indeed correct, and many stocks of fish have declined throughout the world's fisheries.

It has been estimated that the sustainable harvest of marine living resources is probably no more than 100 million metric tonnes a year. This level has been reached over the last few years, and it is believed that around 90–95 per cent of the world's fish stocks are now fished at, or beyond, the maximum sustainable levels.

Commercial fisheries

The commercial catch is made up of around 35 fish species of which six – pollack, mackerel, herring, sardine, cod and anchovy – make up over half of the total landings.

The demersal catch, which includes prawns and shrimp, is dominated by cod, hake and haddock. These species amounted to around 10 per cent of the total world catch of finfish in the late 1980s. In contrast, the pelagic catch is dominated by herrings, anchovies and sardines, all of which are exploited both for human consumption and for the production of oil and fish meal. Several of these large, pelagic fisheries contribute significantly to the total world catch, such that fluctuations in one stock can significantly affect total world landings.

The processes controlling the size of fish populations and patterns of recruitment are not well enough known to regulate the harvest within sustainable limits. The recovery of North Sea fish stocks during World War II led to the assumption that fish stocks were largely regulated by fishing effort. More recently it has been realized that inter-annual variations in ocean characteristics, including variations in current patterns, upwelling and salinity, can affect the survival of juvenile fry and hence the size of the stock.

Sustainable levels of harvest

Following World War II, and during the major period of expansion of fisheries up to the 1970s, world fish catch grew at around 7 per cent per annum. By the 1960s, when the world fish catch had reached 50 million metric tonnes, about half its present level, it was already understood that some of the stocks were being fished at unsustainable levels. The decline in stocks of cod and haddock in the North Atlantic, herring in the North Sea, and salmon in the North Pacific, all attest to the destructive nature of modern harvesting techniques.

As traditional fish stocks declined, the fishing industry developed more sophisticated techniques for locating and catching fish. In some instances these innovations resulted in temporary increases in yield, but overall the increased fishing effort merely depleted stocks further. At the same time commercial fishing turned to less desirable species such as squid and shellfish, and the focus of fishing shifted from the Northern to the Southern Hemisphere.

Most fisheries appear to go through a boom and bust cycle. The Californian sardine fishery of the 1930s is a classic example. Fishing boats and gear were poured into this fishery to such an extent that the fishery became over-capitalized, with too many boats chasing a diminishing supply of fish. The stock finally crashed to extinction in the early 1950s.

▶ **The rapid growth** of European populations in the 19th century produced an increased demand for food. The fishing industry expanded dramatically; in Grimsby in England, for example, the number of fishing trawlers grew from 24 in 1855 to 600 by 1877. Increased competition and declining resources forced fishermen farther offshore and by 1900 the whole of the North Sea was being fished. As fish stocks declined in the North Sea, boats went further afield to the Grand Banks off Canada and into the Arctic Circle.

◀ **The waters** off the Peruvian coast are among the most productive fisheries in the Pacific. The concentration of anchovetta in the productive zone of upwelling off the South American coast makes these waters a suitable target for large-scale industrial fishing. During normal years, larger catches of up to 20 million tonnes are obtained, but in years when an El Niño event occurs, production declines and the fishing industry stagnates.

▲ **The mile-long complex** of canneries and reduction plants that made up Cannery Row was built in the 1920s and 1930s to process the huge quantities of sardine caught by the Monterey fishing fleet which, at its height, had as many as 80 purse seiners. By 1947 the number of canneries and reduction plants had grown to 31. From 1950 onward, the size of the sardine catch dwindled rapidly, and the canneries were eventually forced to close down.

As a consequence of stock declines, a number of governments have taken action to control and, in some instances, reduce the numbers of fishing vessels operating in different fisheries. Some fisheries have been closed, and in others, the numbers of fishing vessels and frequency of fishing have been curtailed in order to reverse stock declines.

A need for control
In the past, when the numbers of fishermen were low and fishing technology was less sophisticated, the concept of open access to fisheries was workable. Unrestricted access to fisheries is no longer feasible in the face of accelerating growth in world populations and the subsequent demand for seafood products. Already the world's harvest of seafood is less than the demand, and this shortfall will increase if stocks continue to be over exploited. Clearly an urgent need exists to manage the oceans' living resources in a more sustainable manner than in the past.

The human use of living marine resources is still at a primitive hunter-gatherer stage of development. Wild stocks are caught and harvested without either sufficient knowledge concerning their biology or the proper management tools to ensure regulation and control of harvest.

◀ **The history of** commercial whaling is one of unmitigated greed and commercialism resulting in the dramatic decline of species such as the blue whale. While most nations observe the moratorium on commercial whaling, some traditional whale hunts continue, and scientific whaling is undertaken by a few countries.

Mariculture

▶ **Oysters,** Ostrea edulis, and mussels have been cultured in Europe at least since Roman times. This picture shows oyster cages lifted above the mud in the shallow intertidal of Cancale in Brittany. Nowadays a wide variety of different oyster species are cultured for food and pearl production in different parts of the world.

The idea of farming the sea is not new. In Southeast Asia the culture of molluscs, crustaceans, fish and seaweeds has been practised for hundreds of years. Today, Southeast Asia produces about two-thirds of the world's aquaculture output, and in some countries in this region up to 60 per cent of the dietary protein comes from "farmed" marine organisms.

Although by 1987, world production of marine products through aquaculture totalled only around 3 million tonnes, or 4 per cent of world marine production, fish farms are becoming an increasingly common sight around the world's coastlines.

Farmed species

The range of organisms cultured varies widely, from the mussel and oyster farms of Europe, through the clam farms of North America, to the seaweed and giant clams cultured in Southeast Asia and the Pacific. Shellfish mariculture, such as that of mussels in Europe, can yield harvests of up to 100 tonnes of organic food per hectare (2.5 acres) per year – considerably higher than that found on land. In part, this high productivity results from the nature of food chains and primary production in ocean waters, where microscopic phytoplankton take the place of larger plants on land. Mussels and other molluscs remove these primary producers from the water, and provided that the productivity of the waters is not affected by other factors such as pollution, the organisms require no feeding, relying instead on natural ocean productivity.

Giant clam mariculture, as recently developed in Australia, the Pacific and Southeast Asia, takes this process one stage further, since the clams contain their own primary producers, or symbiotic algae. The giant clams, therefore, require feeding only during the larval stages. Once they have acquired the symbiotic, algae they

can be placed in shallow water and grow rapidly due to the high productivity of their symbiotic algae rather than being dependent on external food sources.

In the Philippines the milkfish (*Chanos chanos*) is a popular mariculture species. Adults spend most of their lives at sea, but return to shallow coastal waters to spawn. Milkfish fry are collected in shallow waters by fishermen using fine nets, who then sell the fry to brackish water farms where they are maintained in shallow nursery ponds before being transferred to larger, deeper ponds. The small fry feed on the phytoplankton blooms in the ponds, while the larger adults graze on the growth of algae on the bottom of the ponds. The fish are harvested at all growth stages. The tiny fry are fried and served as a side dish and the larger adults are eaten as a main course.

Norway is the world's largest producer of farmed salmon with an estimated yield, in 1990, of 150,000 tonnes. The size of the industry is reflected in the fact that 10 per cent of Norway's workforce is currently employed in fish farming, using the naturally sheltered enclosures of the deep-water fjords which line Norway's coastline. Salmon farming is also of growing importance in Chile, the Faroes and New Zealand. The yellowtail, another important farmed fish species, yielded 160,000 tonnes worldwide in 1987. In Singapore and Hong Kong, a wide variety of groupers and other tropical species are being farmed on an experimental basis, while in the Mediterranean, sea bream and dorade are being farmed experimentally in a range of different net enclosures.

Some 30 different species of seaweed are grown in farms, mainly in Southeast Asia. Here sea mustard, kelp and *Euchuema* are produced both for human consumption and for the production of alginates used for clarifying, gelling and thickening food products as well as a source of food for animals in culture. Korea and China each produce

▲ **Salmon** are farmed in suspended pens, such as those shown at this fish farm in a fjord in Norway.

▼ In the Asian region, different species of seaweed are grown for food and to produce alginates.

more than 250,000 tonnes of seaweed annually. Agar, a gelatinous substance used in health care as a growth medium for bacteria, is derived from seaweed.

The wide diversity of marine organisms with a unique ability to accumulate elements at low concentration in seawater may provide further resources for development in the future. In recent years, screening of marine organisms for anticancer agents and other potential chemicals of importance to industry and medicine has greatly increased. The numbers of useful chemical products extracted from marine organisms is at present hardly tapped.

Problems of sustainability

Although mariculture was, and is still, considered a potential solution to the problems of marine production and meeting global demand for seafood products, it is not without its problems. Intensive pen culture has resulted in severe environmental degradation in a number of areas where the faecal waste and uneaten food result in high rates of bacterial decomposition and deoxygenation of bottom waters. Correct siting of mariculture installations in areas where water flushing reduces or removes this problem is now recognized as a necessary prerequisite for sustainable mariculture.

In addition, most farming of fish, shellfish and some crustaceans relies on specialized centres that produce "spat" (juvenile animals) which are then sent to the farms where they grow to a harvestable size. However, because many of the farms are themselves incapable of producing sufficient juveniles to continue the culture, they have to rely on the unsustainable harvest of wild juveniles. Furthermore, many of the farms are constructed in the very habitat necessary for the supply of individuals to maintain the farm. The clearance of mangrove swamps for prawn or mariculture "farms" destroys the habitat required for the growth and survival of the juvenile prawns on which both the farming industry and the commercial trawl catch of prawns depends.

Offshore Minerals

Mineral deposits in the marine environment derive from three major sources: those, such as oil and gas, which have formed in sedimentary deposits on the continental shelf area; minerals that have been eroded from rocks on the land before accumulating along sedimentary coastlines; and finally minerals, such as those found in calcareous sands, which are formed as the result of chemical and/or biological processes in the ocean itself.

It is only the last 30 years that have seen growing interest in extensive exploitation of the minerals of the inshore and continental shelves. Most of this interest has centred around the rapid expansion of oil and gas extraction. Far less obvious, but both more widespread and of major economic importance, is the recovery of sands and gravels for construction purposes on land. The economic value of such sand and gravel greatly exceeds that of any other mineral type, other than oil and gas, mined from offshore areas.

Sand and gravel

Offshore sand and gravel deposits result from weathering processes occurring on land. The deposits are sorted by the action of currents in such a way that materials in any one location tend to be of a relatively uniform size and density. Gravels extracted from the North Sea include a high proportion of flints weathered from chalk and limestone deposits, which ice sheets have moved into the ocean basin. Subsequent current and water action have removed the finer sediments, leaving graded materials suitable for use in a variety of construction aggregates. In some instances, such as Faxa Bay on the west coast of Iceland, the bank of shell sand is constantly replenished by the action of tidal currents and winter storms. Their action breaks up the shelled molluscs growing on offshore rocky shoals and moves the shell sand materials into the bay.

Pumping sand onto beaches is now widely practised in many places around the world including the east coast of the United States and along the Netherlands coast. Although not strictly mining, such beach replenishment schemes are designed to contribute to coastal protection by replacing sand which has been moved offshore by the action of waves and tidal currents. In the case of the Netherlands this is one component of the complex coastal protection schemes necessary to protect land below sea level. In the United States such beach replenishment is

*▲ **Placer deposits** along the coast such as these diamond deposits in Namibia are mined to produce a variety of precious minerals. Offshore dredging of cassiterite occurs in Southeast Asia to produce tin.*

designed to maintain the beaches used by tourists during the summer which have been eroded over winter.

For many small islands in tropical regions the sole source of sand for construction is the calcareous sand derived from the weathering of coral skeletons, algae and animals such as foraminifera. These deposits collect in lagoon areas, often in close proximity to coral reefs, and their extraction by dredging results in suspension of fine materials which seriously damage the neighbouring reefs. In some atoll states no other source of building material exists other than the coral reefs themselves, and the skeletons of large corals may be mined and used in construction. The largest single offshore mining operation involves the extraction of aragonite sands in the Bahamas. These are exported to the countries surrounding the Caribbean and used in the manufacture of high-grade cement. The use of calcareous sands derived from the shells and skeletons of marine animals for the manufacture of cement is widespread from Iceland to India, and relies on the renewable nature of the resource.

Salt and phosphorite

Salt in solution is an important resource. It represents about two-thirds of all the minerals in solution in sea water, and it can be extracted relatively cheaply in countries with a suitably hot, dry climate. In such areas, salt water is allowed to flood shallow pans where evaporation occurs, leaving deposits of sea salt.

Phosphorite deposits occur in a number of offshore areas, particularly in areas of strong boundary currents and zones of upwelling. Most of the known deposits of phosphorite are geologically fairly old, but some are known to be forming today. Phosphorite is important as a fertilizer and as a chemical agent used in the manufacture of a wide variety of other chemicals.

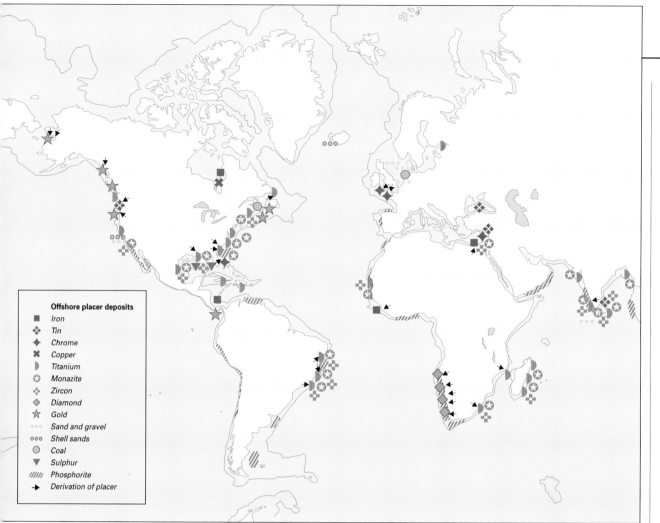

Offshore placer deposits
- ■ Iron
- ◆ Tin
- ♦ Chrome
- ✖ Copper
- ◗ Titanium
- ✪ Monazite
- ✧ Zircon
- ◆ Diamond
- ★ Gold
- ∘∘∘ Sand and gravel
- ooo Shell sands
- ● Coal
- ▼ Sulphur
- ///// Phosphorite
- → Derivation of placer

◀ **Most offshore minerals,** with the exception of oil and gas, do not occur in sufficient high ore grades to warrant their economic production. Nevertheless significant amounts of tin, diamonds, gold and titanium are recovered from beach and offshore placer deposits world-wide. Sulphur is extracted commercially in a number of areas, including the Gulf of Mexico, and the most important mineral resources after oil and gas are sand and gravel dredged from offshore deposits for use in the construction industry.

Precious metals

For most minerals in the sea (with the exception of salt) the economic costs of extraction exceed the value of the materials. Consequently, most offshore mineral deposits can be considered to be reserves rather than resources, in that prevailing circumstances make them uneconomic. One of the best examples were the diamonds mined from placer deposits off the Namibian coast. Despite the high grade of the deposits, the costs of extraction exceeded the returns and most operations were suspended in the 1970s.

Tin, however, is still mined from shallow water in Southeast Asia, particularly in Indonesia and Thailand. The ore is extracted from shallow water deposits by dredging. Unfortunately, such extraction results in extensive resuspension of fine sediments which have considerable impacts on neighbouring reef ecosystems.

▶ **Salt is made** from seawater in many parts of the world with a hot dry climate. The salt pans, such as these at Lanzarote in Spain, are flooded and the water left to evaporate, producing raw sea salt.

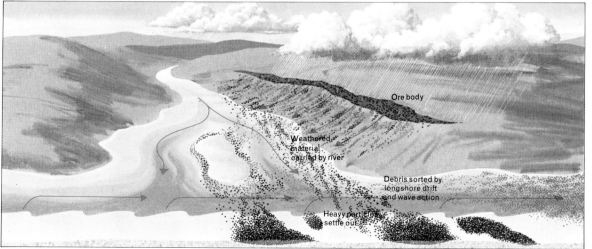

Ore body

Weathered material carried by river

Debris sorted by longshore drift and wave action

Heavy particles settle out

◀ **Most offshore metals** are found as placer deposits. A metal-bearing rock on land is weathered and the debris produced is washed down to the sea by rivers. There it is sorted by the currents, waves and tides so that the heavy metal particles accumulate to form deposits of mineral sand. These typically take the form of beach deposits, but where the sea level has changed they can be found well out on the continental shelf. The sands are lifted by dredgers.

Deep-sea Minerals

The report of the Scientific Results of the Exploring Voyage of HMS *Challenger* (1872–6) records that, "the dredges and trawls yielded immense numbers of more or less circular nodules and botryoidal masses of manganese oxides of large dimensions. To mention all the regions where manganese was observed would take up too much space. . ."

This was the first report of deep-sea mineral deposits and led to the well-known term "manganese nodules". In reality, however, the term polymetallic nodule is perhaps more descriptive since these nodules contain a number of elements in addition to manganese (see table). The nodules are largely composed of oxides of iron and manganese, but also include quantities of titanium, chromium copper, nickel, cobalt and zinc in varying amounts. The composition of the nodules is such that the processes involved in separating the minerals are likely to be complicated since the copper and nickel occur within the manganese oxides rather than as separate minerals.

Until the 1950s, interest in these potential resources was low. The availability of much richer ore deposits on land made the polymetallic nodules, economic exploitation unfeasible. Following World War II, however, the average grade of copper deposits being worked on land fell from around 2.6 per cent in 1900 to 0.7 per cent by 1965, and the nickel content of ores being mined in New Caledonia – which lies some 1600 kilometres (1000 miles) off the east coast of Australia – was down to about 2.8 per cent compared with around 7 per cent in 1900. At that time, projections of future demand for copper and nickel suggested that lower and lower grade ores would need to be mined. For this reason interest in the potential mining and extraction of deep-sea manganese nodules increased from the 1960s onwards.

The diameter of polymetallic nodules ranges from 2.5–5 centimetres (1–2 inches) and their density ranges from a sparse cover to an almost continuous pavement on the seabed. Although the complex physical, chemical and biological processes involved in their formation are still not well understood, extensive surveys of the world's oceans have shown that such nodules are widely distributed throughout the deep ocean basins.

The richest nodules are generally found in areas away from the input of land-derived sediments, and in depths of more than 4000 metres (13,000 feet). The vast majority of ore-grade nodules is now known to occur in the Pacific Basin, with the highest density lying between the Clarion and Clipperton fracture zones, which are found southeast of the Hawaiian islands. Within this region there are an estimated 8–25 billion tonnes of nodules.

Mining the seabed

A number of large international consortia have investigated possible methods of mining manganese nodules. At the present time, however, the costs of such an operation are prohibitive. Mining the seabed will involve two processes: the collection of the nodules from the bottom, and their transport to the surface. Unmanned collecting devices could be used on the seabed for aggregating the nodules, which subsequently could be lifted either by suction or by dredging operations. The use of deep-water trawls is likely to prove too expensive since the energy required to lift the trawl is considerable, and the time spent dropping and retrieving any trawl-like device would be greater than the time it could be used on the bottom actually collecting the nodules. Processing nodules on site or at a submerged processing station would make the mining more efficient, and these possibilities are also being considered.

Vents, brines and metalliferous muds

Long before the polymetallic nodules are being exploited on any large scale, another source of metals on the seabed is likely to prove more economic. Associated with deep-sea vents in areas of mid-ocean ridges are found deposits of metalliferous muds. These contain varying concentrations of zinc, copper, manganese and lead.

Metalliferous muds are formed by seawater permeating the molten oceanic crust. As the seawater filters through the crust, it becomes superheated, dissolving various metals from the molten basalt. As these superheated brines, which may be as hot as 104°C (220°F), are released into the colder water of the deep-ocean floor, precipitation of the dissolved metals takes place and deposits of metal-rich mud form.

Where water movement is restricted, as in the case of deep pits in the axial valley of the Red Sea, the superheated brine – at temperatures of 60°C (140°F) – fills the pits to a depth of around 200 metres (650 feet). The surface of the ocean floor is covered with metal-rich muds reaching between 2 and 25 metres (6 and 80 feet) in depth. The Atlantis II Deep in the Red Sea is of considerable commercial interest since the muds in this region contain as much as 40 per cent of iron, 3.5 per cent of manganese, 2 per cent of zinc and 0.95 per cent of copper.

Deep-sea mining and the Law of the Sea

The United Nations Law of the Sea entered into force in November 1994, 12 years after its signing in 1982. At this time, an International Seabed Mining Authority was estab-

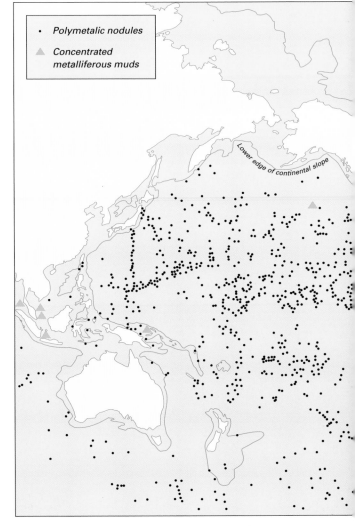

- • Polymetalic nodules
- ▲ Concentrated metalliferous muds

lished under the United Nations to regulate and control the exploitation of deep-sea mineral resources. Part XI of the Convention contains detailed provisions concerning the operation of deep-sea mining, and it is this section of the Convention which has prevented a number of industrialized countries from ratifying. Under this section, the deep ocean bed, and consequently the minerals of the ocean floor beyond the continental shelf, are considered as "the common heritage of mankind".

The International Seabed Mining Authority will become the body responsible for licensing and controlling deep-sea mining. Much of the opposition from the industrialized nations stems from the influence of the few large companies which have developed the technology to exploit these resources and which would prefer to operate under national rather than international jurisdiction. The sediments of the deep ocean are unconsolidated. Their disturbance will cause resuspension of this material which could be potentially damaging to the diverse living communities of the deep ocean. As a consequence, a number of environmental groups have expressed strong reservations about any deep-sea mining operations, and the mining companies fear that international jurisdiction will place too many restrictions on their mode of operation.

COMPOSITION OF MANGANESE NODULES (AIR-DRIED, % BY WEIGHT)

Element	Northeast Pacific Ocean	South Pacific Ocean	West Indian Ocean	East Indian Ocean
Manganese	22.33	16.61	13.56	15.83
Iron	9.44	13.92	15.75	11.31
Nickel	1.080	0.433	0.322	0.512
Cobalt	0.192	0.595	0.358	0.153
Copper	0.627	0.185	0.102	0.330
Lead	0.028	0.073	0.061	0.034
Barium	0.381	0.230	0.146	0.155
Molybdenum	0.047	0.041	0.029	0.031
Vanadium	0.041	0.050	0.051	0.040
Chromium	0.0007	0.0007	0.0020	0.0009
Titanium	0.425	1.007	0.820	0.582

◄ **Most of the world's** deep-sea mineral deposits constitute reserves, rather than resources, since the costs of their extraction are much too great to permit economic exploitation at the present time.

▲ **Manganese nodules** were first discovered by the Challenger *expedition and are now known to occur over vast areas of the seabed. Nodules grow very slowly by deposition of minerals around a small object.*

Oil and Gas

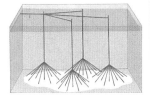

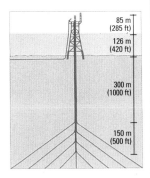

85 m (285 ft)
126 m (420 ft)
300 m (1000 ft)
150 m (500 ft)

▲ **The Forties field** in the North Sea is worked by four production platforms, each of which can drill 27 wells to 3500 m (11,500 ft). Each well is drilled out to an angle to reach over a radius of 2700 m (9000 ft) and tap a wide area of oil-bearing rocks. The wells need to penetrate only to the surface of the oil-bearing strata since water pressure forces oil up the well to the well head.

The world's continental shelf areas are, in relative terms, a more promising area for oil and gas exploitation than the land. Although the continental shelves occupy only about 26 million square kilometres (10 million square miles), approximately 15 million square kilometres (6 million square miles), or about two-thirds, are composed of the sedimentary basins where oil and gas reserves are normally found. Although the total area of the land is much greater, around 150 million square kilometres (60 million square miles), only about a third of that area is likely to yield oil and gas.

Oil and gas are formed by the decomposition of organic material in sedimentary rocks. The geological structures suitable for their accumulation must include permeable rocks, such as sandstones and limestones, through which the oil and gas can freely move as they are formed; and cap rocks, such as shale and salt, which seal the deposits in structures such as anticlines, preventing their escape to the surface or diffusion through surrounding rocks.

Seismic surveys are generally used to determine the structure of the underlying rocks of the continental shelf, and serve as the initial means of identifying likely areas for exploratory drilling. A ship towing a ray of sonic sensors fires a "shot", usually an explosive charge. The sensors detect the reflected waves from the seabed and underlying strata and generate a "sonic picture" of the geological formations. Following the detection of a likely geological structure, exploratory drilling may proceed. However, a very high proportion of the structures investigated prove to contain either only water or gas and oil reserves which are not economic.

Offshore drilling

The first oil to be extracted offshore was from the Summerland field of California in 1896. At this time, wells were drilled from piers that extended up to 250 metres (800 feet) from the shore. During the 1920s and 1930s, wells were drilled in the Baku region of Russia from trestles running out into the Caspian Sea. However, it was the discovery of the large Bolivar oil field in Lake Maracaibo, Venezuela, which really spurred interest in offshore oil reserves. By the 1940s, the first specifically designed offshore steel drilling structure was installed in 7 metres (23 feet) of water in the Gulf of Mexico.

Since that time, the design of offshore oil rigs and production platforms has advanced considerably. The height of these structures has increased as the search for oil has moved further away from the land. By 1975, 77 wells had been drilled in various places around the world in waters of about 200 metres (650 feet) deep. Initially, offshore oil exploration and extraction relied on transferring land-based technology to the ocean environment. However, sea-based rigs were limited in that they not only had to reach the necessary height above the well head for the successful operation of the drilling equipment, but they also required legs and support structures that were long enough to reach the seabed. One of the tallest platforms in operation today is found near Santa Barbara, California. It stands 355 metres (1165 feet) from the seabed to the top of the derrick – a mere 30 metres (100 feet) or so shorter than the Empire State Building in New York.

From about 1953 onwards, rigs were designed that could operate in around 90 metres (300 feet) of water

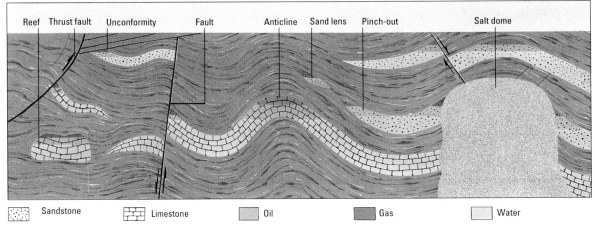

Reef　Thrust fault　Unconformity　Fault　Anticline　Sand lens　Pinch-out　Salt dome

Sandstone　　Limestone　　Oil　　Gas　　Water

▲ **Petroleum** is formed from decomposed organic material trapped in sedimentary rocks. Once formed, it migrates – along with the gas formed at the same time – through permeable strata until it is trapped by an impermeable layer. Here it accumulates to form a reservoir, with the gas and oil floating on the surface of the water. Permeable rock strata such as red sandstones may be capped by impermeable rocks such as anhydrite or shale, serving as traps for the oil and gas.

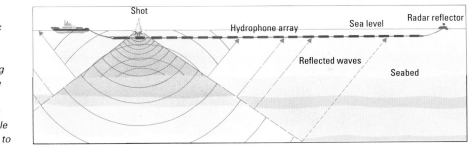

Shot　　　Hydrophone array　　Sea level　　Radar reflector
Reflected waves
Seabed

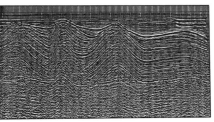

▲ **Geological structures** which might trap oil and gas can be detected by seismic soundings which provide a picture of the strata. When a seismic shot is fired the reflected sounds are picked up by sensors. Different rocks and structures reflect the sound differently, providing a seismic profile such as the one on the left.

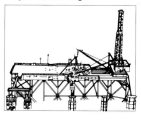

▲ *Template rig*
Shallow/medium water

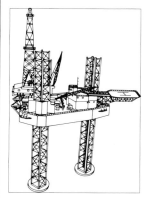

▲ *Semisubmersible rig*
Deep-water operation

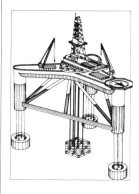

▲ *Jackup rig Medium-water operation*

▲ *Tension-leg rig Deep-water operation*

▲ *Drilling ship General survey*

▲ **A North Sea** *gas platform as seen from a supply vessel. The number of offshore platforms for oil and gas production has increased dramatically world-wide over the last two decades as the demand for petroleum products continues to increase.*

▼ **The height** *of oil production platforms has steadily increased as the search for oil and gas has reached deeper and deeper waters. The illustration below shows the increase in height of production platforms above the seabed from 1966 to 1977.*

through legs lowered into the seabed on which the drilling platform was jacked up above the sea surface. During the 1960s, the semi-submersible rig was designed. This rig was fitted with pontoons that provided buoyancy to a partially submerged structure anchored to the seabed.

Mobile rigs

Although drill ships have been used for some time, they were usually anchored to the seabed in order to maintain position over the well head. More recently, however, the development of dynamically positioned drilling rigs has dispensed with the need for such ships to be anchored to the seabed during exploratory drilling operations. The drilling ships maintain their position over the well head by continuous adjustments of specialized propellers which respond to sonar signals from sources located at known positions around the well head.

The first of these ships was put into service in 1971, and by 1976 the fleet of mobile drilling rigs had increased to 350 worldwide. By the late 1970s, drilling had been carried out in 1000 metres (3300 feet) of water off Thailand, and during the 1980s test wells were drilled in the Maldives and Philippines at depths in excess of 2000 metres (6500 feet).

The rapid increase in offshore oil exploration and exploitation worldwide reflects the dominance of fossil fuels to the economies of the industrialized nations, and in turn the increase in demand for this source of energy. In 1960, for example, 90 per cent of all offshore oil drilling took place in the United States. By the 1970s, however, the American contribution had dropped to approximately a quarter of world offshore production, a proportion which has further declined as increasing numbers of countries around the world discover and exploit their reserves.

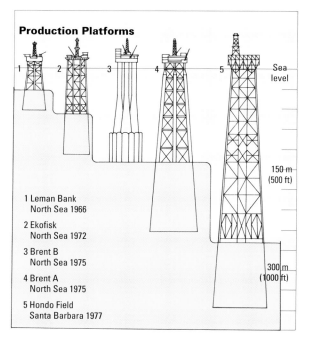

Production Platforms

Sea level

150 m
(500 ft)

300 m
(1000 ft)

1 Leman Bank
North Sea 1966

2 Ekofisk
North Sea 1972

3 Brent B
North Sea 1975

4 Brent A
North Sea 1975

5 Hondo Field
Santa Barbara 1977

Power From the Sea

The idea of harnessing the power of the seas for the production of energy has existed for some time. Yet the efficiency of most systems developed to date is inadequate, making the electricity they generate more expensive than burning fossil fuels. As a result, few operational systems have been developed. However, concern over the continuing use of fossil fuels has led to increased investment in such systems over the last few years.

Ocean power, in all its many forms, is essentially a renewable energy source. The power derives either from the energy of the Sun, which is stored as heat in the oceans, or it is transferred from the atmosphere to the sea in the form of waves. Tidal energy results from the force of gravity from the Sun and Moon acting on the water of the ocean basins.

Wave and tidal power

Wave-power systems are based on the fact that energy is continuously transferred to the ocean from the wind. Tidal systems rely on the ebb and flow of the sea under the influence of gravity and the Earth's rotation. Wave-power systems have the added advantage of providing coastal protection since the energy of the waves is removed and the water passing over or through such systems moves more slowly. No fully operational systems of wave-power generating electricity are yet in operation, although prototypes have been constructed and tested in a number of countries. These prototypes are based on the action of the waves driving huge rocker-shaped structures which pump

▼ **Ocean Thermal Energy Conversion (OTEC)** *is designed to utilize the great temperature difference between the surface and deep waters to drive a closed evaporation condensation cycle based on ammonia or another liquid with comparable thermal properties. Cold water from the ocean depths would be used to condense the vapour, which would be cycled through evaporators utilizing warm surface water, and the cycle would be used to drive turbines linked to banks of generators.*

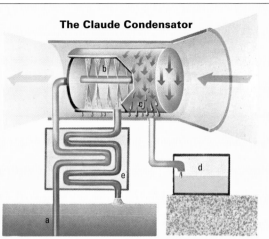

The Claude Condensator

One ingenious suggestion for using the ocean's thermal gradient directly is in the desalination of seawater to produce fresh water in arid regions of the world. Cold deep water (a) is pumped to the surface and used to cool the surface of a spray chamber (b). Moisture then condenses from the warm humid air drawn into the apparatus and is drawn off into tanks (d). The coolant water warmed in the chamber is passed through a heat exchanger (e) so that the temperature is reduced before it is returned to the sea.

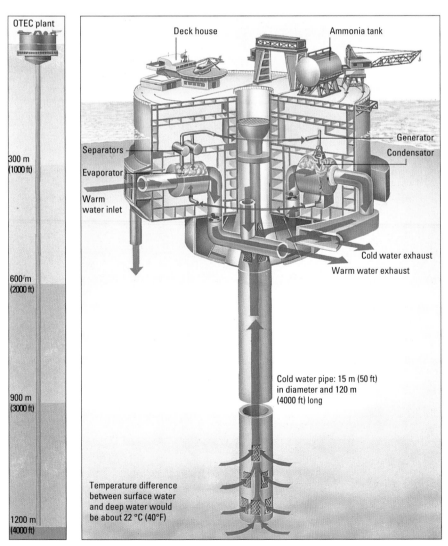

OTEC plant

300 m (1000 ft)

600 m (2000 ft)

900 m (3000 ft)

1200 m (4000 ft)

Deck house

Ammonia tank

Separators

Evaporator

Warm water inlet

Generator

Condensator

Cold water exhaust

Warm water exhaust

Cold water pipe: 15 m (50 ft) in diameter and 120 m (4000 ft) long

Temperature difference between surface water and deep water would be about 22 °C (40°F)

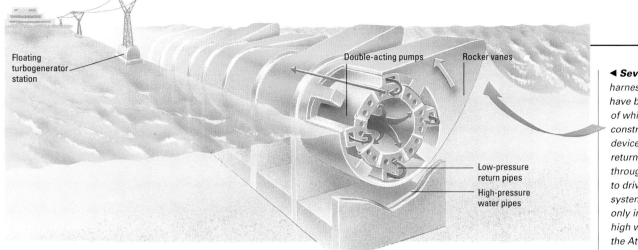

Floating turbogenerator station

Double-acting pumps

Rocker vanes

Low-pressure return pipes

High-pressure water pipes

◄ Several designs for harnessing wave energy have been proposed, one of which involves the construction of huge rocking devices which activate non-return valves, forcing water through small-bore pipes to drive turbines. Such a system would be suitable only in areas of continuous high wave energy such as the Atlantic approaches to Britain.

water through non-return valves to drive the turbines used for generating the electricity.

The first fully operational system harnessing tidal power has been working in the Rance estuary of northern France for several decades. Tidal systems such as this generate electricity through turbines located in a dam spanning the estuary. The turbines are driven by the ebb and flow of water with the tides.

Ocean Thermal Energy Conversion

Some experimental systems such as Ocean Thermal Energy Conversion (OTEC) are being developed. This system depends on the temperature differential between warm, surface water and the considerably cooler water of the deep ocean. The temperature difference may be as much as 20°C (36°F). The cold, deep water is used to condense ammonia or similar fluids which are then passed through evaporators warmed by the surface waters. The cycling of the gas through the condensation cycle is used to drive turbines for electrical power generation. Although the efficiency of such systems is relatively low, alternative uses for this thermal gradient include the Claude Condensator, which can be used for generating freshwater supplies in coastal areas of arid countries.

Other forms of power

While it would be inefficient to return to the days of sailing ships, ocean currents and waves could be harnessed for maritime transport. The deep-water outflow from the Mediterranean, for example, is used by submariners to navigate the straits of Gibraltar with engines turned off. It has been calculated that the wave power which causes a ship to rise and fall at sea is considerably more than that required to propel it. If this energy could be efficiently harnessed, then long-distance transport costs could be reduced. In addition, small, man-made currents could be used to influence the direction of sediment transport by tidal currents. Again more efficient use of such developments would take the place of dredging and expensive beach replenishment schemes, both of which rely on the expenditure of energy by pumping sand from offshore areas onto eroding coastlines.

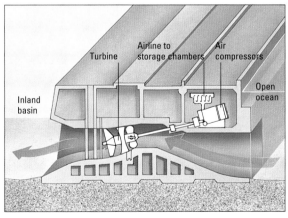

Turbine

Airline to storage chambers

Air compressors

Open ocean

Inland basin

◄ The tidal power station in the Rance estuary in France was the first system to use ocean power. The dam spans the estuary and contains the turbines below the waterline which are driven by the ebb and flow of the tides.

▲ A possible future development of tidal power without the need for construction of a dam involves harnessing the tidal flow to compress air which can be used to drive gas turbine generators.

Marine Pollution

At the time of the United Nations Conference on the Human Environment held in Stockholm in 1972, the greatest threat to the marine environment was perceived to be marine pollution. Waste discharged from ships, as well as urban and industrial effluent from land-based sources, were all seen as contributing to a significant reduction in the health and quality of the marine environment.

Although marine pollution is still considered a major cause of concern, it is now recognized that other human activities, including large-scale commercial fishing and the extensive modification of coastal environments, may be having just as great an impact on the quality of the seas and oceans.

In 1990, a report published by the Joint Group of Experts on the Scientific Aspects of Marine Pollution, which advises the United Nations' agencies, concluded that, "In 1989 man's fingerprint is found everywhere in the ocean. Chemical contamination and litter can be observed from the poles to the tropics and from beaches to abyssal depths. But conditions in the marine environment vary widely."

This group further noted that although the open ocean remains relatively clean, the greatest problems of pollution and contamination are found in coastal areas. Many coastal habitats are being lost irretrievably to the construction of harbours and industrial installations, the development of coastal settlements and cities (including tourist facilities) and an increase in mariculture.

By the time of the United Nations Conference on Environment and Development held in Rio de Janeiro in 1992, land-based pollution was considered to be the major source of pollution in the marine environment. Of even greater importance, however, was the over-riding need to develop more rational management of human uses of the coastal zone and inshore resources. Agenda 21, which was approved by the governments attending the Rio Conference, calls for all coastal states to develop integrated coastal zone management plans by the year 2000.

▲ **The Exxon Valdez** *spill in Alaska had devastating local effects, but chronic marine pollution by oil has declined in recent decades.*

▼ **A small vessel** *dumps jarosite waste off the coast of Australia. Dumping mining waste presents localized pollution problems.*

Chemical and oil pollution

In the recent past, heavy metals such as mercury, cadmium and lead were considered among the most pervasive pollutants. It is now recognized, however, that a number of marine organisms naturally concentrate these elements, and hence high concentrations may not necessarily reflect man-made pollution. Nevertheless pollution by such elements remains a concern in areas of high-industrial discharge. Anti-fouling agents such as organo-tin compounds are known to have major effects on the reproductive biology of shellfish, and their use has been banned by some countries. Chlorinated hydrocarbon pesticides may also be causing problems along tropical coastlines, although the concentrations of these materials has declined in the developed north following control and restriction of their use.

Contaminants enter the sea either through direct discharge or indirectly through rivers and through the atmospheric transport of particles in aerosols and gases. Around 80 per cent of all marine pollution is derived from land-based sources, a further 10 per cent results from marine dumping, and the remaining 10 per cent from maritime operations, such as ship-based discharge of sewage.

Maritime activities have considerably less impact now than in the past, mainly due to the introduction of international conventions limiting the discharge of wastes at sea. Nevertheless maritime accidents, such as the *Exxon Valdez* disaster, may have devastating and highly visible local impacts. Although oil may be considered a highly visible contaminant of the marine environment, particularly following tanker disasters, it is generally of less concern than many other materials. Floating oil is generally less damaging than oil which comes into direct contact with bottom-dwelling organisms, either in the intertidal or subtidal areas. Damage from such accidents is not usually irreversible, although recovery may be slow.

Most of the materials entering the ocean remain in the continental shelf areas, and in semi-enclosed bays and seas where they may be deposited in sediments and resuspended at a later date during storms or dredging operations. In some semi-enclosed areas, such as the North Sea, the build-up of contaminants has reached unacceptably high levels, resulting in algal blooms, toxic red tides and viral deaths of marine mammals.

Plastic and other refuse

The haphazard disposal of plastic material on land and from ships at sea results in the fouling of beaches, and seriously affects marine wildlife, particularly mammals, diving birds and turtles. These animals may become entrapped or tangled in such materials and drown.

A considerable quantity of fishing gear is also lost at sea each year and nylon nets are reported to "ghost" fish, catching fish years after they have been lost.

Nutrients

Present discharges of sewage, both treated and untreated, not only represent a potential health hazard to bathers and seafood consumers, but perhaps more importantly are increasing the rate of primary production in coastal waters. Sewage and agricultural run-off are high in nitrogen and phosphorus which encourage phytoplankton production in coastal waters. These high inputs cause rapid growth, or blooms, of phytoplankton, resulting in unsightly algal scum on tourist beaches. When the algae die and sink to the bottom the resulting bacterial decomposition uses up available dissolved oxygen causing deoxygenation of bottom waters, which in extreme cases will kill fish. Often the species of algae in the blooms produce toxic substances which may be taken up by shellfish, rendering them unfit for human consumption.

Managing the Oceans

▲ *The Great Barrier Reef Marine Park off Australia is managed by a single authority, which regulates the use of the area by fishermen and tourists and controls the numbers of visitors to offshore islands to reduce the impacts of tourism in this unique environment.*

The historical development of the frameworks for managing the oceans and their resources is based on the concept of the freedom of the seas. The concept was articulated by Dutch and English seafarers at the time the Spanish and Portuguese were dividing the New World and the newly discovered oceans into spheres of influence. The idea of free access to the oceans was also applied to the resources of the oceans, which were considered open for use by anyone possessing the means to exploit them.

Such a perspective was acceptable only when the limits of human use did not result in conflict and over-exploitation. When human populations were much lower than today and technologies were not highly developed, the extent to which the global population could affect the world's oceans was limited. However, the rapidly expanding world population, combined with improved technologies for maritime transport, for exploiting the hidden resources of the seabed and for catching fish, have now resulted in levels of use which threaten the long-term sustainability of the ocean environment.

Managing living resources

The history of whaling and many fisheries demonstrates the problems of open access to living marine resources. When a new fish resource is discovered, fishermen invest in boats and gear. If the fishery is not controlled or regulated the numbers of boats increase dramatically until a point is reached when the catch starts to decline. Initially, this may result only in a decline in the average size of fish in the population. However, this can soon be followed by decline in catches. If the fishing effort remains at the same level, complete collapse of the fishery can occur, with subsequent economic and social impacts for the fishing communities concerned.

In addition, where access to the fish resource or fishing grounds is uncontrolled, the monetary economies of the developed world ensure the incentive to over-exploit is high. The argument is: "If I don't catch them, someone else will." The catch is viewed no longer as a long-term resource and vital necessity for subsistence, but as a source of short-term, personal, economic benefit.

The Exclusive Economic Zone

In an attempt to address such problems, the concept of the Exclusive Economic Zone (EEZ) was developed. EEZs try to provide a form of ownership which restricts access to the fish stocks of the EEZ to fishermen of the country concerned or licensed fishermen from other nations. Unfortunately, this does not prevent over-capitalization of the national fishing fleet and hence the problems of over-exploitation of the resource. These problems require urgent attention at all levels and recognition of the fact that many of the management techniques currently in use, including "days in port" and restrictions on gear size and use, have failed to halt the decline of some stocks.

Conflicts of use

Not only is marine exploitation rapidly exceeding sustainable capacity, but the extent to which different human uses of the oceans conflict with one another is also increasing. This is particularly apparent in coastal areas where around 60 per cent of the world's present population live. Two-thirds of all cities with more than 2.5 million inhabitants are located on the coast, often in fertile and productive estuarine and delta areas.

Urban and industrial development conflict with coastal agriculture and fisheries. The discharge of pollutants affects the health and acceptability of marine foodstuffs, both wild and farmed. Tourist development depends on healthy coral reefs, unpolluted beaches and clean water. However, both mariculture farms and the development of ports and harbours for industry conflict with such use.

Resolving these conflicts requires not only a sound scientific basis of information and knowledge, but also the recognition that ocean resources, including space, are not infinite. Many marine pollution problems stem from the view, now no longer valid, that the ocean was so vast that it had an infinite capacity to absorb, dilute and disperse toxic and other waste materials without harmful effects. Regrettably this is not the case. The ocean, although vast, is not well mixed, and the bulk of contaminant materials discharged to coastal areas remains in close proximity to the shore. Here they are recycled between the sediments, the water and the living organisms – often for many decades following their discharge – before degrading into less harmful products. Enclosed bays and semi-enclosed seas, where the rate of water exchange with the open ocean is limited, are particularly vulnerable. Seas such as the Baltic, Black and North Sea have all reached a critical

◄ **Many of the world's** coral reefs are under threat from the growing coastal populations in tropical developing countries which depend on these fragile systems for food. Like rainforests, coral reefs are areas of high species diversity, and indeed they are the most diverse marine ecosystems.

▼ **Marine animals** have no respect for the political and administrative boundaries drawn through their world by politicians and decision-makers. Managing wide-ranging species such as whales and migratory species requires the consent and co-operation of all parties if we are not to lose such species permanently.

stage whereby the present loadings of contaminant materials are at, or even beyond, the level which can cause irreversible damage to the marine ecosystems concerned.

Spatial limits and management units

One of the major problems with developing rational management tools and strategies is the problem of what constitutes the unit to be managed. Administrative boundaries between towns and villages, between privately and publicly owned land and coastal zones rarely reflect the biological or physical boundaries which separate populations of organisms or functional ocean units. In countries such as Australia, for example, the individual states claim rights to the territorial waters, while the Federal Government claims rights over the 200-mile EEZ. The artificial boundary between territorial seas and EEZ has no biological or physical meaning, and may divide stocks or whole ecosystems in a purely arbitrary manner.

National boundaries often cross physical features in such a way that two states may control different coastal areas used by the same population of fish. The feeding and breeding areas of a single species may be controlled and exploited by two different states. Transboundary stocks that cross on migration, or merely exist in an area divided between two owners, cannot be successfully managed without the joint consultation, consent and agreement of both parties.

In recognition of this problem, particularly in relation to pelagic tuna stocks, regional fisheries commissions have been established to provide a forum for negotiation on the fish catches of individual states. Unfortunately, at present, the area of the high seas outside the 200-mile EEZ can be fished by any nation. Consequently, regulating the catch of transboundary stocks by long distance fishing fleets is not possible.

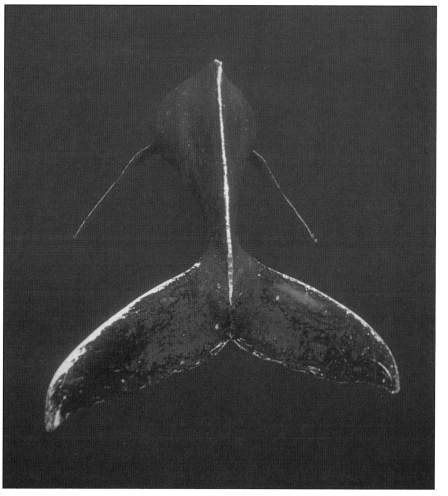

OCEAN ATLAS

◄ **The world's** oceans vary from the freezing southern ocean of Antarctica to the warm tropical Caribbean; from the supersaline Red Sea to the almost freshwater of deltaic coastal regions; some areas are teeming with life, while others are barren. The diversity is enormous.

The World Ocean

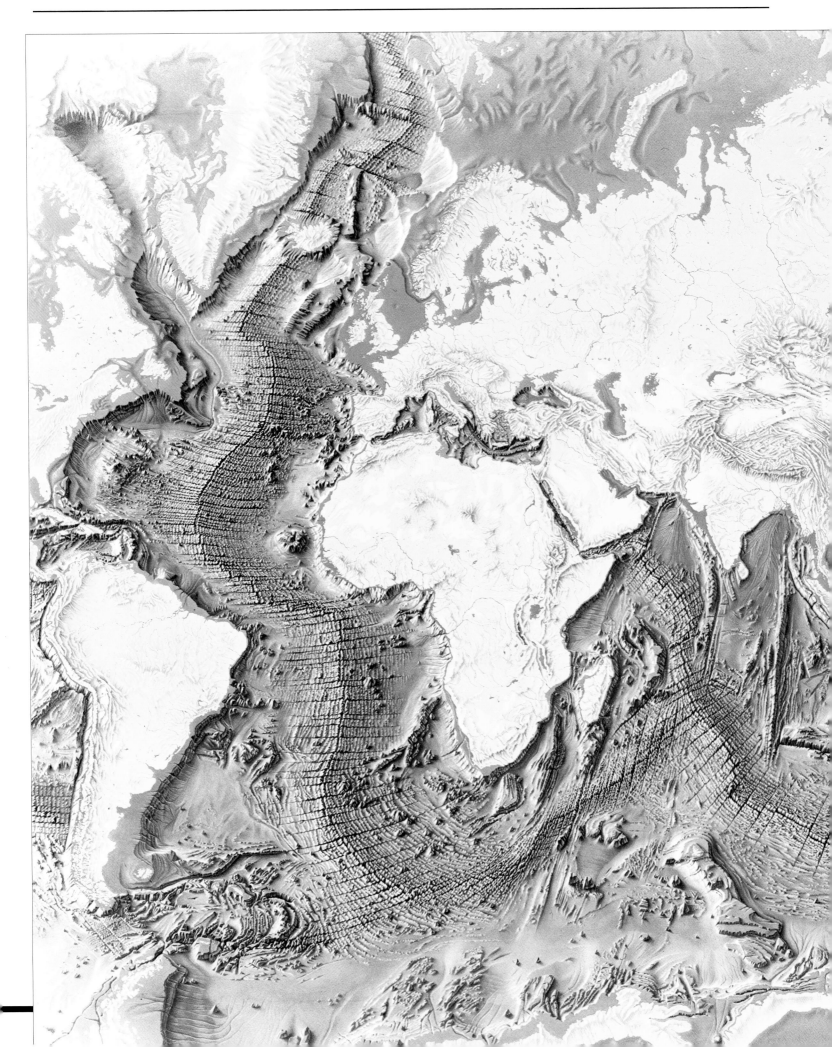

The Atlantic Ocean

▲ **The Atlantic Ocean**
Area: *82,000,000 sq km (31,660,000 sq miles)*
Average depth: *3330 m (10,930 ft)*
Volume: *321,930,000 cu km (77,235,000 cu miles)*
Max. depth: *(South Sandwich Trench) 9144 m (30,000 ft)*

The Atlantic Ocean, although some 84 million square kilometres (32 million square miles) smaller than the Pacific, receives freshwater runoff from a far larger area of land. In fact, an area four times as large as that which drains into the Pacific. The saline content of surface waters in regions of high rainfall and freshwater runoff from land are generally lower than in other areas. Because the Atlantic receives so much freshwater runoff, low-salinity waters dominate the surface of much of the Atlantic.

To the north, in the area of the Greenland Sea, the Atlantic is connected with the almost entirely land-locked Arctic Ocean. It is via this channel that approximately 80 per cent of water exchange from the Arctic takes place. Furthermore, it is here that the cold, high-density Atlantic bottom water forms, known as the North Atlantic Deep Water. This mass of water sinks and spreads out at a depth of around 1500–4000 metres (5000–13,000 feet), creating a south-flowing, deep-water current that continues south to beyond the Equator.

Formation of the Atlantic

The Atlantic first began to form 195–135 million years ago, in what is now the central part of the North Atlantic. North America at this time drifted away from the combined landmass of Africa and South America at a rate of about 3 centimetres (1 inch) a year. By about 150 million years ago, the Central Atlantic had opened to roughly 30 per cent of its present width.

It was not until the Cretaceous period that the South Atlantic began to form as South America and Africa began to drift apart. Finally, the extreme North Atlantic was created as the seafloor between Greenland and the Rockall Plateau began spreading about 60 million years ago.

The Atlantic gyres

Dominating the current pattern of the North Atlantic is the North Atlantic Gyre. An almost circular system of warm, surface-water currents, the North Atlantic Gyre is driven by the atmospheric circulation of the Northeast Trade winds, which blow across the Atlantic between 10° and 30°N; and the Westerlies, found between 40° and 60° N.

In the south, the Antarctic Circumpolar Current carries deep, Antarctic bottom waters into the Atlantic, contributing to the strength of the Benguela Current off the west coast of Africa. Unlike the North Atlantic, the South Atlantic forms a single oceanographic unit. It is also dominated by a warm-water gyre, which again is driven by the atmospheric circulation of the Southeast Trade winds and the Westerlies.

In contrast, the open ocean waters of the North Atlantic are connected in the west to the semi-enclosed Caribbean Sea and Gulf of Mexico, while to the northeast the North Sea is connected via a narrow channel, the Skaggerack, to the Baltic. Further south, the Mediterranean Sea receives water from the North Atlantic through the narrow Straits of Gibraltar. It is in turn linked to the Black Sea via the Dardanelles and to the Red Sea by the Suez Canal which was opened in 1869.

The western Atlantic and Sargasso Sea

The surface waters of the western Atlantic are dominated by high salinities and high temperatures as a consequence of the high evaporation and low inputs of freshwater into the Caribbean Sea. These dense, warm waters eventually cool, becoming denser in the northern North Atlantic, sinking to slide south. This combination of density and salinity differences drives the thermohaline circulation pattern of the North Atlantic.

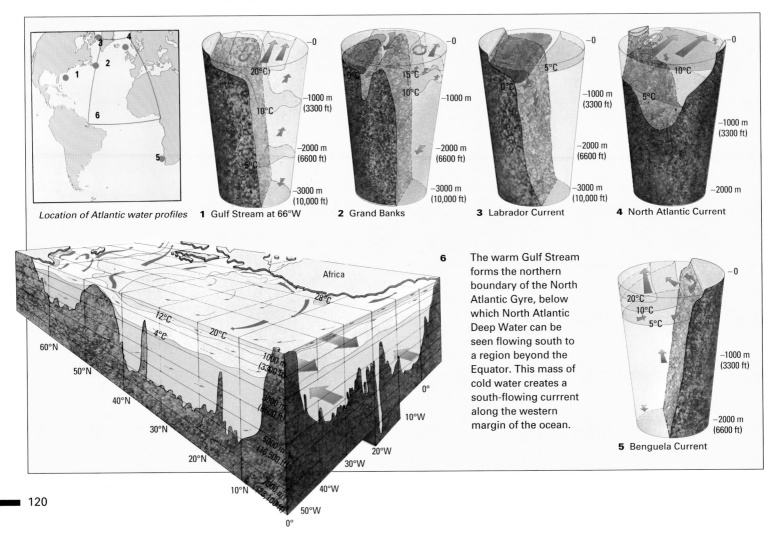

Location of Atlantic water profiles   **1** Gulf Stream at 66°W   **2** Grand Banks   **3** Labrador Current   **4** North Atlantic Current

6 The warm Gulf Stream forms the northern boundary of the North Atlantic Gyre, below which North Atlantic Deep Water can be seen flowing south to a region beyond the Equator. This mass of cold water creates a south-flowing currrent along the western margin of the ocean.

5 Benguela Current

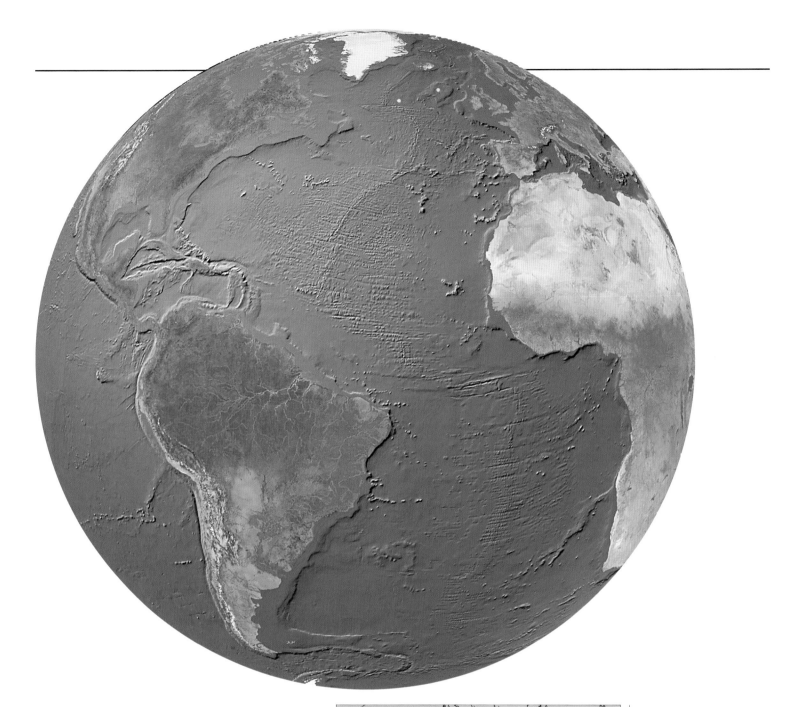

One of the results of of the circular motion of the surface currents in a gyre, is that, towards the centre of the gyre the the surface waters often have lower motion and indeed may be higher than the surrounding waters balancing the Coriolis force of the surrounding currents. As a consequence, the Sargasso Sea has a mean surface level approximately a metre (3 feet) higher than that of the neighbouring coasts. Within this area of relatively still water is found a community based on the floating Sargassum weed, related to the brown seaweeds of temperate shores. The Sargassum forms a food source and hiding place for small crustacea, molluscs, sea anemones and fish such as *Histrio*, and a place of attachment for barnacles and algae. This community differs considerably from that of the surrounding waters, since the large size of the algae compared with the plankton of the open ocean provide a major food resource for herbivorous animals.

Not only does the Sargasso Sea form a home for the permanent seaweed community, but this is also the area to which freshwater eels from North America and western Europe migrate on their annual spawning migrations. The larval eels being carried by the currents of the North Atlantic Gyre northwards along the Atlantic seaboard of the United States and across to western Europe.

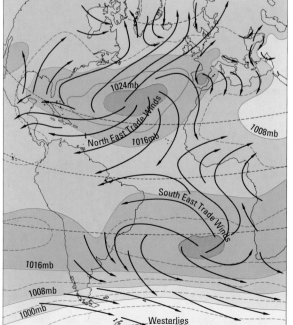

◄ **The atmospheric** circulation of the Atlantic Ocean closely mirrors the surface ocean currents. The Northeast Trade winds drive the Canaries and North Equatorial currents from east to west across the North Atlantic, and the Westerlies drive the Gulf Stream to the north. The Southeast Trade winds in the South Atlantic drive the South Equatorial and Brazil currents, while the westerly winds of the Southern Hemisphere complete the circle of the South Atlantic Gyre.

The Atlantic Ocean Basin

The Atlantic Ocean floor is dominated by an S-shaped, mid-ocean ridge. This vast, underwater mountain range extends from north of Iceland as far south as Bouver Island on the margin of the Southern Ocean. Known as the Mid-Atlantic Ridge, it divides the Atlantic Ocean Basin into two parallel troughs, which in turn are sub-divided by transverse ridges.

Some peaks along the ridge emerge as the islands of the Azores, Ascension Island and Tristan da Cunha, but most of the ridge lies 1.5–3 kilometres (1–2 miles) below the surface.

Along the centre of the ridge runs a deep rift valley, which varies in width between 24 and 48 kilometres (15 and 30 miles). The valley is displaced by east-west transform faults which can be seen as narrow ridges and deep clefts, in some places extending more than 2000 kilometres (1250 miles) from the centre of the ridge.

The Atlantic Ocean is continuing to expand, at a rate of 1–2 centimetres (0.4–0.8 inches) per year, due to the process of seafloor spreading. The volcanic eruption south of Iceland in 1963, which formed the island of Surtsey, was a dramatic example of this process in action.

Passing away from the mid-ocean ridge, towards the continental rise, the two troughs become progressively deeper, and the lateral ridges and clefts become obscured by increasing depths of sediment brought down from the adjacent land and continental shelves by bottom currents.

Sediment eroded from the land and brought down into the coastal ocean flows across the continental shelf and into the ocean basin via deep canyons cut into the face of the continental rise. At the foot of the continental rise the sediment may be redistributed by bottom currents to accumulate in a specific region, such as the Argentinian Rise which lies northeast of the Falkland Islands.

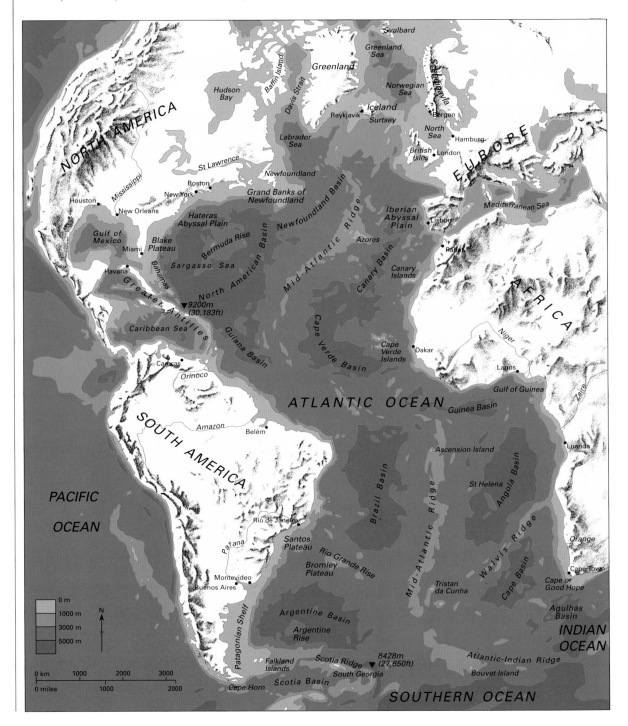

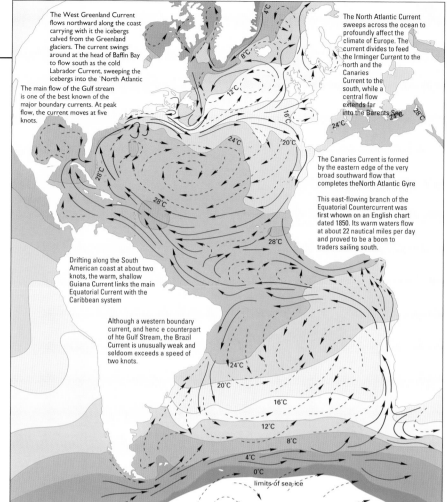

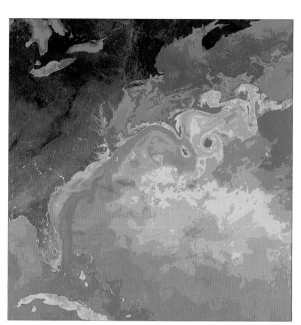

▲ *Meanders and eddies* in the Gulf stream are vividly portrayed in this false-colour satellite image. The Gulf stream appears as a red (warm) river producing warm core (3) and cold core (4) rings.

At the edge of the continental slope land derived sediments may exceed 5 kilometres (3 miles) in thickness and off the North American and North African coasts the sediments date back as far as the Jurassic period (170-160 million years ago). In contrast, the depth of sediment over the younger seafloor is usually less than a kilometre (0.6 miles) deep and consists of pelagic ooze and clay.

Currents and circulation

Perhaps the most familiar current of the North Atlantic is the Gulf Stream – a strong, narrow river of warm water which flows north at more than 130 kilometres (80 miles) a day. It flows along the eastern seaboard of North America before leaving the shore around the latitude of the westerly winds (40–60°N) and crossing the Atlantic as the North Atlantic Current, the northern boundary of the warm-water North Atlantic Gyre. Without the influence of the Gulf Stream the winter climate of western Europe would be much more severe than at present.

A second warm-water gyre lies in the south. The South Atlantic Gyre is formed by the northern portion of the Antarctic Circumpolar Current which flows along the eastern coast of Africa as the Benguela Current before crossing the Atlantic as the Equatorial Current. This current divides at the coast of Latin America with the northern branch sweeping into the Caribbean Sea as the Guiana Current and the southern branch passing southwards along the coast as the weak Brazil Current.

Separating the northern and southern warm-water gyres is the Equatorial Countercurrent which flows on the surface towards the North African coast. This current overlies the substantial Equatorial Undercurrent, a body of water about 200 kilometres (125 miles) wide and which flows east at about 80 kilometres (50 miles) a day at a depth of 100 metres (330 feet).

At higher latitudes the westerly winds drive two cold-water gyres. In the north the subpolar gyre consists of the Irminger Current, the Greenland Current and the North Atlantic Current. In the south a similar cold-water gyre is found in the area of the Weddell Sea.

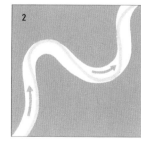
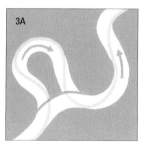

▲ *Growing instability* causes Gulf Stream meanders to increase in amplitude (1–2) until they break away as independent eddies. If the meanders break away on the northern side of the Gulf Stream (3A), then the eddy has a warm core of Sargasso Sea water and rotates anticlockwise (4A). If, however, the meander breaks away to the south (3B), the eddy rotates cyclonically around a cold core of slope water (4B). North-side eddies progress westward to rejoin the

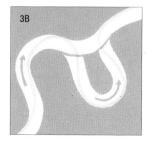
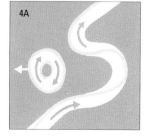
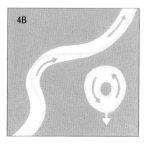

eastward-flowing Gulf Stream off Cape Hatteras in North Carolina, and becoming part of the distinctive northern boundary of the North Atlantic Gyre. Meanwhile south-side eddies make a long, irregular, southerly loop to rejoin the Gulf Stream near Florida. The eddies, which may be anything up to 320 km (200 miles) in diameter and approximately 2 km (1.2 miles) deep, may persist for up to two years. It is thought that these vast eddies play an important role in the overall return flow of the Gulf Stream.

Within map:

The West Greenland Current flows northward along the coast carrying with it the icebergs calved from the Greenland glaciers. The current swings around at the head of Baffin Bay to flow south as the cold Labrador Current, sweeping the icebergs into the 'North Atlantic

The main flow of the Gulf stream is one of the best known of the major boundary currents. At peak flow, the current moves at five knots.

The North Atlantic Current sweeps across the ocean to profoundly affect the climate of Europe. The current divides to feed the Irminger Current to the north and the Canaries Current to the south, while a central flow extends far into the Barents Sea.

The Canaries Current is formed by the eastern edge of the very broad southward flow that completes the North Atlantic Gyre

This east-flowing branch of the Equatorial Countercurrent was first whown on an English chart dated 1850. Its warm waters flow at about 22 nautical miles per day and proved to be a boon to traders sailing south.

Drifting along the South American coast at about two knots, the warm, shallow Guiana Current links the main Equatorial Current with the Caribbean system

Although a western boundary current, and henc e counterpart of hte Gulf Stream, the Brazil Current is unusually weak and seldoom exceeds a speed of two knots.

limits of sea ice

Atlantic Resources

Most of the mineral deposits of the Atlantic, with the exception of oil and gas, are of insufficient economic value to merit commercial exploitation. However, some of the finest diamonds in the world are found on the marine terraces of Oranjemund, off the southwest coast of Africa, and gem-quality diamonds are recovered from raised beaches and placer deposits lying below the tide line.

Production of those minerals which are currently being exploited is largely concentrated in shallow water on the continental shelf close inshore. Petroleum and gas extraction is centred on the wider Caribbean region, the North Sea and along stretches of the West African coast.

Apart from oil and gas, the value of sand and gravel is greater than that of all other minerals combined. Most of the extraction occurs off the coast of northwest Europe, where around 10 million tonnes of gravel are dredged annually by operations in the North Sea. Dredging aragonite sands from the Great Bahamas Bank is also of commercial importance.

Depleted living resources

The majority of the North Atlantic fish catch comes from the continental shelf areas bordering northwestern Europe, eastern Canada and the United States. The North Atlantic is the world's most heavily fished ocean area, and numerous fish stocks are suffering from over-fishing. The North Atlantic salmon fishery, for example, has declined in recent years. Although the salmon spawn in rivers on both sides of the Atlantic, much of their adult life is spent off Greenland, where commercial fishing has resulted in depletion of the stocks. Much of the salmon has now been replaced by farmed salmon, particularly from Scandinavia.

The most important fisheries by weight of catch in the North Atlantic are the pelagic species, dominated by sardine and anchovy. Demersal fish, particularly cod, flounder and plaice in the north, and hake from southern Europe, northern Africa and America are also heavily fished. Finally, although not as large as that of the Pacific, the Atlantic tuna fishery is also significant.

Crustacean fisheries for lobster, prawn and shrimps are widespread. Most of the lobster come from the northeast United States, the Caribbean, northern Brazil and along the South African coast. Penaeid shrimps are fished off the West African coast and in the Caribbean, while crabs form

an important resource off the eastern seaboard of the United States and to a lesser extent in the North Sea.

The rich fisheries of the West African coast are the consequence of high primary production resulting from the upwelling of cold, nutrient-rich waters off Senegal and Zaire. In addition, the high nutrient inputs from the Congo and Zaire drainage basins also contribute to the high productivity of this area.

Farming the sea

Mariculture is widespread in the North Atlantic, with salmon and trout being farmed in Canada, Scandinavia and western Scotland. Additionally, production of Pacific oysters in France exceeded 125,000 tonnes in 1985, while in the same year, Spain produced more than 240,000 tonnes of blue mussel. Mariculture of oysters, clams, mussels and various fish expanded in the United States throughout the 1980s; however, degradation of water quality may limit further growth of this industry.

Northwest Europe		
100 m (330 ft)		
200 m (660 ft)		Sole, flounder
300 m (1000 ft)		Hake
400 m (1300 ft)	Cod, haddock, whiting	Plaice, rays
500 m (1650 ft)	Norway pout, saithe	Blue whiting

North Africa		
100m (330 ft)		Flatfish, surmullets
200 m (660 ft)		Axillary breams, gurnards
300 m (1000 ft)		European hake, Norway pout, horse mackeral
400 m (1300 ft)		European hake
500 m (1650 ft)		Deep-sea shrimps
600 m (2000 ft)		

Demersal fish live on continental shelves, where temperature, light intensity and pressure vary more quickly with depth than regionally. This limits the vertical distribution of some species.

▲ **The world's largest,** single, offshore mining operation occurs on Ocean Cay, an artificial island, created from aragonite sands dredged from the Great Bahamas Bank. Aragonite contains calcium carbonate, and, among other things, is used in the manufacture of cement, glass and animal feed supplements.

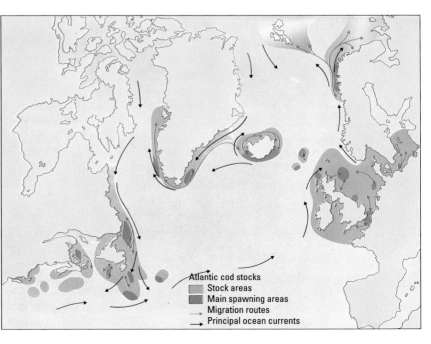

Atlantic cod stocks
- Stock areas
- Main spawning areas
- Migration routes
- Principal ocean currents

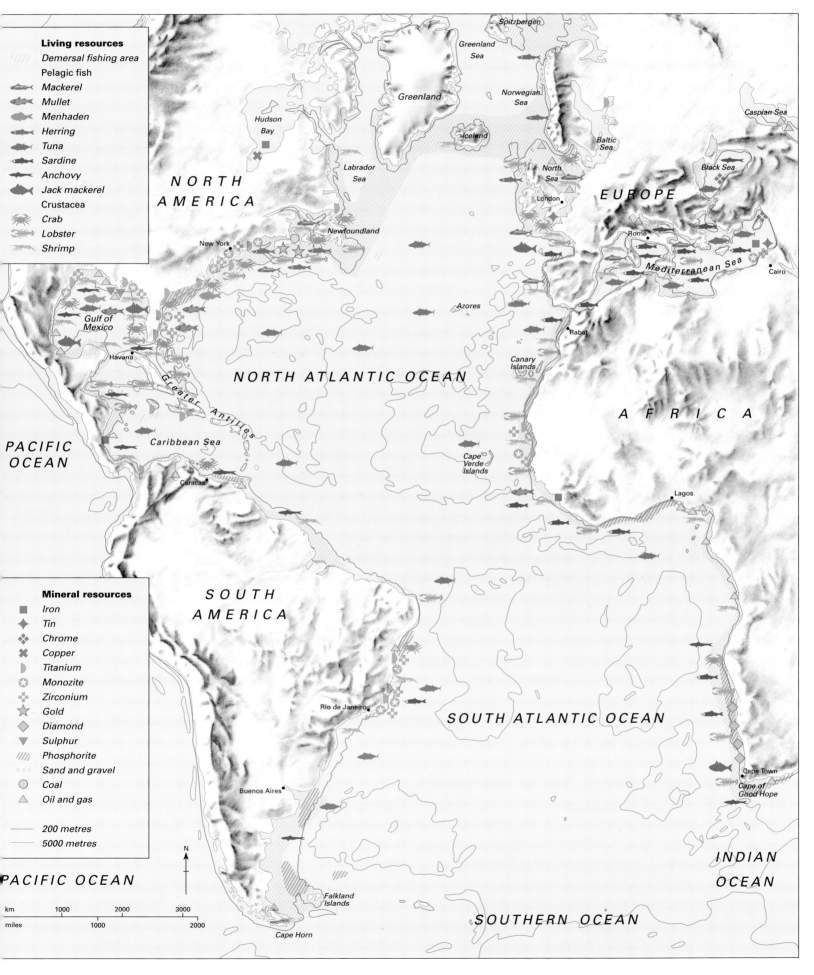

Living resources

- Demersal fishing area
- Pelagic fish
- Mackerel
- Mullet
- Menhaden
- Herring
- Tuna
- Sardine
- Anchovy
- Jack mackerel
- Crustacea
- Crab
- Lobster
- Shrimp

Mineral resources

- Iron
- Tin
- Chrome
- Copper
- Titanium
- Monozite
- Zirconium
- Gold
- Diamond
- Sulphur
- Phosphorite
- Sand and gravel
- Coal
- Oil and gas
- 200 metres
- 5000 metres

NORTH AMERICA

Hudson Bay

Greenland

Spitzbergen

Greenland Sea

Norwegian Sea

Iceland

Baltic Sea

North Sea

London

EUROPE

Caspian Sea

Black Sea

Rome

Labrador Sea

Newfoundland

New York

Mediterranean Sea

Cairo

PACIFIC OCEAN

Gulf of Mexico

Havana

Greater Antilles

Caribbean Sea

Caracas

NORTH ATLANTIC OCEAN

Azores

Rabat

Canary Islands

AFRICA

Cape Verde Islands

Lagos

SOUTH AMERICA

Rio de Janeiro

SOUTH ATLANTIC OCEAN

Buenos Aires

Cape Town

Cape of Good Hope

INDIAN OCEAN

PACIFIC OCEAN

N

km   1000   2000   3000

miles   1000   2000

Falkland Islands

SOUTHERN OCEAN

Cape Horn

The Caribbean Basin

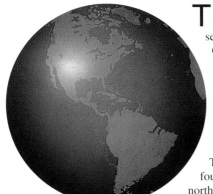

▲ The Gulf of Mexico
Area: 1,543,000 sq km
(598,000 sq miles)
Volume: 2,322,000 cu km
(560,000 cu miles)
Av. depth: 1512 m (4960 ft)
Max. depth: 4029 m
(13,218 ft.)
The Caribbean Sea
Area: 2,640,000 sq km
(1,020,000 sq miles)
Max. depth (Cayman
Trench) 7686 m (25,216 ft)

The wider Caribbean region is located in the southwestern margin of the North Atlantic, but separated from it by the Bahamas, and the Greater and Lesser Antilles. The region, which includes the Caribbean Sea and the Gulf of Mexico, is a relatively young ocean area which only assumed its present form some three million years ago following the closure of the Isthmus of Panama.

Bathymetry

The wider Caribbean ocean floor is divided into four major basins: the Gulf of Mexico to the north, the Yucatan Basin in the centre, and the Colombian and Venezuelan Basins to the south, which, together with the Grenada Trough to the east, form the Caribbean Sea. The floor of the Caribbean Sea is believed to be a fragment of the Pacific Ocean crust cut off at the time that South American joined central America.

To the east, Atlantic ocean crust slides beneath the Lesser Antilles, resulting in volcanic eruptions such as that at Mt Pelée which caused the death of the 30,000 inhabitants on St Pierre, Martinique on 8th May 1902. To the west the Pacific ocean crust slides beneath the landmass of Central America and again violent earthquakes and volcano activity occur in this area. Along the northern edge connecting these two zones is a giant fault along which North America is sliding westwards past the Caribbean plate. A movement of around 3 metres (10 feet) along this fault caused the Guatemala City earthquake in February 1974. To the south along the northern coast of South America the faults are more complex, although the general direction of movement of South America, like North America, is westwards past the Caribbean plate.

Basin divisions

The Aves Swell, which separates the Grenada Trough from the Venezuelan Basin, is composed in part by rocks of continental origin while the Beata Ridge which separates the Venezuelan and Colombian basins is in contrast a section of ocean crust which appears to have been uplifted at the time of the separation of the Caribbean plate from

the Pacific. The Nicaragua Rise, however, is geologically more complex and of uncertain origin.

Circulation

The surface water of the top hundred metres or so of the Caribbean Sea behaves as an extension of the North Atlantic. The Guiana Current and part of the North Equatorial Current flow past St Lucia into the Caribbean and continue westward at a speed of around 32 kilometres (20 miles) per day.

In the western Caribbean Sea the trade winds cause the surface currents to flow northwards away from the South American coast, drawing up colder nutrient-rich water

▲ Sediment load, eddies and current paths show as tone variation in this satellite photograph of Mobile Bay, Alabama. Silt-laden water flowing into the Gulf of Mexico from the Alabama and Tombigbee rivers is swept southwestward by longshore currents.

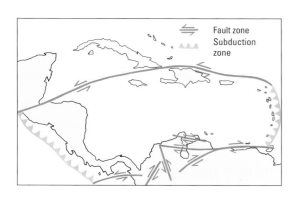

Fault zone
Subduction zone

▲ The elongated plate of the Caribbean Sea is bordered to east and west by subduction zones, where major plates dip beneath the Caribbean Basin. A simple fault zone across the northern margin and a complex fault system in South America absorb lateral movement.

► The circulation of the Caribbean's surface waters is dominated by the warm Gulf Stream. The Gulf Stream flows out of the Gulf of Mexico and then follows the northeastern coast of America, before passing across the Atlantic and on to western Europe.

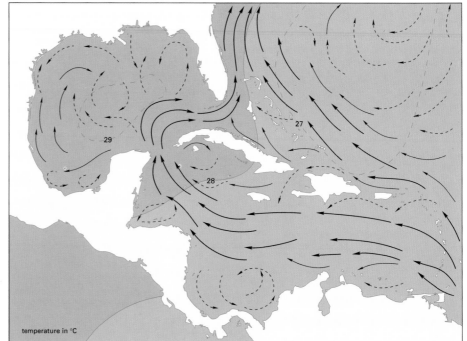

temperature in °C

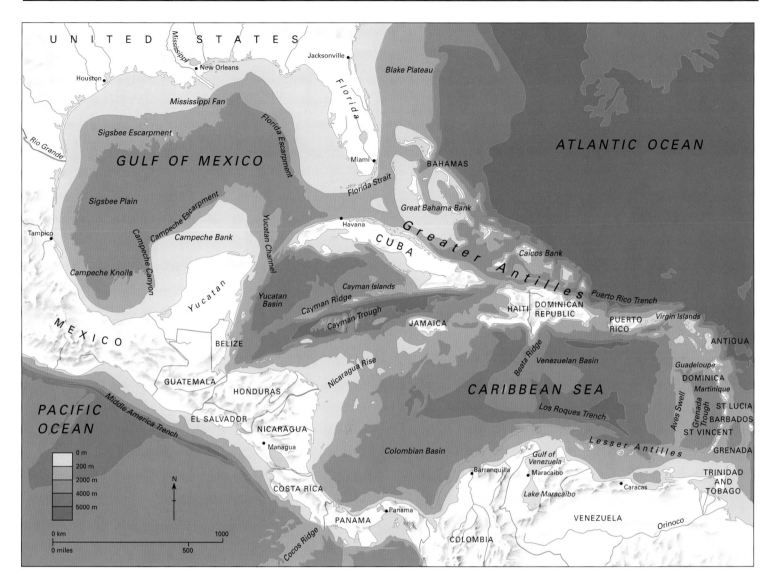

from around 200 metres (650 feet). This upwelling of water supports a rich fishery in the area.

In the Yucatan Basin surface water flows north through the Yucatan Channel into the basin of the Gulf of Mexico where it is forced to the east towards the Straits of Florida. This turn is sharp and the current meanders may become cut off as warm water eddies which drift westwards across the Gulf of Mexico. The heating of the surface waters as they flow through the Caribbean and Gulf of Mexico regions contributes substantially to the warming of the Gulf Stream.

The Caribbean's deeper water

The deep basins of the wider Caribbean region are as deep as much of the North Atlantic, up to 5000 metres (16,400 feet). They are, however, separated from the Atlantic by the island arcs and from each other by the ridges. They contain water at roughly the same temperature – 4.85°C (40.73°F) – as the Atlantic Ocean at a depth of 1590 metres (5250 feet) and it seems likely that this Atlantic water spills over the ridges periodically to form the Caribbean Deep Water.

The separation of the Caribbean from the Atlantic is also demonstrated by the pattern of their tides. In the Caribbean, tidal range is smaller than in the Atlantic and the dominant tide is the diurnal tide giving a single high and low each day. In the Atlantic the dominant tide is the semi-diurnal, giving two high and two low tides each day.

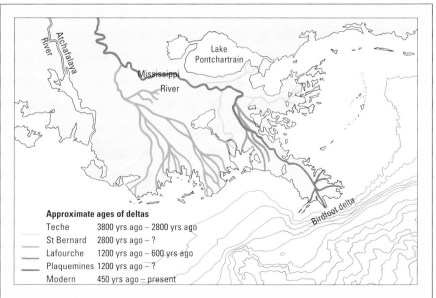

Approximate ages of deltas

Teche	3800 yrs ago – 2800 yrs ago	
St Bernard	2800 yrs ago – ?	
Lafourche	1200 yrs ago – 600 yrs ago	
Plaquemines	1200 yrs ago – ?	
Modern	450 yrs ago – present	

Over thousands of years, the Mississippi River has built a series of deltas across the continental shelf. The deltas formed as sediment build up, discharged by the river. However, flood defences along the banks have increased the speed the water enters the sea, so carrying the sediment to deeper water. As a result the delta is shrinking by about 100 sq km (40 sq miles) each year.

Caribbean Resources

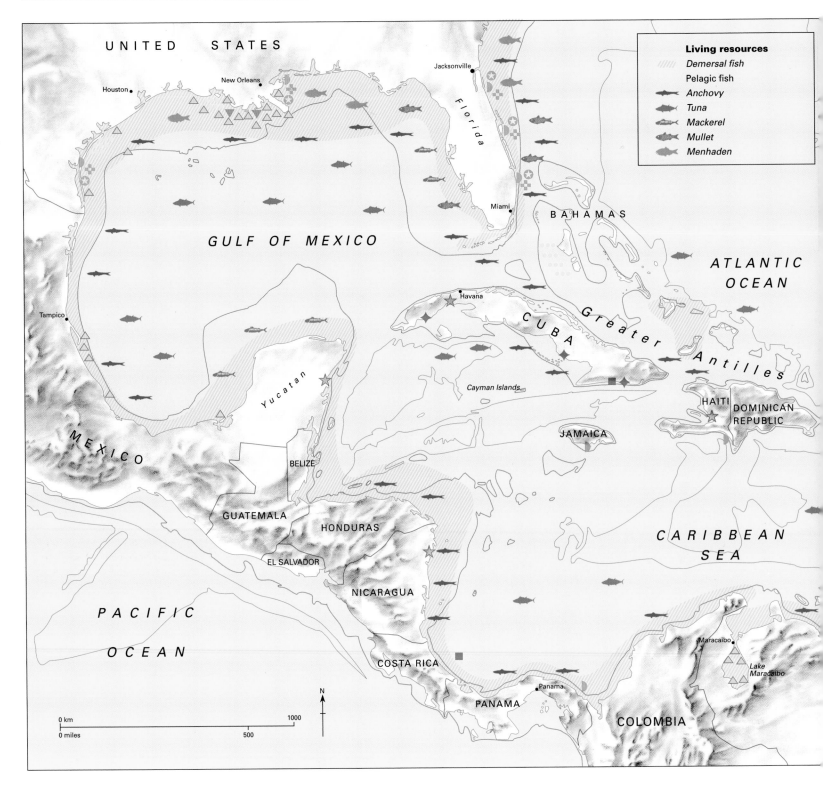

The Caribbean Sea and Gulf of Mexico have a long and complex sedimentary history and this, combined with the restricted water circulation between the basins, has resulted in varied and valuable mineral resources.

Oil and gas
Since the late 1930s, the Gulf of Mexico has become an important centre for the extraction of oil and natural gas. The development of these resources, however, has led to increasing levels of contaminants in the Gulf, with unfortunate consequences for the fisheries of the area. The fisheries that have been worst affected are those that are dependent on the region's extensive mangrove systems.

Important oil and gas fields lie off the coast of Louisiana and the Mexican states of Veracruz and Campeche, as well as Venezuela and the island state of Trinidad and Tobago.

Sulphur deposits
The north and western margins of the Gulf of Mexico contain large evaporite beds of sedimentary rocks, formed when seawater evaporated during earlier geological periods, leaving salt deposits. These evaporite beds contain domes and plugs both of which are formed from volcanic activity during which the magma of the volcano solidifies before it reaches the surface. The domes and plugs contain

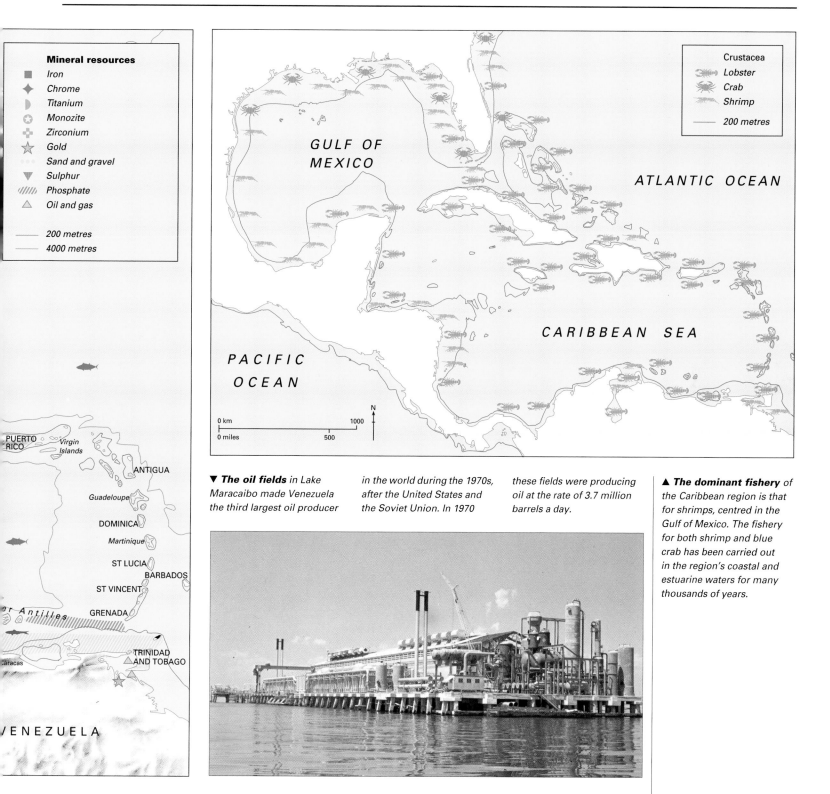

Mineral resources
- ■ Iron
- ◆ Chrome
- ◗ Titanium
- ✪ Monozite
- ❖ Zirconium
- ☆ Gold
- ∙∙∙ Sand and gravel
- ▼ Sulphur
- ▨ Phosphate
- △ Oil and gas

—— 200 metres
—— 4000 metres

GULF OF MEXICO

ATLANTIC OCEAN

PACIFIC OCEAN

CARIBBEAN SEA

0 km 1000
0 miles 500
N

Crustacea
- Lobster
- Crab
- Shrimp
—— 200 metres

PUERTO RICO · Virgin Islands · ANTIGUA · Guadeloupe · DOMINICA · Martinique · ST LUCIA · BARBADOS · ST VINCENT · er Antilles · GRENADA · TRINIDAD AND TOBAGO · Caracas · VENEZUELA

▼ **The oil fields** in Lake Maracaibo made Venezuela the third largest oil producer in the world during the 1970s, after the United States and the Soviet Union. In 1970 these fields were producing oil at the rate of 3.7 million barrels a day.

▲ **The dominant fishery** of the Caribbean region is that for shrimps, centred in the Gulf of Mexico. The fishery for both shrimp and blue crab has been carried out in the region's coastal and estuarine waters for many thousands of years.

significant amounts of salt, potash and magnesium, while the anhydrite beds comprise reserves of sulphur (in the form of calcium sulphate), which occurs as cap rocks over the salt domes. The world's first offshore sulphur mine began operation in 1960, 11 kilometres (7 miles) off the coast of Louisiana. The sulphur is extracted by the Frasch process in which hot water is pumped down the boreholes under pressure causing the sulphur to melt. The sulphur is then pumped out in liquid form.

Placer deposits of monazite, zircon and titanium are known to occur around the northern margins of the Gulf of Mexico, while iron and titanium are present in continental shelf deposits off the coasts of Costa Rica and Colombia.

Placer deposits of chromite, titanium and gold are also present in the shelf sediments off Cuba and Haiti.

Inshore fishing

The Caribbean region is characterized by small-scale, commercial and artisanal fisheries. The fisheries reflect the ecological complexity of the region, which contains some 14 per cent of the world's coral reefs.

Considerable areas of fringing mangrove forest, salt marshes and sea grass beds contribute to the inshore productivity, and as a result of this diversity of habitat, the fisheries are characterized by a high diversity of species in the catch, including groupers, snappers, jacks, flying fish and tuna.

The Arctic Basin

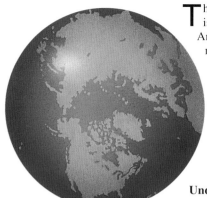

The Arctic Ocean is the world's smallest ocean. It is surrounded by Eurasia, Greenland and North America, and is on average approximately 1000 metres (3300 feet) deep – although in places it can reach six times this depth. There are four major basins separated by three oceanic ridges of which the largest, the Lomonosov Ridge, extends northwest to southeast for 1750 kilometres (1100 miles), and rises 3000 metres (10,000 feet) above the Pole Abyssal Plain to reach within 1100 metres (3600 feet) of the surface.

▲ The Arctic Ocean
Area: *12,173,000 sq km (4,700,000 sq miles)*
Average depth: *990 m (3250 ft)*
Max. depth: *(Pole Abyssal Plain) 4600 m (15,091 ft)*

Underwater plains

Running parallel to the Lomonosov Ridge is the Arctic Mid-Ocean Ridge, which is offset from the line of the Atlantic Mid-Ocean Ridge. Between the two lies the Pole Abyssal Plain where in certain places the sea can reach depths in excess of 4500 metres (15,000 feet). On the Russian side lies the shallower Barents Abyssal Plain, which has an average depth of around 3500 metres (11,500 feet). The Barents Abyssal Plain is separated from the shallow Kora and Barents seas by the island chain that includes Zevernya Zempla, Franz Josef Land and Spitzbergen islands.

On the Canadian side of the mid-ocean Lomonosov Ridge lie the Fletcher and Wrangal abyssal plains. These plains are separated from the Canadian and Mendeleyev abyssal depths by the Alpha Ridge, which is believed to be an inactive section of an old mid-oceanic ridge. The Canadian Abyssal Plain, which has an average depth of around 3600 metres (12,000 feet), is by far the largest of the Arctic sub-basins.

Shallow seas

The continental shelf regions of the Arctic are one of the oceans' most unusual features. While the continental shelf off the Canadian and Alaskan coasts is of average width, that is generally between 50 and 125 kilometres (30 and 75 miles) wide, the continental shelf off the north Asian coast shelf is considerably wider, extending out for more than 1600 kilometres (1000 miles) at its widest, and

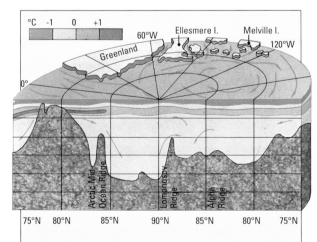

The almost entirely enclosed Arctic Ocean has two primary surface-current systems: a well-defined, clockwise-rotating gyre that occupies most of the Canadian side of the ocean basin, while a more direct transarctic current flow sweeps across from the Chukchi Sea towards the Greenland Sea. The deeper, warmer layer flows in from the North Atlantic.

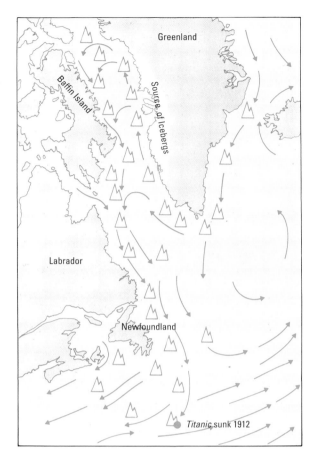

▲ Treacherous conditions in the North Atlantic, southeast of the Newfoundland Banks, are caused by a combination of fog, due to humid winds blowing over the meeting of warm and cold currents, and icebergs carried out of the Davis Strait by the Labrador Current. Newly calved icebergs typically weigh 1.5 million tonnes and stand 80 m (260 ft) out of the water extending to more than 370 m (1200 ft) below the surface. By the time they reach the Atlantic they have decreased in size to a mere 150,000 tonnes.

nowhere is it less than about 480 kilometres (300 miles) wide. This extensive shelf area is divided by small island chains into the Chukchi, East Siberian and Laptev shallow sea areas.

Arctic circulation and icebergs

Compared to the world's other oceans, the Arctic Ocean is virtually enclosed. The vast majority of water movement, about 80 per cent, flows via the Greenland Sea through a narrow gap between Greenland and Svalbard, the only deep-water connection to the world ocean. The remaining 20 per cent passes through the shallow Bering Straits into the Pacific.

Two per cent of the water leaving the Arctic Ocean does so in the form of icebergs. The icebergs are calved from the Greenland ice cap and carried into the North Atlantic by the Labrador Current which flows between Greenland and Baffin Island. The Arctic ice bergs come in a variety of colours and shapes. The colours can vary from an almost pure white, through green-blue shades to browns and blacks; the different colours are dependent on the amount and type of soil and/or debris that is mixed in with the ice. The majority of Arctic icebergs are of the "glacier" type. Such icebergs usually stand about 80 metres (260 feet) out of the water and over 1000 metres

(3300 feet) long. The other common type of iceberg is of the sort found in the Antarctic. Known as "tabular" icebergs, they tend to have sheer sides and a distinctive flattened top. Finally, there is a type of iceberg that comes only from the ice shelves at North Ellesmere Island and North Greenland. Made up of very old ice, "ice islands" although small, being 5 metres (17 feet) high, are 200 metres (660 feet) thick.

Of greater danger to shipping, however, is superstructure icing. When a ship's superstructure becomes covered in ice, the centre of gravity becomes higher so making the ship less stable and in danger of capsizing.

Arctic sea ice
During the winter, sea ice extends over an area of 15 million square kilometres (6 million square miles) of the Arctic Ocean. Its extent shrinks to around half that during the summer. This ice cover affects the surface circulation in the Arctic by restricting heat exchange between the ocean and the atmosphere. Water directly below the ice is much more variable in temperature and density than is normal for ocean surface waters.

The generally thin, cold, surface waters overlie a deep layer of slightly more saline water which enters the basin from the North Atlantic. This mid-water layer may extend from about 180 metres (600 feet) below the surface to 900 metres (3000 feet) below, becoming colder and more dense as it passes across the deep ocean basins.

On the Canadian side of the Arctic, a clockwise gyre is seen, while a more direct flow occurs in a broad arc across the shallower, continental shelf bordering the landmass of the Asian side.

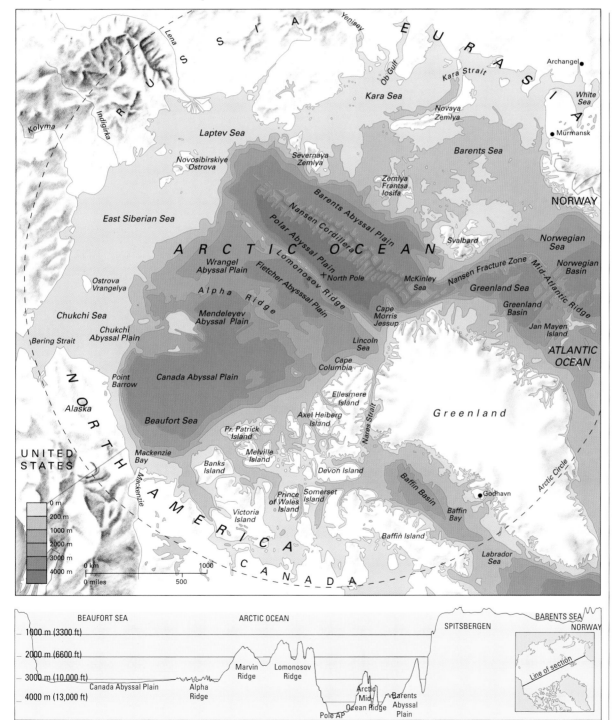

Arctic Resources

Despite the inhospitable conditions of the Arctic region, the broad sedimentary basins of the Arctic Ocean represent suitable geological formations for the accumulation of oil and gas deposits. Anhydrite deposits are known to occur in the Laptev Sea and off the Canadian coast. In the Beaufort Sea large oil deposits have been discovered and gas reserves are extensive in the region of Melville Island.

Oil extraction

Oil was discovered on the Arctic coast of Alaska in 1968, and nine years later the 1270 kilometre (800 mile) long trans-Arctic pipeline was constructed. The pipeline carries crude oil south from Prudhoe Bay on Alaska's Arctic coast. On the Russian coast, known oil and gas reserves have not yet been exploited.

Problems of oil and gas extraction include the extreme cold, gale force winds, and constantly shifting sea ice. In a number of instances artificial islands have been constructed by dredging and pumping seabed gravel and sand to provide a base for drilling rigs. Land-based pipelines have proved both technically and economically more practical than early ideas concerning the possibility of submarine tankers carrying crude oil under the Arctic ice sheet.

Given the low density of human populations surrounding the Arctic Ocean and the inhospitable conditions, mineral extraction from the Arctic remains at a relatively low level at present.

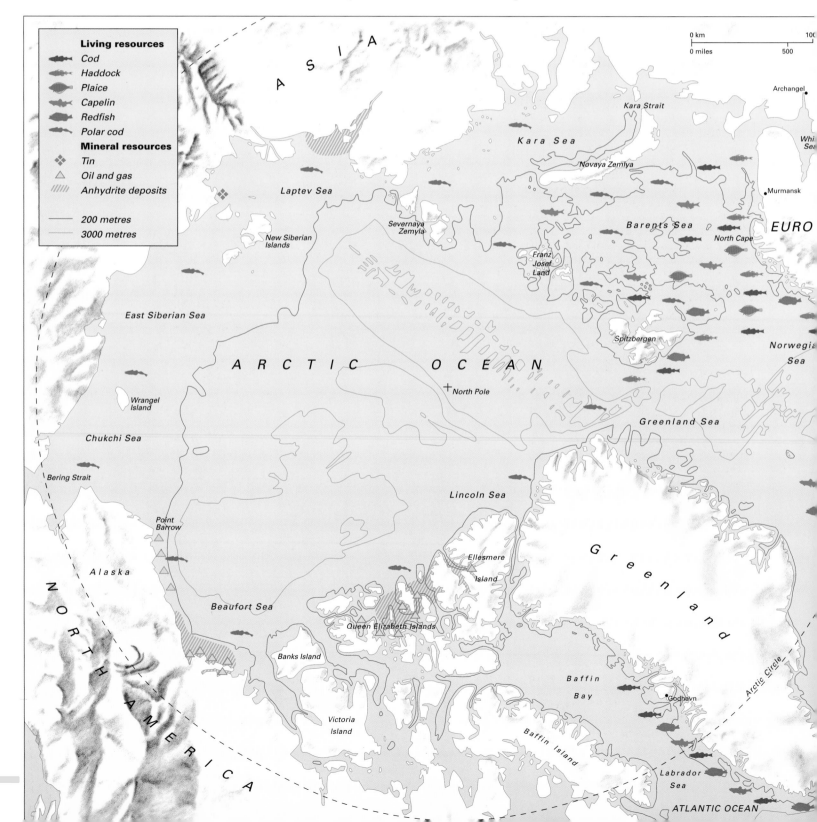

▲ **The Inuit** of the far north have a close relationship with their environment, taking enough to provide them with food and clothing, but never overexploiting.

◀ **Thousands of Harp seal** were killed annually during the 1970s and 1980s to satisfy the international market in attractive animal furs. However, recently, partly due to the anti-fur campaign, the annual culling of Harp seals has now been halted altogether.

▼ **Ice-breakers** plough their way through the sea ice of the Arctic. The ice breakers' primary role is to keep sea lanes open, however, they also perform an important role undertaking research in this inhospitable region.

Living resources

The living resources of the Arctic are dependent on a system of primary production which is highly seasonal, with little or no photosynthesis occurring during the winter. Total primary production is only around one tenth of that in temperate ocean areas.

During February, when light reaches the Arctic, small phytoplankton, mainly diatoms, increase their productivity. Located on the undersurface of the ice and in areas of open water, by March they have formed a yellow-brown layer within the bottom 30 centimetres (12 inches) of ice. The plankton are fed on directly by the Arctic cod, the only commercially important species found in the central Arctic Basin. On the shallow continental shelves the fisheries are dominated by Arctic char and capelin.

There are two main communities of fish in the Arctic. First, the cod, haddock and plaice which inhabit the warmer waters of the West Spitzbergen and North Cape currents, and second, the Arctic cod and capelin, which inhabit colder water areas. In response to long-term climatic fluctuations the distribution and abundance of these two communities of fish alter, reflecting changes in the current systems and ocean circulation.

Fishing and hunting

Although cod have been caught on their spawning grounds around the Lofoten Islands since the 12th century, they were not fished from the nearby Barents Sea until the 1920s, when the temperature of the North Atlantic reached a peak. During the 1960s and 1970s, capelin became more important in Arctic catches, the shift reflecting a period of colder climate.

By the early 1970s, rising fuel costs and restrictions placed by the Canadians on the North Atlantic fishery resulted in an increase in fishing effort in the Barents Sea, such that by 1975, a catch quota of 810,000 tonnes was imposed by the Northeast Atlantic Fisheries Commission on the Arctic-Norwegian cod stocks.

All the major fisheries of the Barents Sea are seasonal, with cod, haddock and redfish being taken between February and September, and capelin being fished over the winter.

The Arctic attracted commercial hunting expeditions particularly for fur seals from as early as the 16th century. Commercial hunting of marine mammals has now been stopped, although the indigenous peoples of North America, Siberia and Greenland continue to hunt marine mammals and birds for subsistence use. The Inuit of the far north take only what they need, and continue to have little impact on the species in the region.

The North Sea Basin

▲ During the Permian
period some 250 million
years ago, the North Sea
was a desert plain bounded
by mountains. Inland seas
and salt lakes were present
and the rocks formed at this
time were predominantly
sandstones and anhydrite,
important for trapping oil
and gas.

▲ The upper Cretaceous
period, 100 million years ago
was a time when much of
the lowlands of the Northern
Hemisphere, including the
North Sea Basin, were
flooded by shallow seas.
Thick deposits of chalk
were laid down during
this period.

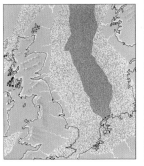

▲ During the Tertiary
period the North Sea Basin
had assumed a shape very
similar to that of today and
thick deposits of sediments
including muds and clays
were deposited at this time.
The sediments were the
result of run-off from rivers.

The North Sea is a large ocean region formed by the inundation of an extensive area of the continental crust. This basin has been subsiding throughout its long history, and the rather featureless bottom topography conceals a deep sedimentary basin containing more than 6100 metres (20,000 feet) of sediments deposited over the last 250 million years.

Running close to the coast of Norway lies the Norwegian Trough. The trough extends from the North Atlantic to the mouth of the Skagerrak in northern Denmark, where it reaches a depth of 700 metres (2300 feet). Although this trough is an obvious feature of the bottom topography of the North Sea, a much deeper trough lies hidden beneath the sedimentary deposits which cover the majority of the North Sea bottom. It is in this trough that the vast majority of North Sea oil and gas deposits are to be found.

To the west of Ireland a line of deep ocean troughs runs roughly north to south, marking the edge of the European continental shelf. The submarine topography of the North Sea is similar to that of the dry land which existed at the end of the last ice age.

Past ice ages
During the ice ages, ice sheets covered much of the British Isles and flowed into the surrounding basins, extending as far south as the Celtic Sea in the west and the Thames Estuary in the east. Additionally, ice sheets extended from the Scandinavian landmass into what is now the North Sea. The well-known "Banks" of the North Sea, such as Dogger, Fisher and Jutland banks, are in fact glacial moraines – piles of boulder clay pushed out in front of extending ice sheets. The Fladen Ground was probably covered by an inland lake surrounded on three sides by the edges of the ice sheet.

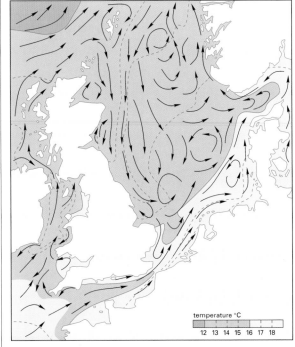

temperature °C
12 13 14 15 16 17 18

▲ Current flow in the North Sea is generated by a combination of the tides, the prevailing winds and density differences in the water masses. The pattern varies considerably on a local and seasonal basis, and local coastal configuration plays a dominant role in determining the direction of inshore currents.

As sea level rose with the melting of the ice and warming of the global climate, waves and currents immediately commenced the redistribution of these unconsolidated sediments, a process that continues today. The finest sediments are moved northwards in suspension, and are eventually deposited in the deeper parts of the northern North Sea. Sand is constantly being moved along the seabed by tidal currents, particularly in areas where strong currents coincide with large storm waves. The approaches to the ports of Liverpool, London and Hamburg are among the many areas where sand has accumulated. Dredging to remove the sand and allow access to shipping has, over the years, cost millions of dollars.

The Straits of Dover is also an area of sand deposition with long sandbanks being deposited parallel to the strong currents which flow through the straits. The Godwin Sands and Norfolk Banks are a hazard for shipping, and the shipping lanes to the major ports of Europe require regular surveying since the sand is constantly shifting.

Circulation
Water movement in the North Sea is generated by the tides, the prevailing winds and density differences between different water masses. In general, the pattern is highly variable both locally and on a seasonal basis. Water enters the North Sea from the north between the Orkney and Shetland islands, and flows down the eastern coast of Scotland and England. At the same time, warmer water from the North Atlantic Current flows into the North Sea through the English Channel. The inflow of warm water ensures that the area remains ice free throughout the year. The warm-water current flows along the coast of the Netherlands to the south of the North Sea. The combination of the two prevailing currents sets up a generally anticlockwise circulation in the North Sea.

▼ Different species of plankton are usually characteristic of different water masses, a phenomenon that is true of most of the world's seas and oceans. In the North Sea, for example, three different species of Arrow worm of the genus Sagitta can be used to clearly distinguish between oceanic, mixed and coastal water masses.

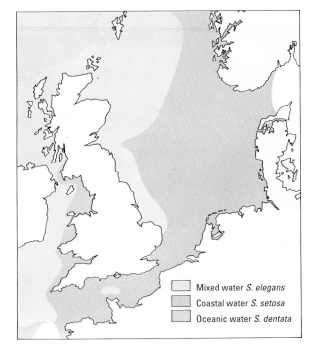

☐ Mixed water *S. elegans*
☐ Coastal water *S. setosa*
☐ Oceanic water *S. dentata*

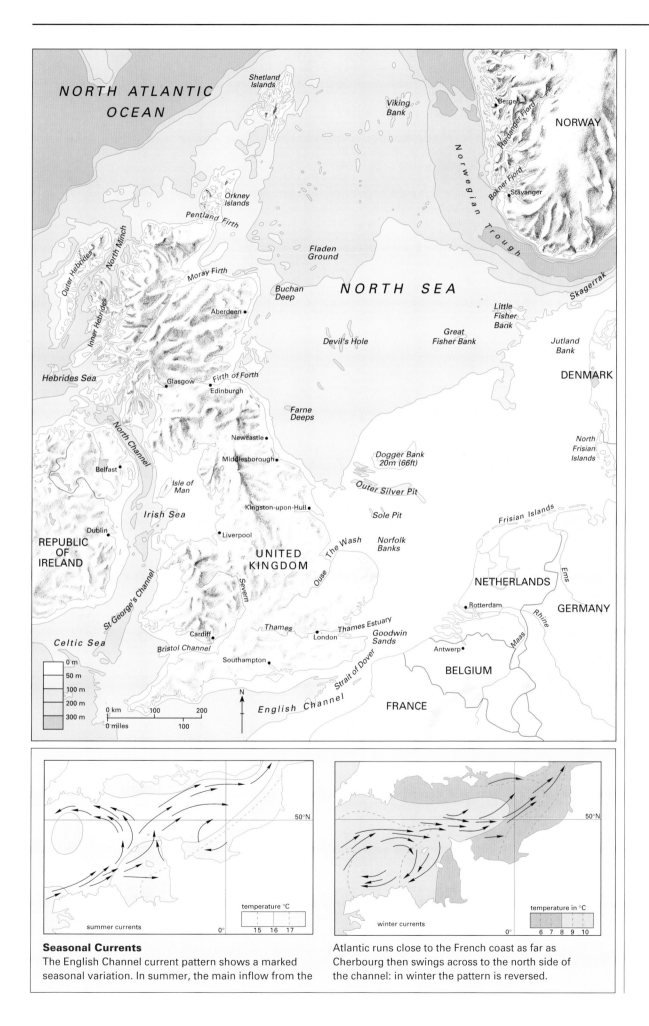

NORTH ATLANTIC OCEAN

Shetland Islands

Viking Bank

NORWAY

Bergen

Hardanger Fjord

Orkney Islands

Pentland Firth

Norwegian Trough

Bokner Fjord

Stavanger

Fladen Ground

Moray Firth

Buchan Deep

NORTH SEA

Skagerrak

Outer Hebrides

North Minch

Aberdeen

Devil's Hole

Great Fisher Bank

Little Fisher Bank

Inner Hebrides

Jutland Bank

DENMARK

Hebrides Sea

Glasgow

Firth of Forth

Edinburgh

Farne Deeps

North Channel

Newcastle

Middlesborough

North Frisian Islands

Belfast

Isle of Man

Dogger Bank 20m (66ft)

Outer Silver Pit

Irish Sea

Kingston-upon-Hull

Sole Pit

Frisian Islands

Dublin

Liverpool

The Wash

Norfolk Banks

REPUBLIC OF IRELAND

UNITED KINGDOM

Ouse

NETHERLANDS

Ems

GERMANY

Severn

Rotterdam

Rhine

St George's Channel

Thames

Thames Estuary

London

Goodwin Sands

Maas

Celtic Sea

Cardiff

Bristol Channel

Southampton

Antwerp

BELGIUM

Strait of Dover

N

English Channel

FRANCE

0 m
50 m
100 m
200 m
300 m

0 km   100   200
0 miles   100

Seasonal Currents

The English Channel current pattern shows a marked seasonal variation. In summer, the main inflow from the

temperature °C

summer currents   0°   15   16   17

temperature in °C

winter currents   0°   6 7 8 9 10

Atlantic runs close to the French coast as far as Cherbourg then swings across to the north side of the channel: in winter the pattern is reversed.

North Sea Resources

North Sea oil and gas extraction has grown considerably since the early 1970s. Oil production is greatest in the Shetland Basin, where water depths are up to 200 metres (660 feet). Gas extraction from the Southern Bight, on the other hand, occurs in shallower waters at depths of less than 50 metres (170 feet).

By 1989, there were 149 platforms operating in the North Sea, of which 92 were British, 36 Dutch and the remainder Norwegian, Danish and German. Some 8000 kilometres (5000 miles) of pipeline have been laid, and investment in the industry to date exceeds US$75 billion. Presently, 50,000 people earn a direct living from this industry, which produces more than 150 million tonnes annually. To date some 20 per cent of the reserves have been exploited and additional reserves are now known to be present in deeper water off the Faroe Islands.

Although the industry represents a vast economic resource, its extraction brings environmental problems. Recent data suggest that the benthic communities of the region display increasing diversity away from oil rigs, reflecting the influence of low levels of chronic pollution.

Sand and gravel

In addition to oil and gas, significant extraction of sand and gravel takes place in the North Sea. In fact, more sand and gravel is extracted annually from marine deposits in the North Sea area than anywhere else in the world. Although the large quantities extracted reflect in part depletion of land-based sources, the materials recovered are well sorted, uniform and of high quality, requiring little processing before use.

Most of the dredgers operating in the North Sea work in water which is less than 35 metres (115 feet) deep, and relatively close to shore, often resulting in coastal erosion of neighbouring land. Gravels are dredged off the British coast and contain a high proportion of flint and quartzite derived from the weathering of the Cretaceous chalks during the Tertiary period (65 to 2 million years ago).

Living resources

Fishing in the North Sea extends back to the Dark Ages between 500 and 1000 AD. The major component of the catch was herring, which were caught by drift nets at night. Following the Napoleonic Wars, and the growth of urban populations in western Europe, fishing intensity greatly increased, with the first steam trawler being launched in 1881. More recently, the introduction of purse seines and mid-water trawls greatly increased the fishing pressure to such an extent that North Sea herring has been virtually exhausted over the last 100 years. In 1977 a total ban on herring fishing in the North Sea was introduced.

Of the flatfish caught in the North Sea, plaice are the most important. The major spawning area is in the south, off the coasts of Holland and Belgium. The eggs drift with the current a few kilometres each day towards Denmark. The hatchling plaice are round bodied, only assuming the flattened shape of the adult at between four and six weeks of age. Although the southern spawning area is the most important, spawning also occurs off the east coast of England, between Flamborough Head and the Dogger Bank and in the German Bight area.

Overfishing

The North Sea fishery has shown a number of major changes in the last few decades, with catches of cod rising steadily since 1955, and very large catches of haddock occurring in the 1960s. Similarly, industrial fisheries (those species used for fish meal, such as Norway pout),

▲ **Since the decline** of the North Sea fishing fleets, many of the once thriving ports, such as Kingston-upon-Hull, have been redeveloped for shopping and entertainment.

▼ **The North Sea** has nutrient-rich water with high plankton production. As such the region provides a good breeding site for many species of commercially important fish species.

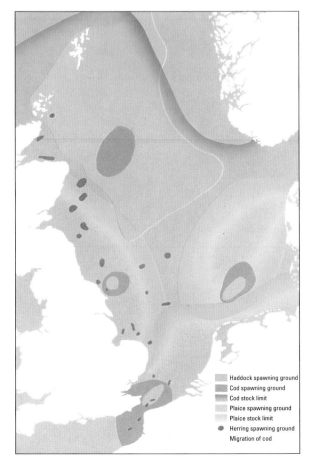

Haddock spawning ground
Cod spawning ground
Cod stock limit
Plaice spawning ground
Plaice stock limit
Herring spawning ground
Migration of cod

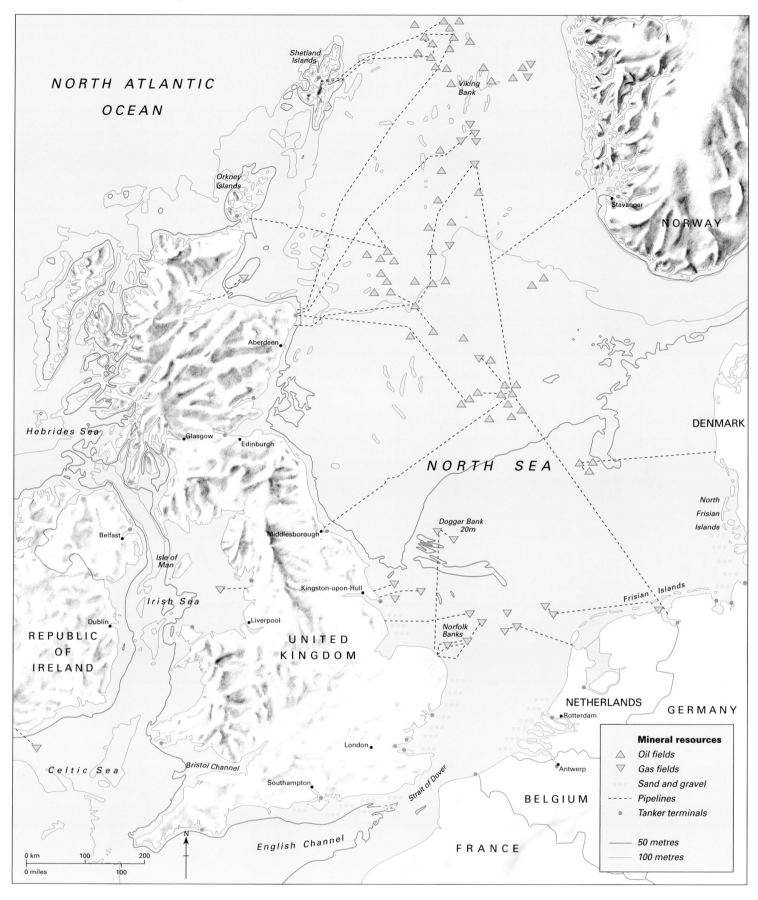

also increased during the same period. Catches of herring and mackerel halved over the same period. The high intensity of fishing has undoubtedly affected fish stocks in the area. However, the fluctuations in stocks also reflect changes in oceanic conditions and circulation patterns, which are still not well understood.

Modern trawlers using heavy bottom equipment damage seabed habitats, affecting both commercial and non-commercial species. Furthermore, although the European quota system, based on the total catch of fish allowable, was introduced as a measure to reverse stock depletion, stocks of many commercial species, particularly those of mackerel, haddock, whiting, saithe and Norway pout are still in decline.

Salmon farming

Farming of Atlantic salmon is practised in Scottish sea lochs and Scandinavian fjords. Salmon eggs are incubated in trays and one-year-old smolts are acclimatized to natural conditions over three weeks before being put into the sea in net cages. Natural water movement oxygenates the water and removes wastes, but high-intensity fish farming can cause environmental problems, such as anoxia (depletion of oxygen) of the bottom water.

The Mediterranean Basin

he Mediterranean Sea lies in a 4000 kilometre- (2500 mile-) long depression running from the coast of Israel, Lebanon and Syria in the east to the narrow Straits of Gibraltar in the west, which represent the only natural connection between the Mediterranean and the world ocean. Although it covers an area of around 2.54 million square kilometres (1 million square miles), this inland sea is much shallower than most ocean areas. The Mediterranean is believed to be a small remnant of the once pantropical Tethys Sea.

▲ The Mediterranean Sea
Area: 2,966,000 sq km
(1,145,000 sq miles)
Average depth: 1500 m
(4921 ft)
Max. depth: (Hellenic Trough) 5092 m (16,706 ft)
Evaporation roughly three times precipitation and run-off

Mediterranean basins

The narrow Strait of Sicily divides the Mediterranean into two distinct basins. The western basin has a broad, smooth abyssal plain that is covered with sediments dating from around 25 million years ago. In contrast, the eastern basin is divided by the Mediterranean ridge system which consists of folded and uplifted sediments compressed by the movement of Africa northwards towards Eurasia. Sediments in this basin date back as much as 70 million years and compression and uplift are still continuing in this area.

In the western basin, the Balearic and Tyrrhenian sub-basins are separated by the island chain of Corsica and Sardinia, both of which have rotated away from the southern coast of France and Spain. In contrast to the relatively flat profile of the bottom topography of the Balearic Basin, the Tyrrhenian sub-basin, which reaches depths of 3600 metres (12,000 feet) – the deepest areas of the western basin – is dominated by ridges, sea mounts and an arc of active volcanoes including Vesuvius, Etna and Stromboli.

In the eastern Mediterranean Basin, between the Mediterranean Ridge System and Greece and Turkey, lie the deepest parts of the modern Mediterranean Sea, where depths of 5000 metres (16,000 feet) are found. The Ionian Abyssal Plain is the largest abyssal area in the eastern section of the basin, and is again divided by a series of ridges and sea mounts.

In contrast to many other ocean areas, the continental shelf is narrow. The greatest extent of shallow coastal waters in the western basin are associated with the deltas of the Rhône and Ebro rivers. More extensive shelf areas are seen off the North African coast and in the Adriatic and Aegean seas, while the extensive delta of the Nile has contributed sediments to a considerable area of the eastern basin and ridge system.

Man-made connections

In addition to its natural connection with the Atlantic via the Straits of Gibraltar, the eastern Mediterranean Basin is now connected with the Indian Ocean via the Suez Canal

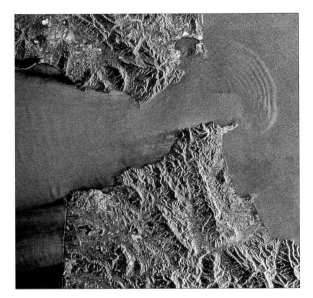

▲ This satellite picture shows waves at the Straits of Gibraltar produced by tidal flow through the narrow straits. Surface water generally flows into the Mediterranean from the Atlantic while at depths of around 80 m (280 ft) the flow is reversed leaving the Mediterranean via the same route.

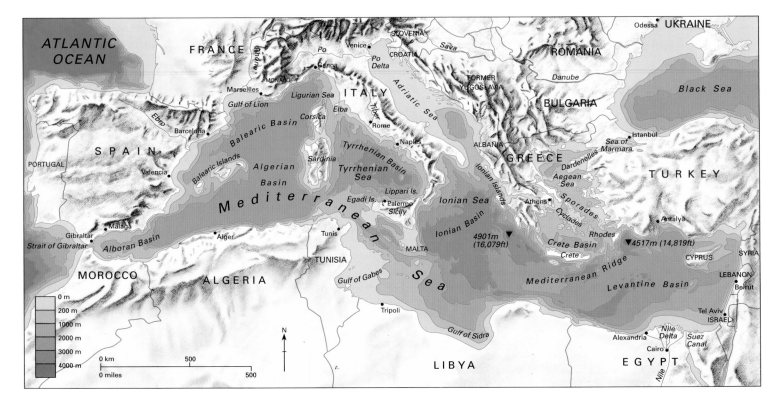

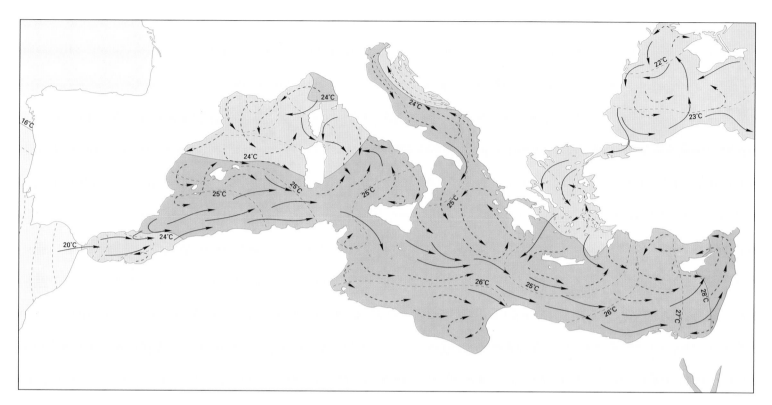

which was opened in 1869. Although water exchange via the canal is insignificant, a number of Red Sea species have invaded the Mediterranean Basin via the canal.

The inland Black Sea is connected to the eastern Mediterranean Basin via the Dardanelles, the Sea of Marmara and the Aegean Sea, and plankton studies have revealed that zooplankton species characteristic of the Aegean are spreading into the Black Sea. This may result from the damming of rivers which feed the Black Sea, thus reducing the flow of water out through the Dardanelles and increasing the influx of Aegean water.

Circulation

Although Atlantic surface water enters the Mediterranean through the Straits of Gibraltar, there is a significant out-flow of more saline, denser Mediterranean water via the same route. According to legend, this deep-water outflow was harnessed by the Phoenicians, who lowered their sails several fathoms into the water to harness this current in order to enter the Atlantic against the prevailing winds. Today, this strong under current is used by submarines which can turn their engines off and pass silently through the straits into the Atlantic.

The Atlantic water entering the basin passes east-wards, becoming progressively more saline as water vapour evaporates from the sea surface. Salinity may reach as much as 39.5 parts per thousand in the eastern Mediterranean. During the summer, the warming of the surface waters leads to the development of a marked ther-mocline at around 20–40 metres (66–130 feet), which sep-arates the warmer, high-salinity water from the cooler water below. During winter, dry winds also remove water vapour, increasing surface salinities and at the same time cooling the surface water which becomes more dense and sinks. These winter water masses flow westwards towards the Straits of Gibraltar or sink to form the dense bottom water of the eastern basin.

The Mediterranean loses by evaporation almost three times as much water as it receives from rainfall and run-off from the surrounding land. However, this imbalance is

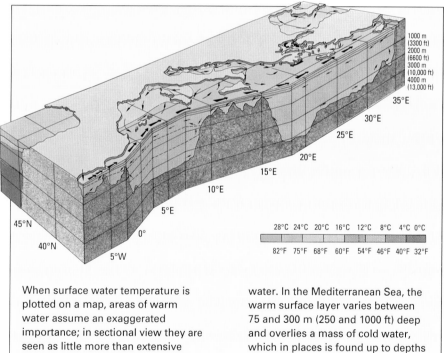

When surface water temperature is plotted on a map, areas of warm water assume an exaggerated importance; in sectional view they are seen as little more than extensive shallow pools of relatively warm water. In the Mediterranean Sea, the warm surface layer varies between 75 and 300 m (250 and 1000 ft) deep and overlies a mass of cold water, which in places is found up to depths of 4500 m (15,000 ft).

compensated for by the inflow of Atlantic water. After passing through the Straits of Gibraltar, Atlantic water flows eastwards, hugging the North African coast, creating the only well-defined current in the Mediterranean. The current feeds the counterclockwise circulation patterns of the western basin, the Adriatic and Ionian seas. Water exchange with the Atlantic is, however, limited and the turnover time of the Mediterranean is estimated at 150 years. Tidal range in the Mediterranean is low, around 30 centimetres (12 inches), hence water exchange between semi-enclosed coastal bays is slight, resulting in the build up of pollutants in close proximity to the shore.

Mediterranean Resources

Seismic studies of the Mediterranean record the presence of a widespread stratum of evaporites, which drilling studies have dated to be around 5 million years old. Formed from the evaporation of seawater, this formation may be as much as 915 metres (3000 feet) thick, and consists mainly of halite or rock salt.

The evaporite beds represent a substantial potential resource of rock salt, sulphur and potash, all of which are currently only exploited where they outcrop on Sicily and other Mediterranean islands. These buried seabed resources are conservatively estimated at 1 million cubic kilometres (240,000 cubic miles) in volume.

The Mediterranean region also has considerable oil and gas reserves, but to date these are only exploited in offshore shelf areas that overlie geological structures extending out from the land. Although the Mediterranean has no significant deposits of polymetallic nodules, manganese- and iron-rich deposits exist, derived from hydrothermal solutions emanating from recently active submarine volcanoes.

Tourism

Perhaps one of the most valuable resources of this area is the combination of warm dry summers and the Mediterranean Sea itself. Together they provide the basis for an extensive tourist industry. With over 70 per cent of the world's tourists visiting the Mediterranean region each year, tourism contributes significantly to the economy of Mediterranean countries.

Living resources

The Mediterranean contains some 500 species of fish, of which around 120 are fished commercially. Although most fisheries are small-scale, artisanal ventures supplying fresh fish to local markets, the high market price and high seasonal demand resulting from the several million tourists who visit the Mediterranean each year, encourage overfishing. At present the fisheries which are most heavily exploited are those along the southern coast of Europe, where there are severely depleted stocks of hake, sole and red mullet. The present harvest of over 2 million tonnes of fish per annum is well in excess of the estimated sustainable yield of 1.1–1.4 million tonnes.

In contrast, fisheries yields are increasing in some areas as a consequence of high inputs of nutrients from

▲ **Tuna** have been traditionally caught in the Mediterranean for centuries. This picture shows the last traditional tuna catch held in Sicily in 1989. Fishermen drive the fish into shallow channels with nets having previously been placed on the seabottom. As the nets are raised, teams of fishermen haul the massive fish out into boats using long poles with sharp, grasping hooks. Like many marine organisms tuna concentrate heavy metals and Mediterranean tuna have high levels of mercury in the flesh.

Posidonia

The sea grass *Posidonia* forms dense stands in sandy areas of the Mediterranean. The species provides an important habitat for marine organisms and as a nursery area for species of certain animal. Its long fronds also trap sediment and its creeping growth form helps to stabilize soft bottom habitats.

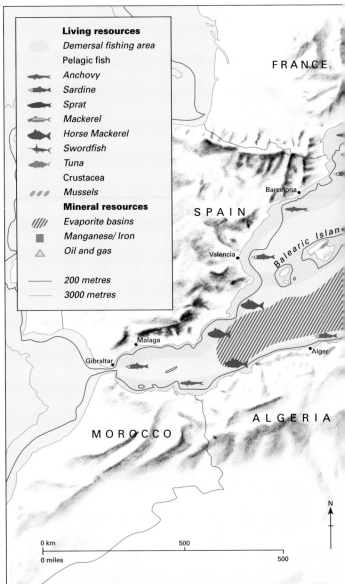

Living resources
- Demersal fishing area
- Pelagic fish
- Anchovy
- Sardine
- Sprat
- Mackerel
- Horse Mackerel
- Swordfish
- Tuna
- Crustacea
- Mussels

Mineral resources
- Evaporite basins
- Manganese/ Iron
- Oil and gas

— 200 metres
— 3000 metres

FRANCE

SPAIN

Barcelona

Balearic Islan

Valencia

Malaga

Gibraltar

Alger

MOROCCO

ALGERIA

0 km   500
0 miles   500

N

sewage. The sewage encourages phytoplankton production, the basis of the marine food web. Over half a billion tonnes of sewage, 80 per cent of which is untreated, are discharged annually to the Mediterranean.

In general, however, phytoplankton production in the Mediterranean is limited. The exchange of nutrients from the cold, bottom water with the surface is limited, and there are few rivers bringing nutrients into the sea. The construction of the Aswan Dam and various barrages on the Nile River has significantly reduced the input of nutrients into the eastern basin, and may have contributed to the collapse of the sardine fishery in the delta region. In addition, while the fisheries along the North African coastline are not yet over-exploited, fishing intensity is also increasing in this area.

Mariculture

Mariculture in the area has considerable potential, with up to 10,000 square kilometres (4000 square miles) of coastal lagoons providing considerable potential for future mariculture of finfish and shellfish. The most important species in the Mediterranean at the moment is the Mediterranean mussel principally produced by Italy.

Other important species include the mullets, gilt head bream and sea bass, all of which are raised in enclosed lagoons using techniques which date back to Roman times. The culture of species such as sea bream in cages suspended from anchored vessels is being experimentally tried, and large sections of the Adriatic coastline are now lined with net cages suspended from flotation rings.

Tourism

Tourism is a major source of revenue to the countries surrounding the Mediterranean. Unfortunately, the infrastructure cannot support this density of visitors, and the resulting environmental degradation can be severe. Sewage, discharged directly into the sea without treatment results in enhanced phytoplankton growth, algal blooms and in some semi-enclosed bays, deoxygenation of the water and fish kills.

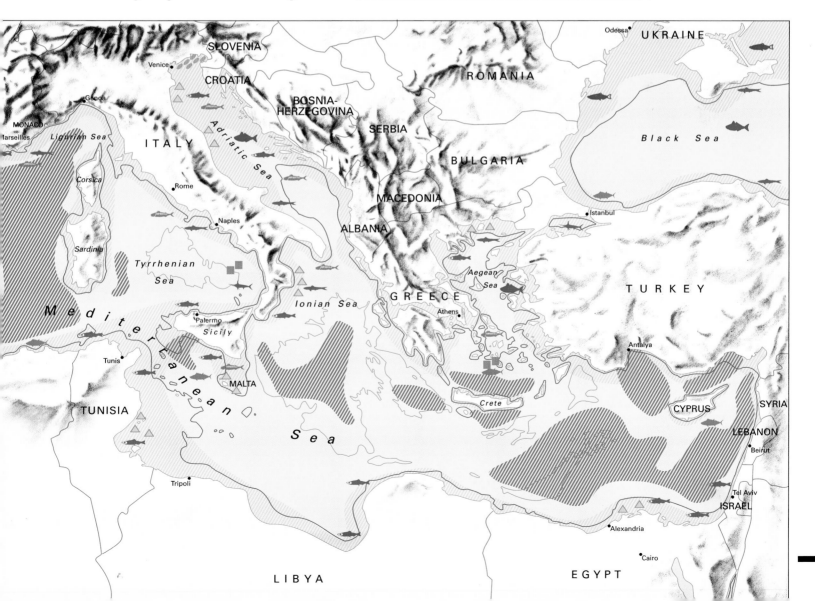

The Indian Ocean

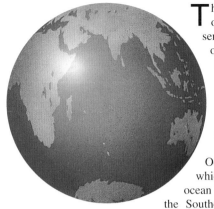

▲ The Indian Ocean
Area: *73,600,000 sq km*
(31,660,000 sq miles)
Average depth: *3890 m*
(12,760 ft)
Volume: *292,131,000 cu km*
(70,086,000 cu miles)
Max. depth: *(Java Trench)*
7450m (24,442 ft)

The Indian Ocean is the world's third largest ocean (after the Pacific and Atlantic) and represents approximately 20 per cent of the total area of the world's oceans. The floor of the Indian Ocean is dominated by the mid-ocean ridge, which forms an inverted Y-shape, the western arm of which runs round the southern tip of Africa to join the Mid-Atlantic Ridge. The eastern arm passes south of Australia to connect with the East Pacific Rise.

The surface ocean conditions of the Indian Ocean are dominated by the Monsoon winds which reverse direction seasonally. However, this ocean basin is open to the South and connects with the Southern Oceans, such that cold, deep Antarctic waters penetrate a considerable distance to the north, in the deeper ocean basins. Storms generated in the turbulent atmosphere of high latitudes result in long-distance swell transmission northwards into the western Indian Ocean, causing periodic flooding in island countries such as Sri Lanka and the Maldives.

Geological history

The geological history of the Indian Ocean is quite complex. Its formation commenced with the breakup of Gondwanaland, when the African continent separated from Antarctica and Australia between 140 and 130 million years ago. Seventy million years ago, India lay south

of the Equator. Over time the continent moved northwards to collide with the Eurasian continental plate, resulting in the formation of the Hindu Kush and Himalayas.

The separation of Australia from Antarctica occurred more recently, around 50 million years ago. In its present form the Indian Ocean Basin is relatively young, a mere 36 million years old.

Some of the bottom features of the Indian Ocean, notably the Ninety East Ridge, were originally at or near sea level. This ridge, which is over 2720 kilometres (1700 miles) long, appears to have formed from a single volcanic source, and at its northern and oldest end, the ridge is more than 1.5 kilometres (1 mile) below sea level. In addition, several of the relatively shallow submarine plateaux appear to be continental fragments left behind during seafloor spreading which have subsided as the larger continental landmasses drifted apart.

Atolls of the Indian Ocean

One of the features of the Indian Ocean, like the Pacific, is the presence of atolls, ring-like structures of living coral reef – indeed the word *atoll* comes from the Indian Ocean being the Dhivehi (the language of the Maldives) word for "place". The living reef supports small sandy islets and grows on the consolidated limestone laid down by previous reef communities which lies on top of submarine mountain ranges and volcanoes. The major atoll island groups of the Indian Ocean include the Maldives and the

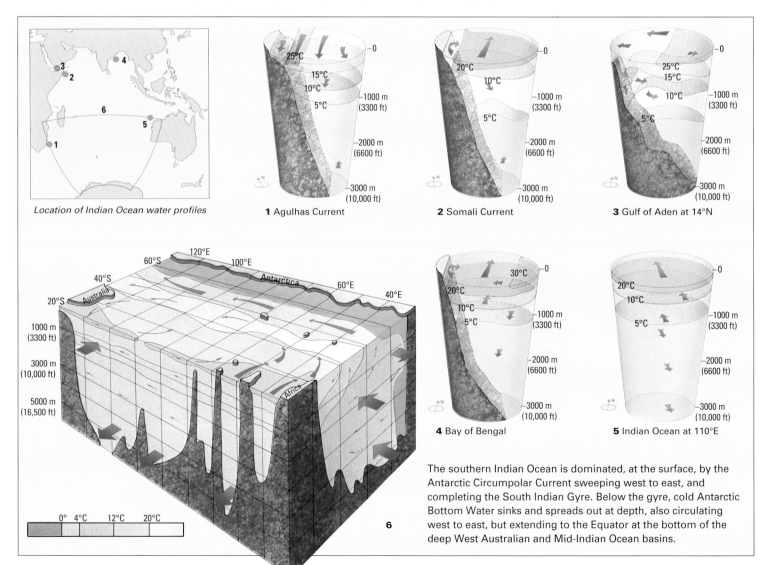

Location of Indian Ocean water profiles

1 Agulhas Current

2 Somali Current

3 Gulf of Aden at 14°N

4 Bay of Bengal

5 Indian Ocean at 110°E

6

The southern Indian Ocean is dominated, at the surface, by the Antarctic Circumpolar Current sweeping west to east, and completing the South Indian Gyre. Below the gyre, cold Antarctic Bottom Water sinks and spreads out at depth, also circulating west to east, but extending to the Equator at the bottom of the deep West Australian and Mid-Indian Ocean basins.

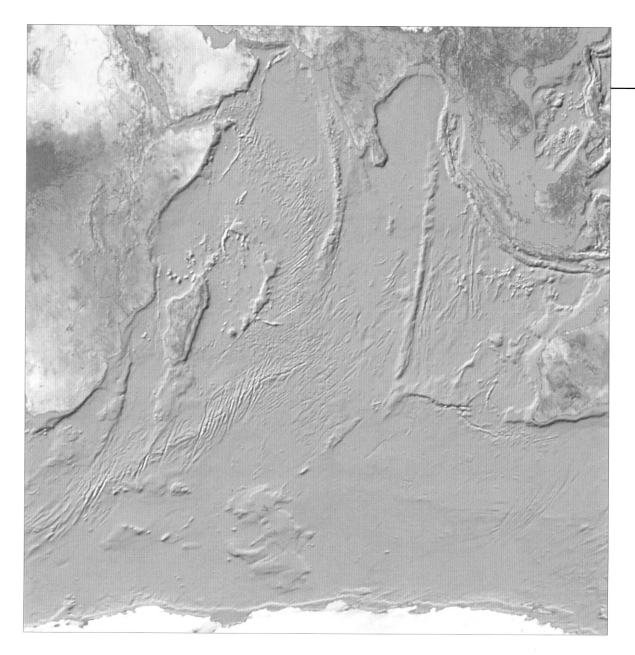

Seychelles and in the former case the formation of the atolls differs considerably from those atolls found in the Pacific Ocean region.

In the Pacific a typical atoll consists of a ring like structure close to the sea surface with a central lagoon lined with coral sand. The lagoon rarely contains emergent structures, although patch reefs and microatolls may occur in some lagoons. In contrast the atoll lagoons of the Maldives contain a wide variety of structures including faros, micro-atolls, and patch reefs. Faros, a structure which is typical of Maldivian lagoons, consist of mini-atolls, small ring-shaped structures often supporting a small island and with a shallow lagoon fringed by living coral reef. The lagoon of a faro is often shallower than that of the main lagoon, and many of the island toursit resorts for which this country is well known have been construct-ed on islands sitting on faros inside the shelter of the main atoll formation.

Bottom sediments

The first sediments deposited on the Indian Ocean seafloor as it formed were peats and low-grade coals, characteristic of shallow-water conditions. The shallow submarine plateaux are generally covered by submarine calcareous oozes, while the deeper, ocean basins contain only red-dish-brown clays or silicaceous oozes.

Two of the world's largest river systems, the Indus and Ganges-Brahmaputra, empty into the Indian Ocean, and have built enormous fans of sediment brought down from erosion of the Himalayas. The Bengal fan is the world's largest, containing an estimated volume of 5 million cubic kilometres (1.2 million cubic miles) of sediments. The large volumes of freshwater entering the Bay of Bengal result in lowered surface salinity, around 34 parts per thousand, compared with 36 parts per thousand in the Arabian Sea.

The Indo-West Pacific

The Indian Ocean covers an enormous area and is con-nected to the east with the Pacific Ocean Basin via the Southeast Asian archipelagos. As a tropical ocean area it is a centre of marine biodiversity with numerous species of coral, fish and other marine organism being found in the shallow waters of the East African coast, and peninsula India. In general terms the diversity of organisms decreas-es north and south of the Equator although over 200 species of coral are recorded from the North and Central Red Sea. The relative isolation of the tropical Indian Ocean from the Pacific, and its total isolation from the tropical Atlantic, has allowed the evolution of distinct shallow-water species in the western Indian Ocean Basin. Similar isolation of the central Pacific has resulted in the evolution of endemic species in the Pacific Ocean as well. Between these two ocean basins lies the Indo-West Pacific a biogeographic region with the highest shallow water marine species biodiversity in the world, having accumu-lated species from both the central Pacific to the east and the Indian Ocean to the west.

The Indian Ocean Basin

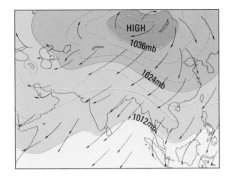

▲ ▶ *Northeast Monsoon: strong high pressure over northern Asia causes an overall outflow of air that drives the surface currents to the southwest.*

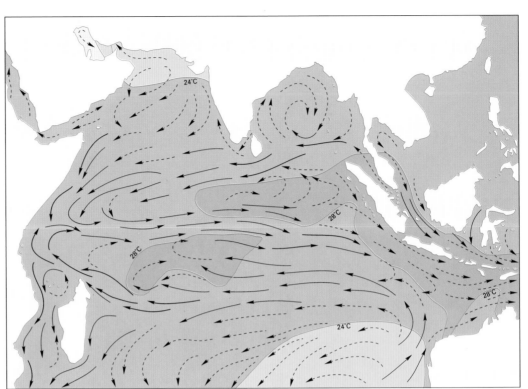

From the junction of the two mid-ocean ridges, which link the Indian Ocean Ridge with the Atlantic and Pacific ridge systems, the mid-ocean ridge runs northwards before swinging west as the Carlsbad Ridge to join the rift system of the Red Sea. At its point of entry into the Red Sea, Africa and the Arabian Peninsula have been actively spreading apart for the last 25 million years.

Off the southeastern African coast lie the Argulhas Plateau, Mozambique Ridge, and Madagascar and Mascarene plateaux. These are aseismic structures supporting islands, and presumed to be fragments of continental crust. The Chagos Laccadive Ridge off the west coast of India is a stable ridge supporting an extensive chain of atoll islands. To the east of this ridge, the seafloor drops to the Chagos Trench. Further east still, lying south of the Indonesian island arc is the only major trench in the Indian Ocean, the Java Trench. This trench reaches depths of 7300 metres (24,400 feet) and marks the subduction zone where the Australian plate slides below the Eurasian plate, a process that has been active for only 2 million years.

Related volcanic activity in the Indonesian region has resulted in the build up of extensive deposits of ash and volcanic sediments on the seafloor in this region. The well-known eruption of Krakatoa in 1883 resulted in about 17 cubic kilometres (4 cubic miles) of mountainous island erupting in a series of four explosions, the loudest of which was heard in Australia 4800 kilometres (3000 miles) away.

Circulation

The northern Indian Ocean circulation is unique in that its surface currents reverse twice a year under the influence of the monsoon winds.

From November to April, the Northeast Monsoon generates the North Equatorial Current. This current carries water across the Indian Ocean towards the African coast. Here the current turns south, forming a western boundary current which flows along the coast of Somalia towards the Equator. At the Equator, it joins the South Equatorial Current and flows east as the Equatorial Countercurrent.

In the region of the Indonesian island arc this current divides, part flowing north to rejoin the North Equatorial Current and part continuing to flow eastwards as the Java Coastal Current.

In April, the Northeast Monsoon ceases and is replaced by the Southwest Monsoon winds which generate a north-flowing current off the Somali coast. By July, the Monsoon Current flows east. After reaching the Indonesian Islands, the entire watermass turns south and returns westwards as part of a much stronger South Equatorial Current which in turn feeds the Somali current.

Upwelling

During the Southwest Monsoon not only are the currents stronger, but significant areas of coastal upwelling occur off the Arabian Peninsula and Somalia where warm surface water is replaced inshore by upwelled colder water from between 100 and 200 metres (330 and 660 feet).

South of the Equator, the Indian Ocean circulation is dominated by the influence of the westerlies and trade winds which drive an anti-clockwise gyre of warm water. The South Equatorial Current flows towards Africa where the part which does not feed the Somali current passes south between Africa and Madagascar and enters the Arghulas Current. This is the strongest western boundary current in the Southern Hemisphere and flows at a rate of 180 kilometres (112 miles) per day along the edge of the southern African continental shelf.

On reaching the southern tip of Africa the Arghulas current turns eastward and occasionally eddies may be cut off and travel westwards into the Atlantic. The return flow of the gyre results from the relatively weak and poorly defined West Australia Current and no area of upwelling is associated with this return.

Most of the water below 1000 metres (3300 feet) in the Indian Ocean originates either as North Atlantic Deep Water or as Antarctic Bottom Water, while in the North Indian Ocean warm saline water from the Gulf is found at about 300 metres (1000 feet) and below this even more saline water which originates from the Red Sea.

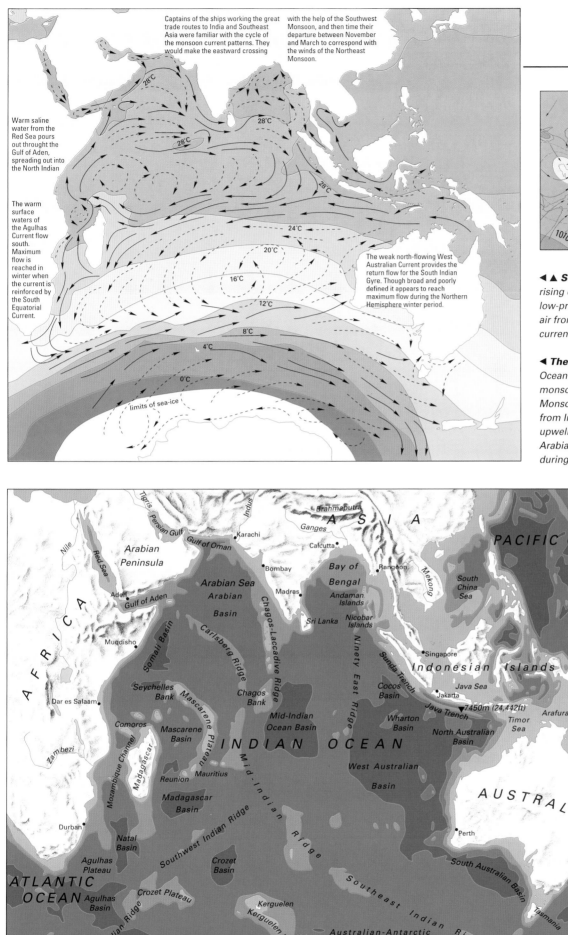

Captains of the ships working the great trade routes to India and Southeast Asia were familiar with the cycle of the monsoon current patterns. They would make the eastward crossing with the help of the Southwest Monsoon, and then time their departure between November and March to correspond with the winds of the Northeast Monsoon.

Warm saline water from the Red Sea pours out throught the Gulf of Aden, spreading out into the North Indian

The warm surface waters of the Agulhas Current flow south. Maximum flow is reached in winter when the current is reinforced by the South Equatorial Current.

The weak north-flowing West Australian Current provides the return flow for the South Indian Gyre. Though broad and poorly defined it appears to reach maximum flow during the Northern Hemisphere winter period.

28°C
28°C
28°C
28°C
24°C
20°C
16°C
12°C
8°C
4°C
0°C
limits of sea-ice

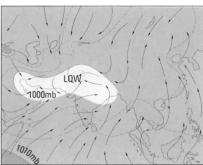

LOW
1000mb
1010mb

◄ ▲ *Southwest Monsoon:* heated air rising over the continent in summer creates a low-pressure region. This causes an inflow of air from the south, driving the surface currents.

◄ *The currents* in the northern Indian Ocean are reversed according to the monsoon seasons. During the Northeast Monsoon the surface waters are driven away from India towards Africa, and the Somali upwelling is suppressed. Water from the Arabian Sea enters the Gulf and Red Sea during this monsoon season.

Indian Ocean Resources

The narrow extent of much of the continental shelf around the Indian Ocean, particularly along the east African coast, has resulted in much lower fisheries production than in either the Atlantic or the Pacific oceans. The reversal of the surface currents in the North Indian Ocean and the shutting off of the upwelling along the western margins of the basin during the northwest monsoon season are also thought to contribute to this relatively low productivity.

Mariculture

Much of the capture fisheries production on the East African coast and around the oceanic islands is from inshore, artisanal fisheries associated with coral reef areas. These subsistence fisheries are important in providing protein resources to coastal populations in the developing countries surrounding the Indian Ocean Basin. As a consequence of the lower fisheries production, mariculture has been extensively developed in the East Asian region. The recent growth in mariculture of finfish, shellfish and seaweed, particularly in the Indonesian region, reflects the importance of marine protein to both the subsistence and export sectors of the economy of these countries.

Larger-scale commercial fisheries are either directed towards pelagic resources such as tuna, sailfish and marlin or towards penaeid shrimp resources trawled from offshore areas closely associated with the large mangrove stands which support the juvenile shrimps during the early stages of their lifecycle. The world's largest stands of mangroves occur in the Sundarbans area which is found near the Bay of Bengal. Similar large areas in the countries surrounding the South China Sea and in Indonesia have been cleared to provide fuelwood, land for rice cultivation and mariculture.

In the Maldives the tuna fishery is unique in being based, not on large-scale purse seiners or long-line vessels, but on a mechanization and expansion of the traditional dhoni fleet. Surface-swimming tunas are caught by pole and line, and transported fresh or on ice to land-based freezer and canning facilities, producing a dolphin friendly product. In the past tuna was cooked to remove the oil, salted and sundried to produce "Maldive Fish" exported to southern India and Sri Lanka where it was traded for rice.

Expansion of tuna production in the Indian Ocean is possible, particularly in the case of the deeper-swimming tunas, which are not caught by surface pole and line, but with long line and purse seines. Significant large-scale tuna fisheries are currently operating in the western Indian Ocean based in the Seychelles and Mauritius.

Marine-based tourism

For many of the smaller island nations tuna and tourism are the sole resources which can be used to generate export income for development. Mass tourism based on the coral beaches and warm waters of the countries surrounding the Indian Ocean has grown considerably over the last two decades. The economy of countries dependent on tourism is extremely fragile. During the Gulf War, for example, tourist arrivals in the Maldives dropped to below half the normal numbers and many resorts closed.

An additional problem is the environmental impacts of tourism. In the Maldives, Seychelles and other smaller islands of the Indian Ocean, coral reefs and their associated white sand beaches form the main attraction for tourists from the Northern Hemisphere. The direct, damaging impacts of large numbers of people swimming, diving and collecting on coral reefs has resulted in significant degradation in the environment on which the industry depends.

◀ **For many small islands** in the Indian Ocean tourism is a source of hard currency earnings. Visitors from Europe come to enjoy the warm sea and coral sand and in countries such as the Seychelles and Maldives whole islands may be converted to tourist resorts.

Cassiterite Dredging
Cassiterite, a major source of tin, was probably one of the first minerals to be exploited on the seabed. Dredgers, like the one shown, scoop ore-containing sediment from the seabed and transfer it to plants on land, where the tin is extracted. The mining takes place mainly off the shores of Myanmar, Thailand and Indonesia.

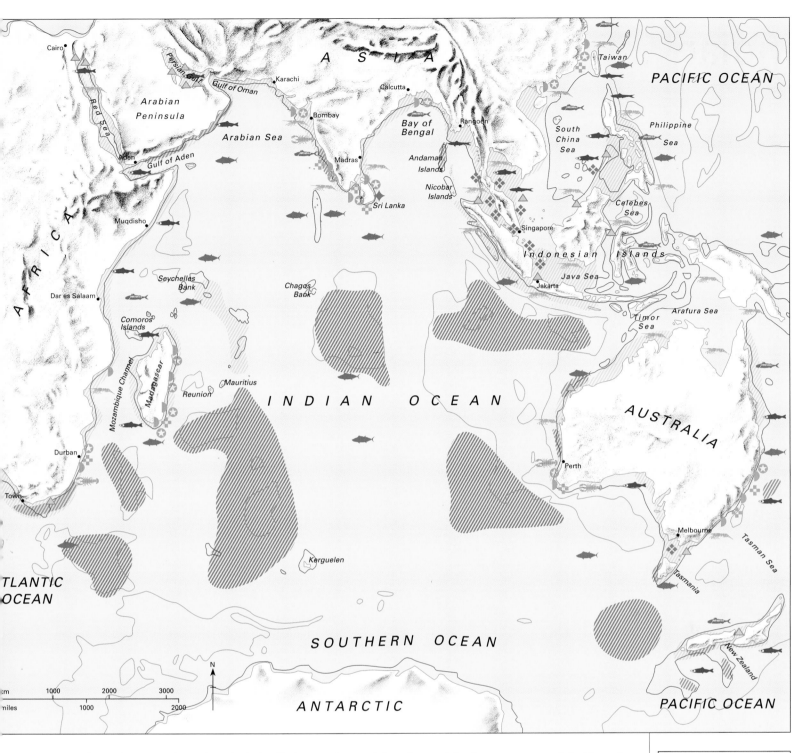

Indirect impacts of tourism include discharge of sewage to reef areas which encourages algal growth, increased beach erosion resulting from construction of hotels on the unstable beach margin, and the increased use of corals for building holiday complexes. In order to supply Male', for example, the capital of the Maldives, every year somewhere in the region of 2000 cubic metres (70,600 cubic feet) of living coral is harvested. At this present rate, in 20 years or so there will be no suitable coral to continue the development. The building of protective sea-walls, groynes and piers, which alter the local patterns of current circulation and sand movement, not only degrades the neighbouring reef environment, but may also threaten the resorts themselves. The countries of the region are taking active steps in environmental management, designed to develop eco-tourism which minimizes the damage to reef ecosystems caused by past unsustainable practices.

Non-living resources

Large areas of the Indian Ocean deep seafloor have been identified as containing significant reserves of manganese nodules, particularly in the southern basins. Phosphate nodules of potential value as a source of fertilizers have been located on the Argulhas Plateau.

Placer deposits have been actively exploited around the margins of the Indian Ocean for over 70 years. Of these deposits, tin in the form of cassiterite is perhaps the most significant. This is mined around 8 kilometres (5 miles) offshore from Myanmar, Thailand and Indonesia. Around a quarter of the world's production of tin came from these deposits during the 1970s. Monazite, ilmenite, rutile and zircon are mined from beach sands along the shores of Kerala state in southern India, while sands rich in ilmenite, rutile, zircon and magnetite are mined in northern Sri Lanka. In eastern South Africa glauconite, rich in potash, is mined for fertilizer production.

Although, after critical evaluation of the sedimentary basins along its land margins, together with geophysical and drilling data from the deep ocean areas initially led to increased interest in hydrocarbon exploitation in the Indian Ocean, oil and gas production from offshore reserves is still largely confined to the northern Arabian Sea, although significant offshore reserves have recently been exploited in the Indonesian region.

Living resources
Demersal fishing area
Pelagic fish
Anchovy
Sardine
Tuna
Mackerel
Crustacea
Shrimp
Lobster
Mineral resources
Tin
Chrome
Titanium
Monozite
Zirconium
Phosphorite
Manganese nodules
Sand and gravel
Oil and gas

200 metres
5000 metres

The Red Sea

The waters surrounding the Arabian Peninsula and its neighbouring countries form a distinctive sub-region in the northern Indian Ocean. The region includes the Red Sea and the Gulf of Aden to the west and south, the Persian Gulf and Gulf of Oman to the north and east, and the Arabian Sea, which receives high-salinity water from all these sources.

▲ The Red Sea
Area: *438,000 sq km (169,000 sq miles)*
Width: *145–306 km (90–190 miles)*
Maximum depth: *(axial trough) 2920 m (9580 ft)*
Evaporation: *more than 200 cm/year (80 in/year)*

Formation of the Red Sea

The Red Sea represents an ocean in the early stages of formation, and is in reality a rift valley which has been flooded. The Red Sea, with its associated arm, the Gulf of Aqaba, is comparatively deep, reaching in excess of 2000 metres (6500 feet) in some areas. In contrast the Gulf of Suez is quite shallow.

The bottom topography of the Red Sea is dominated by the wide, broad, smooth continental shelf, and an axial trough which is itself split by an even deeper axial valley almost 25 kilometres (15 miles) wide. The axial valley was formed by relatively recent spreading of the seafloor, and the bottom shows volcanic features including recent lava flows.

On the continental shelf areas, the thick layers of sediment are of uniform thickness and overlie a layer of anhydrite and salt – minerals which are normally associated with evaporite basins of shallow depth. These deposits are around 5 million years old and show that at that time the Red Sea was an area of high evaporation. Their absence from the axial valley further demonstrates that the valley is a younger geological feature. The uniform thickness of the sediment layers above the evaporite layers suggests that the spreading did not occur at that time, because had seafloor spreading been occurring continuously, then the sediments would have been thicker towards the margin, decreasing in thickness towards the axial valley.

Seafloor spreading

On the basis of this evidence it seems likely, therefore, that two periods of seafloor spreading have occurred. The first is thought to have occurred between 20 and 30 million years ago, and resulted in the formation of the present continental shelves. This was followed by an inactive period during which sediments were deposited and an evaporite basin was formed. The second, more recent onset of seafloor spreading commenced around 2 million years ago and has resulted in the formation of the axial valley and the mid-ocean ridge.

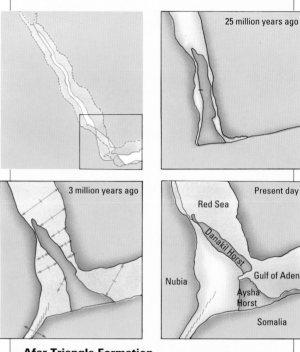

Afar Triangle Formation
About 25 million years ago the Yemen fitted between Somalia and ancient Nubia, which started moving apart from spreading centres at either side of a crust which now forms the Danakil Horst. The area to the southwest subsequently became the Afar Triangle.

The Red Sea is widening at a rate of around 1.25 centimetres (0.5 inches) per year. Although it is today barely 320 kilometres (200 miles) wide at its widest point, if seafloor spreading continues at its present rate, then in 200 million years the Red Sea would be the same width as the present-day Atlantic.

The Afar Triangle

The Afar Triangle represents an unusual piece of coastal geomorphology. It is found at the triple-point junction of the Red Sea, the Gulf of Aden and the East African Rift systems. About 25 million years ago, the tip of the Arabian Peninsula fitted snugly between Nubia and Somalia. However, two centres of spreading developed on either side of what is now the Donakil Rise and, following

◄ The Afar Triangle was formed when Nubia and Somalia started to be split apart by two spreading centres. A small piece of crust the Danakil Horst lies between them and the Afar Triangle was formed to the southwest.

► The Afar Triangle was once below sea level as indicated by ash rings formed from fragments of volcanic glass which erupted underwater. The desert surface is covered with thick deposits of salt formed from evaporation of saline water.

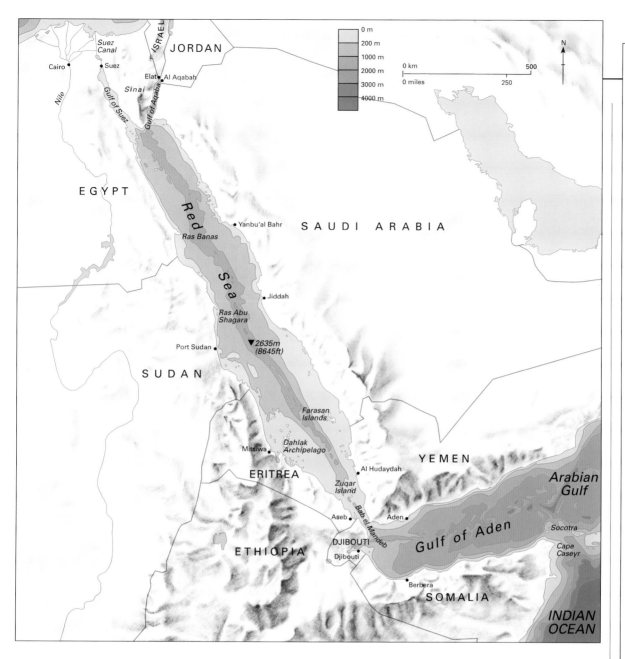

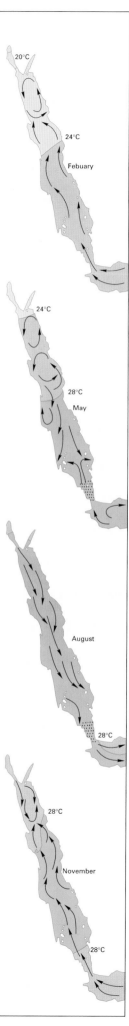

uplift, the oceanic crust of the Afar Triangle has become dry land. The land surface of the Afar Triangle is inhospitable desert. Its unique surface features of old, volcanic cones faced with shattered volcanic glass and thick deposits of salts are seen nowhere else on Earth, although a number of the volcanic cones are similar in form to submarine guyots or dormant, deep-sea volcanoes.

Circulation

The waters of the Red Sea are relatively clear, and the coastline is fringed with extensive coral reef systems. Circulation in the Red Sea Basin is generally wind driven, and during the northwest monsoon the flow is towards the Gulf of Suez. During the southeast monsoon, however, the current reverses, passing down towards the narrow straits of Bab el Mandab. Exchange of water through these straits with the neighbouring Arabian Sea is dominated by the tidal currents.

The reversal of currents is similar to that which occurs in the Arabian Sea, resulting in upwelling which occurs off the coast of Somalia and Oman under the influence of the southwest monsoon winds. These pass parallel to the coast and cause the surface waters to flow away from the coast drawing up colder, nutrient-rich waters from beneath. During the northeast monsoon season, the cold waters become trapped below the warmer surface waters and the subsequent lowered nutrient supply results in reduced primary production.

▼ **This satellite picture** shows the Gulf of Suez and Gulf of Aqaba separated by the mountainous Sinai Peninsula. These two gulfs appear to have been formed during different periods of spreading of the Red Sea.

► **Current flow** in the Red Sea changes, setting to the northwest from November to March, and southeasterly during the remainder of the year; the switch reflects the seasonal changes in currents in the Arabian Sea.

Red Sea Resources

The clear waters of the Red Sea and low, freshwater run-off from the surrounding arid lands have resulted in the development of extensive coral reef systems along both shores. The coral reefs extend for a distance of nearly 2000 kilometres (1250 miles) along the shores, and dominate the marine environment of the area.

Living resources

Over 350 species of corals have been recorded in the Red Sea region, which has a much higher diversity of marine organisms than either the Persian Gulf or the Arabian Sea.

The diversity of corals is greater even than that of the Caribbean Sea region and equal to the highest recorded in the Indian Ocean as a whole. An estimated 6 per cent of the species are found only in the Red Sea, and as much as 90 per cent of some groups of coral reef fish are unique to the area.

This diverse fish fauna includes important food fish, such as snappers, grouper and parrotfish, which are exploited by small-scale artisanal fishermen. Since the density of human populations along the coastlines of the Red Sea is generally not great, artisanal fishing pressure is

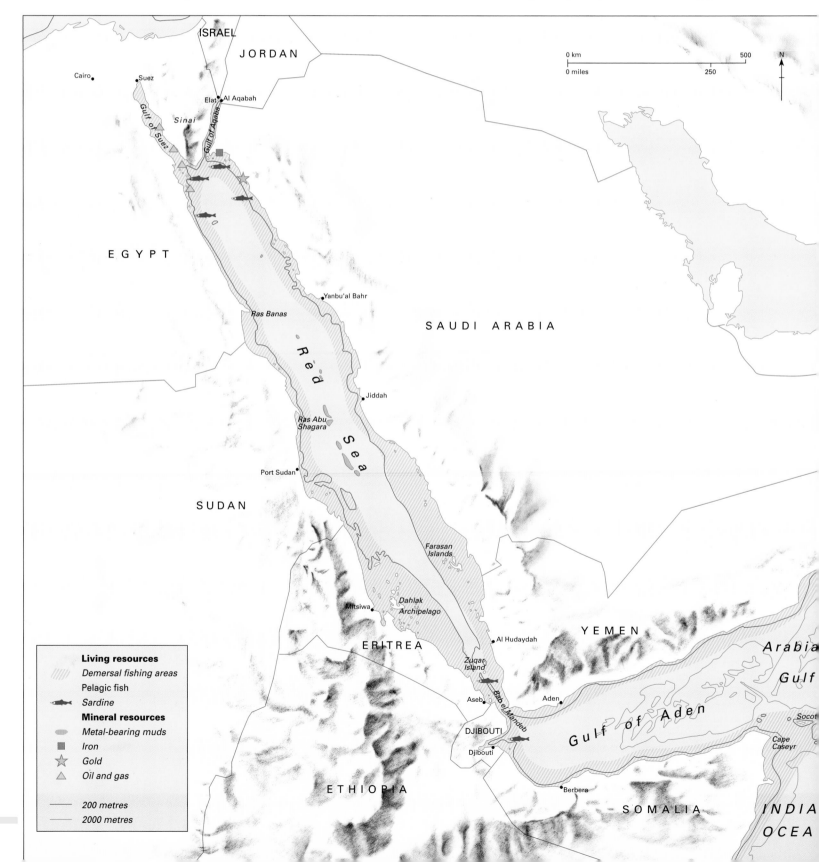

Living resources
- ⫽⫽ Demersal fishing areas
- Pelagic fish
- 🐟 Sardine

Mineral resources
- Metal-bearing muds
- ■ Iron
- ☆ Gold
- △ Oil and gas

— 200 metres
— 2000 metres

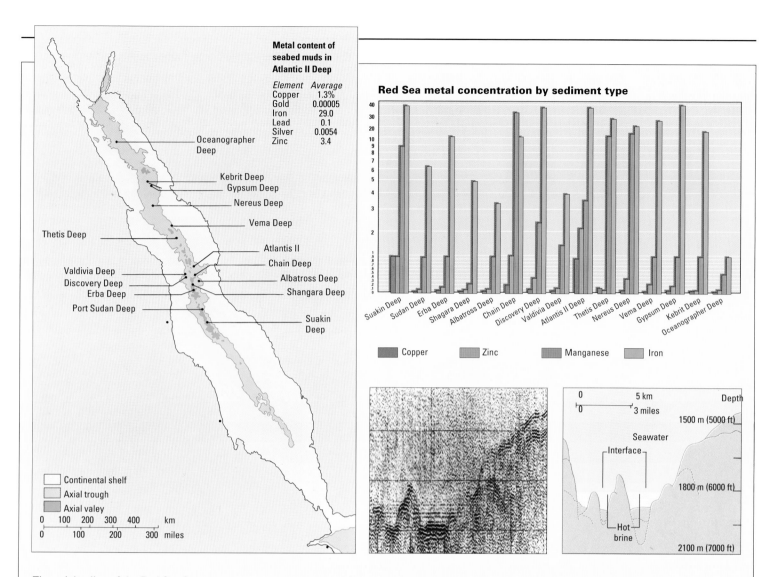

Metal content of seabed muds in Atlantic II Deep

Element	Average
Copper	1.3%
Gold	0.00005
Iron	29.0
Lead	0.1
Silver	0.0054
Zinc	3.4

Red Sea metal concentration by sediment type

Suakin Deep, Sudan Deep, Erba Deep, Shagara Deep, Albatross Deep, Chain Deep, Discovery Deep, Valdivia Deep, Atlantis II Deep, Thetis Deep, Nereus Deep, Vema Deep, Gypsum Deep, Kebrit Deep, Oceanographer Deep

Copper ■ Zinc ■ Manganese ■ Iron

Oceanographer Deep
Kebrit Deep
Gypsum Deep
Nereus Deep
Vema Deep
Thetis Deep
Atlantis II
Chain Deep
Valdivia Deep
Discovery Deep
Erba Deep
Port Sudan Deep
Albatross Deep
Shangara Deep
Suakin Deep

□ Continental shelf
▨ Axial trough
▨ Axial valey

0 100 200 300 400 km
0 100 200 300 miles

0   5 km
0   3 miles   Depth

1500 m (5000 ft)

Seawater
—Interface—

1800 m (6000 ft)

—Hot
 brine

2100 m (7000 ft)

The axial valley of the Red Sea features a number of "pits" over 1800 metres (6000 feet) deep that contain hot brines, which have emerged from the seabed and are associated with the creation of new crustal rocks through volcanic activity. The brine pools have temperatures of up to 60°C (140°F), suggesting that the brine is as hot as 104°C (219°F) when its is extruded. Calculations indicate that the production of brine in some areas is 200 times greater than that of Old Faithful Geyser in Yellowstone National Park in Wyoming in the United States. The brines are rich in potentially valuable metals and these are subsequently deposited in the muds of the seafloor. The interface between normal seawater and brine shows clearly on echo-sounder tracers.

generally low. To the south of the region and in the north around Ghardaqa, small commercial trawl fisheries for demersal species have been established.

The most important fishery of the area is, however, a pelagic fishery for sardine in the northern reaches of the Red Sea and Gulf of Suez. The sardines are caught at night by means of large, circular nets. The nets are used in conjunction with lights, which attract the fish to the surface above the suspended net. The nets can then be raised, so trapping the fish.

A few mangrove stands are found on the landward side of the coral reefs, but these are not well developed, and indeed the coastal trade in mangrove poles from East Africa northwards into this region continues even today.

Metalliferous muds

Metallic muds and brines were first discovered in the Red Sea region in 1963, and are associated with the volcanic activity of the axial rift valley. The submarine rocks of the region contain significant concentrations of zinc, copper, manganese and lead, and it is believed that over time the saline waters of the Red Sea permeate through the fractured submarine rocks. As the hot brine trickles through the rocks, it leaches the metals. The hot brine, which may be as high as 104°C (219°F) when extruded, is trapped in deep pits more than 1850 metres (6000 feet) deep. The metals are then brought to the seabed by the process of convection, where, as the brine mixes with the much cooler Red Sea water the metals are precipitated out.

Cost of extraction

Although the metal concentrations may be high, the costs of recovery are such that economic extraction is not possible given current metal prices and available reserves. Of the 15 hot brines discovered in the Red Sea region, the Atlantis II Deep is believed to contain the highest concentrations of metals, thereby making it the most viable in economic terms.

The deep falls within the EEZ of Sudan and Saudi Arabia, and to control exploitation a Red Sea Commission has been put in place. In contrast to the Persian Gulf, oil and gas reserves are not important in this region.

The desert surface of the Afar Triangle is covered in places by thick deposits of salt formed from the evaporation of pools of saline water following the formation of dry land in an area previously below sea level.

The Persian Gulf

In contrast to the Red Sea, the Persian Gulf is extremely shallow, with the majority of its seafloor lying only 100 metres (330 feet) or less below the surface. In addition, unlike the Red Sea, the Persian Gulf represents an area of active subduction where the Arabian Peninsula is sliding beneath the Asian continental plate. As a consequence, the northeastern shore drops rapidly into a trough, while the opposite shore shelves more gently.

The Arabian Peninsula is one of the world's smallest continental plates. Oceanic crust is formed in the Red Sea rift valley, and is pushing the Arabian plate northeastwards (where it is lost beneath the Asian plate in the Persian Gulf Subduction Zone). The Zagros Mountains, which lie close to the northeastern shore, represent the folded margin of the Asian plate, which is being compressed as the Arabian plate slides beneath it. The sediments that are being deformed at the foot of the Zagros Mountains include deposits of salt, pushed up into domes, which give rise to the extensive oilfields of the region.

The continued movement of the Arabian plate may ultimately result in the complete closure of the Persian Gulf, a process which could take only a few tens of thousands of years.

Circulation

Due to the shallow nature of the Persian Gulf Basin, the waters are well mixed, and nutrient inputs from the neighbouring land mean that primary productivity is higher than in the Red Sea. The diversity of species, however, is much lower than in the Red Sea and Arabian Gulf, and the coral reefs are not as well developed. However, the Gulf's stands of mangroves and seagrass meadows are more extensive than in the Red Sea.

Surface circulation is driven by the winds. A reversal of surface currents reflects the dominant monsoon wind patterns seen in this area (as in the Arabian and Red seas). Warm, high-salinity water passes out of the Gulf into the Arabian Sea during the southwest monsoon. Since the water is not as highly saline as that of the Red Sea, it disperses out into the Arabian Sea at depths of only around 185 metres (600 feet).

The temperatures of water in the Gulf are generally high, reaching as much as 33°C (91°F). The corals in this semi-enclosed basin have adapted to the harsh conditions, and are capable of withstanding temperatures which would result in bleaching and mortality on reefs in more open and cooler water environments.

The Gulf War

During the Gulf War extensive oil spills caused considerable pollution of the marine environment of the Gulf resulting in the death of numerous seabirds such as this cormorant and threatening vitally important desalination plants in neighbouring Saudi Arabia. Oil fires in Kuwait caused considerable air pollution and blackened the skies of the region for several months. Contrary to expectations, however, the damage to the marine ecosystems was less than expected and the oil fires were brought under control more quickly than originally anticipated. However, the effect of the air-bourne oil pollution to soldiers on the ground has not been accurately determined and may be responsible for some respiratory disorders.

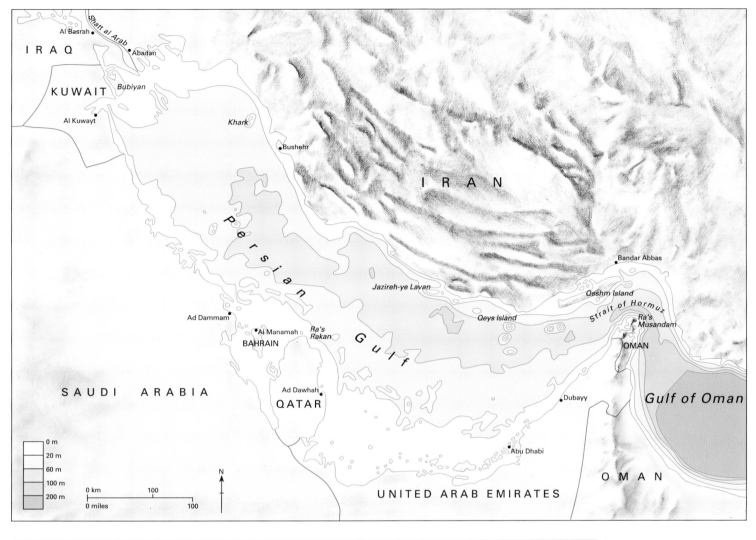

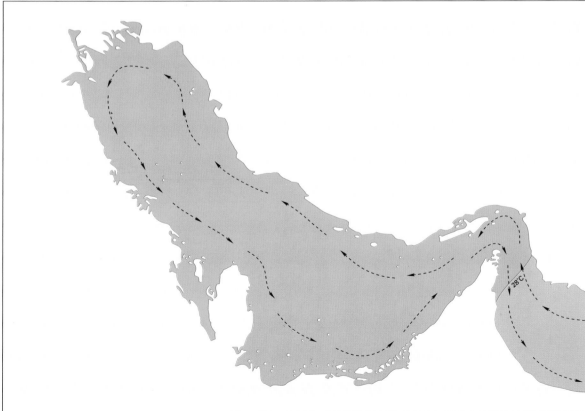

Persian Gulf Resources

The Persian Gulf is synonymous with oil, and indeed the mineral resources of this area are both rich and diverse. What is perhaps less well known is that, as a consequence of the inputs of nutrients from the Zagros Mountains and the Gulf's shallow depth which results in well-mixed waters, the biological productivity of this area is also much higher than that of the Red Sea.

Non-living resources

Although the region's first oil well was sunk in 500 BC, at Shush in Iran, the history of modern oil exploration and exploitation is a mere 100 years old. Modern exploration began in the 1890s, but was relatively unsuccessful until the discovery of the Naft-i-Shah field in Iran in 1923, which commenced production in 1935. Today, this area remains one of the major oil-producing regions in the world, using a number of land-based and offshore fields.

Sedimentary rocks have been deposited in the Persian Gulf region for nearly 280 million years. They consist of a sequence of permeable limestones, inter-bedded with organic-rich layers and evaporites. During the Tertiary period (between 65 and 2 million years ago), these deposits were folded and pushed upwards by the movement of the Arabian continental plate towards the Asian plate. Anticlines in these sediments became traps for the oil that was produced from the decomposition of organic matter in the sedimentary sequence over tens of thousands of years. Such oil reservoirs are found in the Jurassic limestones and dolomites of Saudi Arabia, Cretaceous sands and limestones to the northeast, and in Tertiary limestones and sometimes reef structures in the Iranian foothills. A comparatively small field in Triassic rocks lies in Iraq. Gas fields are located in a number of areas in the foothills of the Zagros Mountains.

◄ *Bahraini pearl* divers, free dive from a dhow to collect the pearl-bearing oysters for which the region was well known even in Roman times. Thousands of oysters had to be collected to reap a handful of pearls, and for this reason pearls were a very valuable commodity. Today, however, the farming of cultured pearls has dispensed with the need for harvesting natural pearls. Cultured pearls are made by inserting a tiny pellet made from nacre or mussel shell into an oyster.

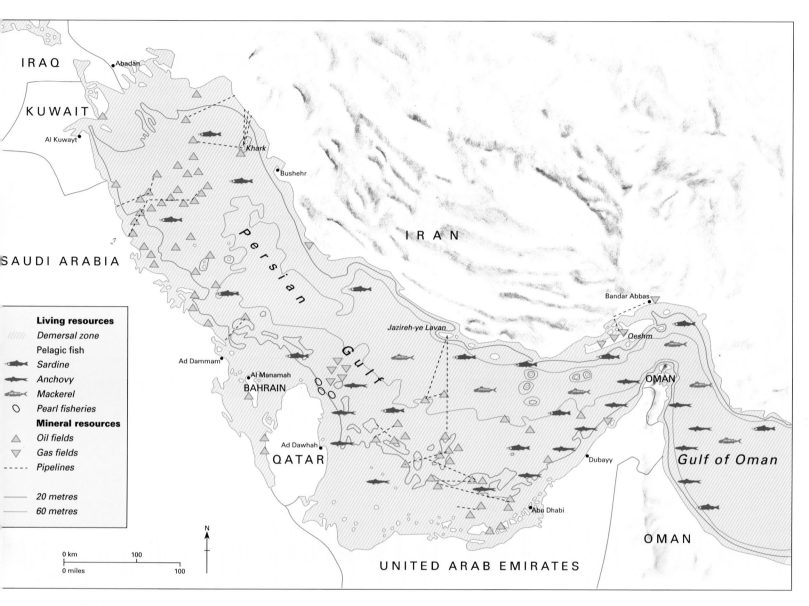

Living resources

- ///// Demersal zone
- Pelagic fish
- Sardine
- Anchovy
- Mackerel
- ○ Pearl fisheries

Mineral resources

- △ Oil fields
- ▽ Gas fields
- - - - Pipelines

—— 20 metres
—— 60 metres

0 km   100
0 miles   100

N

Oil from both the inland and offshore fields is pumped through pipelines, which in the case of Das Island in Abu Dhabi is 90 kilometres (55 miles) long, to land-based terminals. Once at the terminal, the oil and gas are separated and the crude oil refined or pumped to tankers at offshore loading berths.

Following the hostilities in the Gulf during the 1991 Gulf War, around 5 million barrels of oil were released into the sea. Apart from threatening the coastal infrastructure, including desalination plants, this action not only caused the deaths of many seabirds and other marine animals, but also resulted in the degradation of reef areas where the heavier oil fractions covered fragile reef surfaces. Recovery of reefs in the area was, however, more rapid than had been anticipated.

Living resources
Historical references to the rich pearl oyster communities of Bahrain date back to the Assyrians over 2000 years ago. The invasion of the area by the Portuguese in 1522, was, in part, a desire to gain control over this resource. Exploitation continued until the 1930s through collection of oysters by free divers who also collected freshwater in leather bottles from submarine springs.

At Masirah Island on the coast of Oman there is an important fishery for green turtle, with over 1000 harvest-ed in the seagrass meadows around the island annually. The island is also an important rookery for loggerhead and other turtle species, with an estimated 30,000 animals nesting there annually.

Commercial fisheries in the Persian Gulf are based on pelagic species, including sardine, anchovy, mackerel and barracuda, and some trawl fisheries based on demersal species. In general, smaller-scale artisanal fisheries target a much wider range of demersal species.

Trade
The prosperous trading civilization of Dilmun encompassed what is now Bahrain and Saudi Arabia, and flourished some 4000–5000 years ago. Taking advantage of the seasonal reversal of winds, mariners could traverse the entire northern Indian Ocean Basin, passing down the Arabian and East African coast in search of wood and charcoal. Arab traders carried valuable cowry shells from the Maldives to northern India. They also travelled round the tip of India into Southeast Asia, where they established an extensive trading empire dealing in Chinese porcelain, spices, grains, dried fish, ointments and slaves. This trading network of sultanates and trading posts reached as far as New Guinea to the southeast, and the Chinese mainland in the north, and predated the appearance of Europeans in this area.

The Pacific Ocean

The Pacific Ocean is the world's largest ocean, although throughout its history it has been decreasing in size as a consequence of the opening of the Atlantic and Indian oceans. The margins of the Pacific Ocean are, therefore, active areas of subduction, with associated deep-ocean trenches around its perimeter and intense volcanic activity associated with the Pacific "Rim of Fire".

The Marianas Trench, where the Pacific plate plunges beneath the Philippine plate at a rate of more than 11 centimetres (4 inches) a year, and the Tonga Trench, north of New Zealand, reach depths of over 10.5 kilometres (6.5 miles), more than twice the average depth of the ocean. In addition to having a rapid rate of subduction at the plate margins, the Pacific plate also has one of the most active mid-ocean ridges, the East Pacific Rise. On this mid-ocean ridge new oceanic crust is being formed at a rate of around 15 centimetres (6.5 inches) per year.

Eastern Pacific

The floor of the eastern half of the Pacific Basin is comparatively simple in structure. Its topography is dominated by the East Pacific Rise and two less active spreading ridges: the Galapagos, which has a spreading centre near the Equator, and the Chile Rise, which lies southeast of the basin. The slope of the seafloor is generally away from the crest of these oceanic ridges.

Along the western coast of North America, the seafloor deepens from east to west, and increases in age as one passes from Baja California to Hawaii. Baja California is moving away from the rest of North America, and this movement results from a branch of the East Pacific Rise. In the remainder of the northeastern Pacific this mid-ocean ridge has, with the exception of a few remnants off the coast of Washington, been consumed beneath the continental margin. Thus the mid-ocean ridge which created the floor of the northeastern Pacific is no longer visible, having disappeared beneath the North American landmass within the last 30 million years.

Western Pacific

The floor of the western half of the Pacific has a more complex structure. In part this reflects the separation of the Australian plate from the Antarctic landmass some 55 million years ago, and the shift in the direction of movement of the Pacific plate. The Australian plate is continuing to move northwards past the East Indies, which further compresses the oceanic crust at its northern margin. This western region of the ocean floor is also marked by paired active trenches, separated by a number of both active and inactive trenches. The Philippine Sea, for example, has been created in three distinct phases, each marked by the formation of a separate basin, between the trenches.

The oldest oceanic crust is located in the western Pacific, where some areas of the ocean floor are between 100 and 135 million years old. Its structure has been complicated by subsequent volcanic activity, and chains of volcanoes such as those forming the Gilbert and Ellice islands, and the Emperor seamounts, are widespread. The western Pacific also has several large volcanic plateaux, including the Shatsky Rise, the Solomon Plateau and the Manihiki Plateau. These have apparently been formed by extensive lava flows similar to the ones that created continental features such as the Deccan Plateau in India.

▲ **The Pacific Ocean**
Area: 166,000,000 sq km (64,000,000 sq miles)
Average depth: 4280 m (14,050 ft)
Max. depth: (Mariana Trench) 11,022 m (36,161 ft)

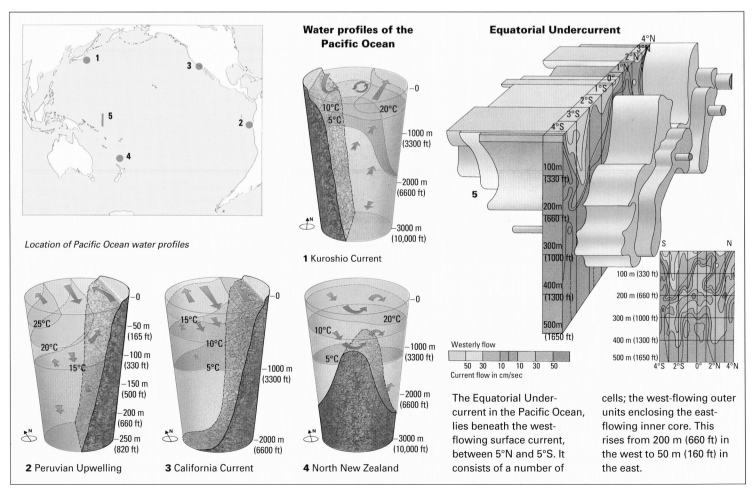

Location of Pacific Ocean water profiles

Water profiles of the Pacific Ocean

1 Kuroshio Current

2 Peruvian Upwelling

3 California Current

4 North New Zealand

Equatorial Undercurrent

Westerly flow
50 30 10 10 30 50
Current flow in cm/sec

The Equatorial Undercurrent in the Pacific Ocean, lies beneath the west-flowing surface current, between 5°N and 5°S. It consists of a number of cells; the west-flowing outer units enclosing the east-flowing inner core. This rises from 200 m (660 ft) in the west to 50 m (160 ft) in the east.

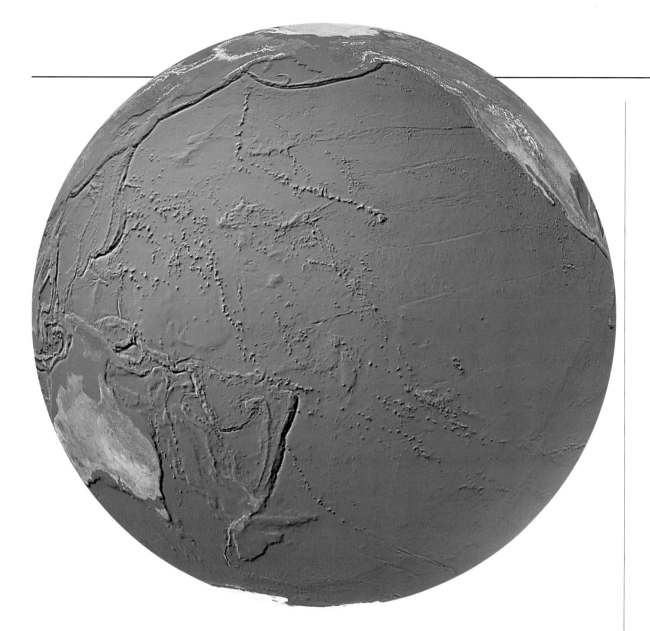

▼ **One hundred million years ago** the Pacific was much larger than it is at present. The ocean floor was formed of four plates separated by spreading centres connected to the mid-ocean ridge of the Indian Ocean, the activity of which was causing India to drift northwards away from Antarctica. Trenches and subduction zones, consuming oceanic crust, surrounded the ocean at this time.

▼ **Between 80 and 60 million years** ago seafloor spreading opened the Tasman Sea between Australia and New Zealand and subsequently, Australia split from Antarctica and began to drift northwards. In the North Pacific the Kula plate was consumed except for the small fragment forming the Bering Sea. By 27 million years ago (below) the ridge between the Pacific and Farallon split into the Gorda, Cocos and Nazca plates.

▼ **Deep-drilled cores** provide scientists with information regarding the age of the oceanic crust which indicate that the oldest part of this basin lies to the west. The original spreading centres have largely been consumed beneath the ocean margin, and the ocean basin is continuing to get smaller as Australia continues to drift north. However, its progress may be impeded as it crashes past the East Indies.

Lost plates of the Pacific

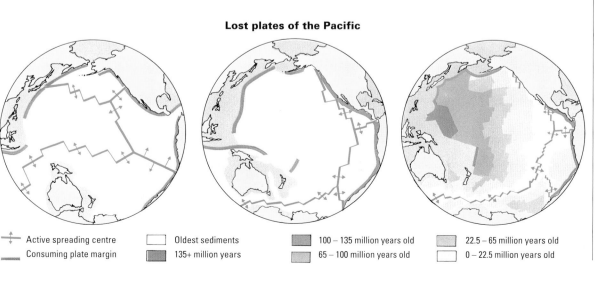

┿ Active spreading centre	☐ Oldest sediments	▨ 100 – 135 million years old	▨ 22.5 – 65 million years old
━ Consuming plate margin	▨ 135+ million years	▨ 65 – 100 million years old	☐ 0 – 22.5 million years old

The Pacific Ocean Basin

The Pacific Ocean covers over a third of the surface of the globe and contains some 724 million cubic kilometres (174 million cubic miles) of water. In area, the Pacific is twice the size of the Atlantic and due to its greater depth, contains more than twice as much water.

The eastern half of the ocean basin is characterized by a comparatively smooth, bottom topography sloping gently away from the North American coastline. In contrast the western half has a rugged surface with numerous trenches, both active and inactive, and volcanic island arcs, crumpled and deformed by the northward movement of the Australian continental plate.

Pacific islands

The Pacific Ocean Basin is characterized by numerous island chains running in a general northwest-southeast direction. The oldest of these island chains show a greater tendency towards a north-south orientation than those which are less than 40 million years old. This difference in orientation indicates a shift in the direction of movement of the Pacific Ocean crust, around 40 million years ago.

In addition to the true oceanic islands, which have never been part of a larger landmass, islands such as New Zealand, represent small fragments of continental crust which have become separated from the larger continental landmasses. The oceanic islands are all volcanic, and in the Pacific they are of two distinct types. Along the landward side of subduction zones in the western Pacific are curved chains or arcs of islands such as the Kurils, Bonins and Marianas. These islands were formed by the explosive eruptions of andesite lavas from the volcanoes on top of the subduction zones, where oceanic crust is being consumed beneath the continental plate margin.

In contrast, the straighter island chains of the mid-Pacific are formed of basaltic lavas which erupt much less violently. These islands are apparently formed over hot-spots in the oceanic crust, where intermittent eruption of lava from a source in the deep mantle occurs. These hot-spots do not erupt continuously and the island may move with the crust away from the hot-spot. Thus the oldest islands lie to the north of the Hawaiian chain.

Beyond the emergent islands, the chains may continue beneath the sea surface as submerged seamounts, or guyots. The Emperor seamount chain (which runs north-south) represents a continuation of the younger Hawaiian Island chain.

Circulation

The two main gyres of the Pacific Basin are separated by a more complex system of equatorial currents and countercurrents than the Atlantic. In addition, anomalous patterns of circulation can occur during El Niño years.

In the Northern Hemisphere, the warm North Equatorial Current carries water some 14,500 kilometres (9000 miles) across the Pacific, forming the longest westward-flowing current in the world's oceans. At the western end, the current turns north and is intensified into the narrow western boundary current, the Kuroshio, which flows northwards at speeds in excess of 145 kilometres (90 miles) a day. In the latitude of Japan, the Kuroshio meets the cold, south-flowing Oyashio Current, and the two are deflected away from the coast to meander eastwards across the North Pacific. To the north of this warm-water gyre lies a cold, subpolar gyre formed by the Alaskan and Aleutian currents in the east, and the Oyashio Current in the west.

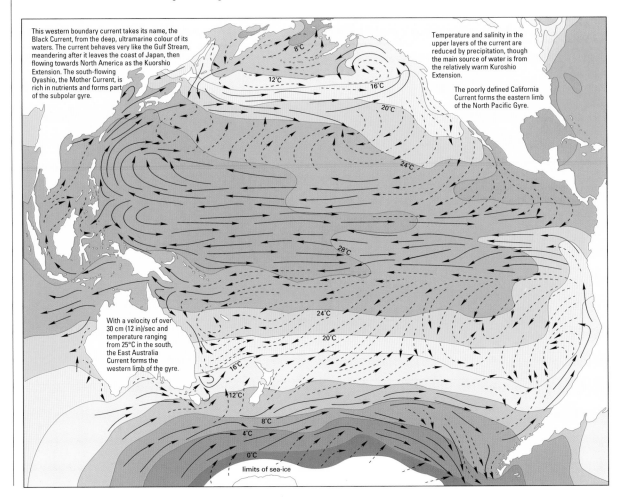

This western boundary current takes its name, the Black Current, from the deep, ultramarine colour of its waters. The current behaves very like the Gulf Stream, meandering after it leaves the coast of Japan, then flowing towards North America as the Kuorshio Extension. The south-flowing Oyashio, the Mother Current, is rich in nutrients and forms part of the subpolar gyre.

Temperature and salinity in the upper layers of the current are reduced by precipitation, though the main source of water is from the relatively warm Kuroshio Extension.

The poorly defined California Current forms the eastern limb of the North Pacific Gyre.

With a velocity of over 30 cm (12 in)/sec and temperature ranging from 25°C in the south, the East Australia Current forms the western limb of the gyre.

limits of sea-ice

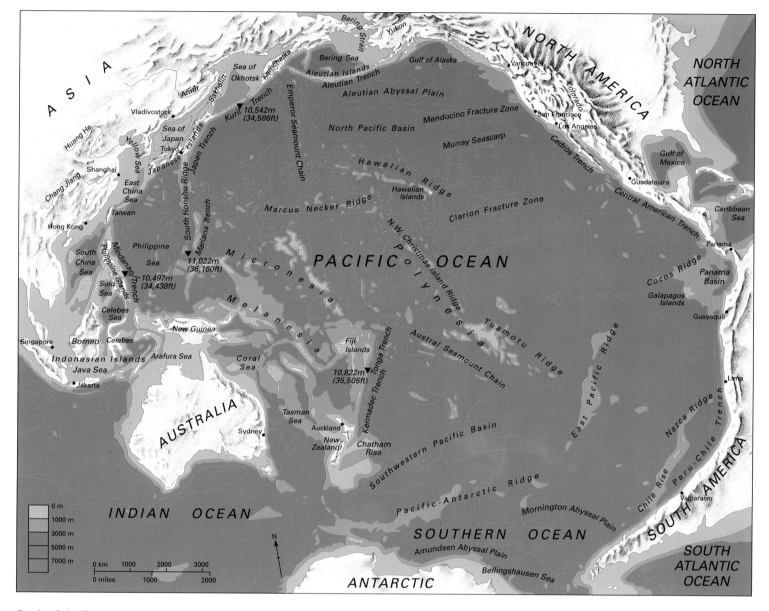

South of the Equator, a second major gyre is formed by the South Equatorial Current, the East Australia Current and the Humboldt Current. The Humboldt Current is stronger than most eastern-boundary currents due to the northward deflection of the westerly winds as they meet the Andes. The western boundary current, called the East Australia Current, turns westwards at around the latitude of Sydney and passes to the north of the North Island of New Zealand. The Antarctic Circumpolar Current is driven by the westerly winds and has a surface speed of only 19 kilometres (12 miles) per day. The total volume of water carried, however, amounts to more than 165 million tonnes per day, greater than the volume transported by any other current in the world's ocean. This current reaches to depths of 3 kilometres (2 miles), hence the immense volumes of water moved.

Countercurrents

In the region of the Equator, the flow is complicated by the presence of the North and South Equatorial countercurrents which result in the Equatorial Undercurrent. This flows east at speeds of more than 145 kilometres (90 miles) per day. The main body of the current lies around 200 metres (660 feet) below the surface in the west, rising to around 50 metres (160 feet) in the east.

Less than 40 m. y. old
Islands
Seamounts

More than 40 m. y. old
Islands
Seamounts

▲ *Younger chains* of islands run northwest-southeast, in contrast to those older than 40 million years, which run north-south. This change indicates a shift in movement of the Pacific plate around that time.

Pacific Ocean Resources

For many isolated, insular nations of the central Pacific, living marine resources are the sole source of export income for development. The potential of deep-sea mineral resources has yet to yield economic benefit.

Non-living resources

Polymetallic nodules are widely distributed throughout the deep Pacific. By 1974, 100 years after their initial discovery, it had been established that a broad belt of seafloor – over an area of 2.15 million square kilometres (1.35 million square miles) – between Hawaii and Mexico was densely paved with such nodules. Despite their relative abundance, mining the nodules remains uneconomic.

Similarly, deposits of phosphate minerals in extensive areas off the western coasts of North and South America, and on submarine plateaux regions near New Zealand and Australia, cannot be economically extracted at present.

Near-shore, many mineral placer deposits are found on the continental shelves surrounding the Pacific. Gold has been mined from the beaches of Alaska for many years, and platinum is known to occur in the same region. Tin is mined extensively in the Southeast Asian region, and mineral sands containing titanium, chromium and zirconium are present along the north American continental margin. In the western Pacific, extensive offshore deposits of iron ore have been mined by the Japanese for many years.

Living resources

In excess of 40 per cent of the world's harvest of finfish comes from the Pacific, and the bulk of this is composed

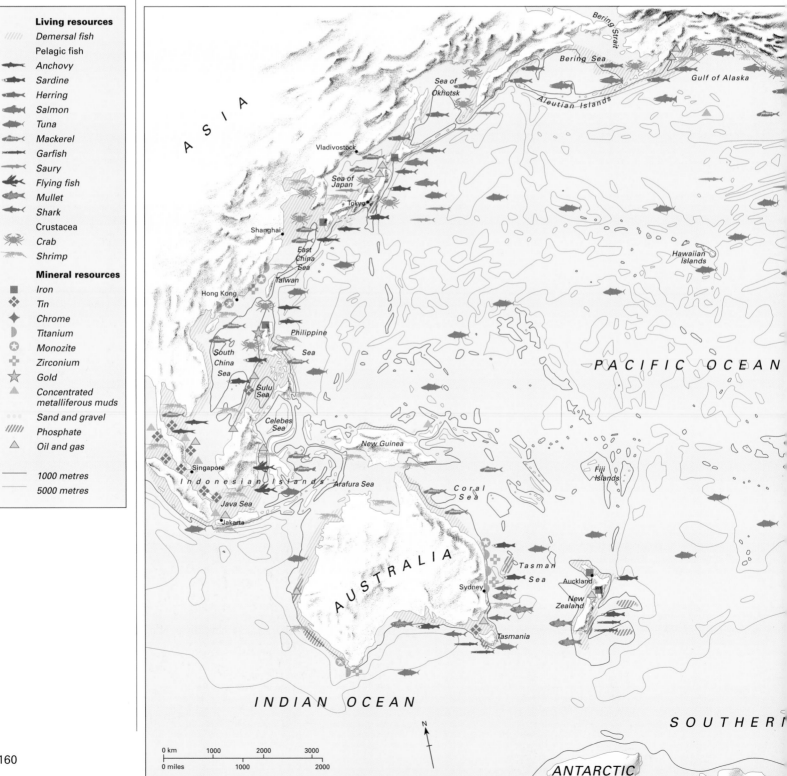

Living resources

- ////// Demersal fish
- Pelagic fish
- Anchovy
- Sardine
- Herring
- Salmon
- Tuna
- Mackerel
- Garfish
- Saury
- Flying fish
- Mullet
- Shark
- Crustacea
- Crab
- Shrimp

Mineral resources

- ■ Iron
- ◆ Tin
- ◆ Chrome
- ▷ Titanium
- ◉ Monozite
- ✦ Zirconium
- ★ Gold
- ▲ Concentrated metalliferous muds
- ∙∙∙ Sand and gravel
- ////// Phosphate
- △ Oil and gas

- —— 1000 metres
- —— 5000 metres

0 km 1000 2000 3000

0 miles 1000 2000

N

of herring-like fishes, including the Japanese sardine and Peruvian anchovy. Between 1969 and 1971, the Peruvian anchovy fishery had risen to almost 30 per cent by weight of the total fishery of the Pacific, one-sixth of the world catch of fish. The harvest of anchovy, which was almost exclusively for the manufacture of fishmeal, had exceeded 10 million tonnes in the peak years. But by 1991, the catch dropped by over half to around 4 million tonnes.

The Pacific tuna fishery is based on purse seine fisheries in the eastern Pacific and long liners in the west and central Pacific regions. These commercial fisheries target the deeper-swimming species of albacore, bigeye and yellowfin tuna, and the catch is frequently processed at sea. The recent introduction of extensive drift nets, several tens of kilometres in length, is a cause of concern to environ-mental groups and the smaller nations of the Pacific for whom tuna represent the sole, exportable resource. These drift nets, used in the Pacific by long-distance fishing fleets from the East Asian region, have been termed the "wall of death", since they result in the entanglement and drowning of turtles, marine mammals and a wide variety of non-resource species. A regional convention has been drafted to ban their use in the South Pacific.

Another high-value species is the North Pacific salmon, taken generally inshore using gill nets, traps and weirs – although they are also fished on the high seas by Japanese fishermen using gill nets and purse seines. Other demersal fisheries include the Alaskan pollack fishery in the subarctic Pacific.

Artisanal fisheries

In the island nations of the central Pacific, the dominant fisheries are small-scale artisanal fisheries based on the wide diversity of species associated with the coral reef ecosystems which surround these islands. The enormous diversity of species in these catches presents problems of processing and marketing outside these countries, where the unknown characteristics of the species result in low international market acceptability.

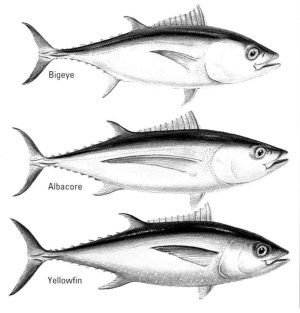

◄ **Tuna** are fast-swimming fish found throughout the tropical and subtropical waters of the Atlantic, Indian and Pacific oceans. The most important fisheries of the six species that occur in the Pacific are the yellowfin, albacore and bigeye.

▼ **The six species** of Pacific salmon migrate from their spawning grounds in the rivers of Asia and North America to the rich feeding grounds of the North Pacific. After periods ranging from six months to five years at sea, the adults return to the rivers where they were born to spawn, making journeys of thousands of kilometres.

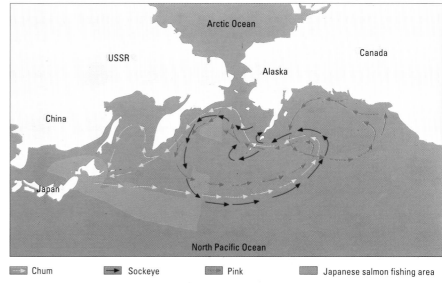

Chum Sockeye Pink Japanese salmon fishing area

Pacific El Niño

The importance of the air-sea interaction is nowhere more dramatically demonstrated than in the Southern Pacific Ocean. Periodic weakening of the major wind patterns in this area leads to alterations in the surface currents in the southern gyre, suppression of the Peruvian upwelling, and changes to biological productivity, sea level and rainfall patterns.

These changes in the southern Pacific have considerable effects on the patterns of rainfall and weather in Australia and Southeast Asia, where monsoon rainfall is reduced during El Niño years. The phenomenon is called El Niño (the child) since it occurs in Latin America around Christmas and forms part of the ENSO (El Niño, Southern Oscillation) pattern of climate in the Southern Hemisphere. The impacts of such events can be traced in patterns of coastal dune formation in eastern South America and may be implicated in changed rainfall patterns as far away as Africa.

The Southern Gyre

Under normal conditions, the South Pacific Gyre dominates water circulation in the southern Pacific Ocean. Surface water flows eastwards under the influence of the trade winds as the South Equatorial Current, passes down the eastern coast of Australia (as the East Australia Current) and turns eastwards at the latitude of Sydney under the influence of the westerly winds. The Antarctic Circumpolar Current then moves eastwards towards Latin America where a major branch, the Humboldt Current, travels northwards along the coast of Chile towards Peru.

▶ *These two satellite images show the differences in sea surface temperature during El Niño and normal years. Blue indicates cool water 0°–12°C (32°–54°F); green intermediate temperatures 13°–24°C (55°–76°F) and yellow, red and purple colours indicate temperatures between 25° and 30°C (77° and 86°F). The image at the top for January 1984 shows the normal pattern of sea surface temperature with a pool of warm water in the western Pacific (1) and a tongue of cooler water penetrating into the Pacific from the west coast of Latin America (2). An image from January 1983 during the height of an El Niño event shows the warm waters of the western Pacific reaching across the surface of the ocean and suppressing the upwelling off the Peruvian coast.*

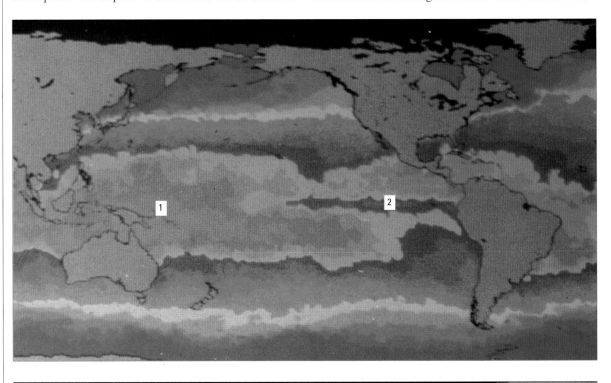

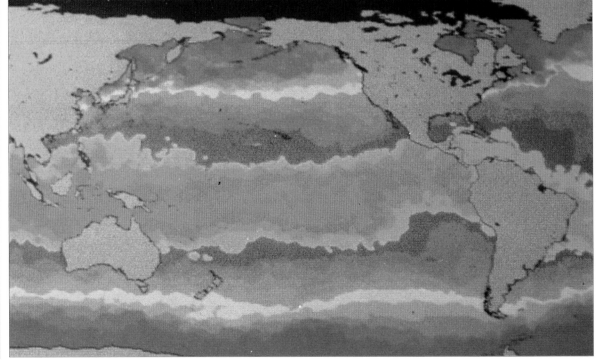

The high Andes block the passage of the westerlies, deflecting them northwards, which not only adds to the speed of the Humboldt Current, but also causes the surface waters to move away from the coast. As they do so they form part of the eastward-moving South Equatorial Current, and subsequently cold, nutrient-rich water is drawn to the surface.

The Peruvian upwelling

Off the coast of Peru, this upwelled, cold, nutrient-rich water stimulates the production of phytoplankton and a characteristic community of comparatively large, multi-cellular phytoplankton develops. The large size of these primary producers when compared with the smaller, single-celled diatoms, characteristic of nutrient-poor, open ocean water means that they can be eaten directly by small fishes, the anchovetta.

The anchovetta themselves form the food of larger predatory fishes and numerous seabirds, whose breeding colonies on the Latin American coast have resulted in the deposition of guano, mined during the last century as phosphate fertilizer. In addition, the anchovetta were extensively fished from small purse seiners, and the catch processed to produce fish meal for domestic animal and poultry feed. The high-fisheries production of the Peruvian upwelling reflects both the high rates of primary production in the area, and the efficiency of the food chain, showing a reduced number of linkages between the primary producers and the top predator, in this case man.

Physical changes

During an El Niño event, upwelling ceases and the biological productivity of the area collapses, since the nutrients are no longer brought to the surface. Extensive research on this phenomenon during the last decade has shown that the surface currents of the Southern Gyre are weaker during El Niño years, and the eastward-moving South Equatorial Current becomes dominated by the westward-flowing Equatorial Countercurrent.

During their 15,500 kilometre- (9000 mile-) journey across the Pacific Ocean, the surface waters of the South

◀ **Anchovetta and sardine** are unloaded at a fishmeal and oil plant on the Peruvian coast. Annually around 20 million tonnes of these small fish are caught, however, during the El Niño year of 1982–3, the anchovetta catch was reduced to one six-hundredth of this total.

Equatorial Current become progressively warmer, leading to the formation of a pool of warm water in the western Pacific. In addition, the movement of the water results in higher sea levels in the western Pacific region, and lower sea levels along the Latin American coast. When the flow weakens or is reversed, the water levels change on each side of the Pacific Basin. During El Niño years, mean sea level in the western Pacific may be as much as 14 centimetres (5.5 inches) below the normal level, while along the Latin American coast it may be as much as 50 centimetres (20 inches) higher than normal. The surface waters in the zone of upwelling are now dominated by the warm waters from the western Pacific, upwelling ceases and the biological productivity of the region collapses.

Economic effects

During the El Niño events in 1972–3 and again in 1982–3, the catch of anchovetta was reduced to one-sixth and one-six-hundredth of the normal catch, respectively. As the fish stocks decline, the birds which depend on them for food also die in large numbers. Following the 1972–3 El Niño, the populations of cormorants, boobies and brown pelicans declined to around 6 million birds, from an estimated number of 30 million in 1950. Following the 1982–3 event a further decline reduced their numbers to a level of only 300,000.

The economic consequences of El Niño are severe. Raised water levels along the Latin American coast increase the frequency of malaria, while inland flooding results in disruption of agriculture and the destruction of coastal infrastructure. Economic impacts also occur elsewhere, with agriculture being adversely affected in the semi-arid regions of Australia, and subsistence food production being reduced in Southeast Asia.

Teleconnections

The El Niño phenomenon appears to be related to changes in atmospheric circulation, and in particular to a weakening of the westerly winds which drive the southern arm of the South Pacific Gyre. The consequences of such a major change in the air–sea interaction are not confined to the Pacific Basin. These interactions appear to be linked to other changes in atmospheric and oceanic circulation in the Southern Hemisphere. The results of these changes are reduced rainfall in Australia, weakening of the Indian monsoons, changes in the Kyushio Current off Japan, shifts in wind patterns in the South Atlantic, with consequent changes in the orientation of coastal dunes, and droughts in Southeast Asia.

TOGA and WOCE

The occurrence of El Niño Southern Oscillation events is not regular, and worryingly there is some evidence to suggest that the frequency of the event may have increased during the last few decades.

Due to the widespread changes to the local climate and sea level conditions that El Niño brings to certain areas of the tropics, and the subsequent substantial economic losses suffered by some countries, two programmes have been set up to monitor this phenomenon with a view to ultimately predicting its occurrence. The international, Tropical Ocean Global Atmosphere programme (TOGA) and the WOCE, World Ocean Circulation Experiment, have been commissioned to collect data on ocean circulation and the interaction between the oceans and the atmosphere, with a view to modelling how the system works.

While scientists are now able to predict the onset of an El Niño phenomenon in the short term, long-term predictions of the event are still not possible.

The Southern Ocean

The Antarctic, or Southern Ocean, is normally taken as meaning the area of ocean which lies south of the Antarctic convergence zone, or between 50° and 55°S. This convergence zone represents the invisible boundary where the cold, surface waters of the Antarctic Ocean meet the warmer waters of the sub-Antarctic.

The southern ocean forms a great expanse of unbroken ocean. This vast area of water is important in the formation of cold water which flows at the surface and at depth into the Atlantic, Pacific and Indian ocean basins, and is replaced by southward-flowing intermediate, warmer water from sub-Antarctic areas.

The currents of the Antarctic Ocean are driven by the Roaring Forties, the Furious Fifties and the Shrieking Sixties, and its southern boundary is the ice and rocks of the Antarctic coastline.

Almost the entire landmass of Antarctica lies within the Antarctic circle, at the centre of which lies the South Pole. Two deep indentations disrupt the circular outline of this continent: the Weddell Sea, which faces the South Atlantic, and the Ross Sea, which faces out towards the South Pacific Ocean.

The Antarctic ice cap

Even during the Antarctic summer, this inhospitable continent is fringed with 3 million square kilometres (1.2 million square miles) of ice-covered sea. During winter, the ice extends to cover an area of some 20 million square kilometres (8 million square miles).

The continent itself supports an ice cap which, over time, has accumulated to an average depth of 2000 metres (6500 feet) and which extends to over 3000 metres (10,000 feet) in some areas. It has been calculated that if all this ice were to melt, then the global level of the sea would rise by 60 metres (200 feet). When released from this enormous weight of ice, the entire landmass of Antarctica would rise by between 200 and 300 metres (650 and 1000 feet).

The Antarctic ice sheet contains over 90 per cent of the world's ice, and around 70 per cent of the planet's freshwater. Glaciers move seawards at numerous points to feed the floating ice shelves which border the continent. These ice shelves extend out several hundred kilometres to a submerged depth of between 150 and 200 metres (500 and 650 feet), and cover the head of the Ross Sea and the inland and western shores of the Weddell Sea.

Life in the frozen seas

Despite the icy seas and frozen wastes of the Antarctic continent, the southern ocean is home to a wide variety of seabirds, seals and whales. These marine animals depend on the biological productivity of the microscopic phytoplankton and the kelp which fringe the shores of the continent. The estimated primary productivity of phytoplankton in the surface waters of the Antarctic Ocean is 610 million tonnes a year. This vast community of phytoplankton is, in turn, eaten by zooplankton of which half are krill, which form the major food source for fish, squid, whales and some seabirds.

Although most seabirds are found in the subantarctic region, a number, including three species of penguin, breed on the Antarctic landmass. The breeding colonies of seabirds are densely packed since the extent of ice-free land is limited to the narrow coastal fringe.

Antarctic seals, of which there are six species, are also abundant. There is, for example, an estimated population

▲ **The emperor penguins** are the largest of all the penguins. After pairing, the female lays one egg, and returns to the sea, leaving the incubation to the male.

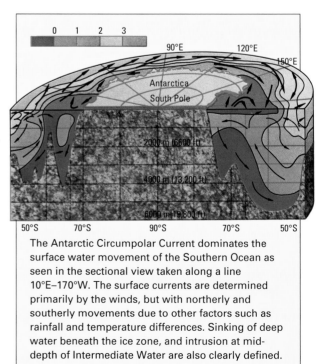

The Antarctic Circumpolar Current dominates the surface water movement of the Southern Ocean as seen in the sectional view taken along a line 10°E–170°W. The surface currents are determined primarily by the winds, but with northerly and southerly movements due to other factors such as rainfall and temperature differences. Sinking of deep water beneath the ice zone, and intrusion at mid-depth of Intermediate Water are also clearly defined.

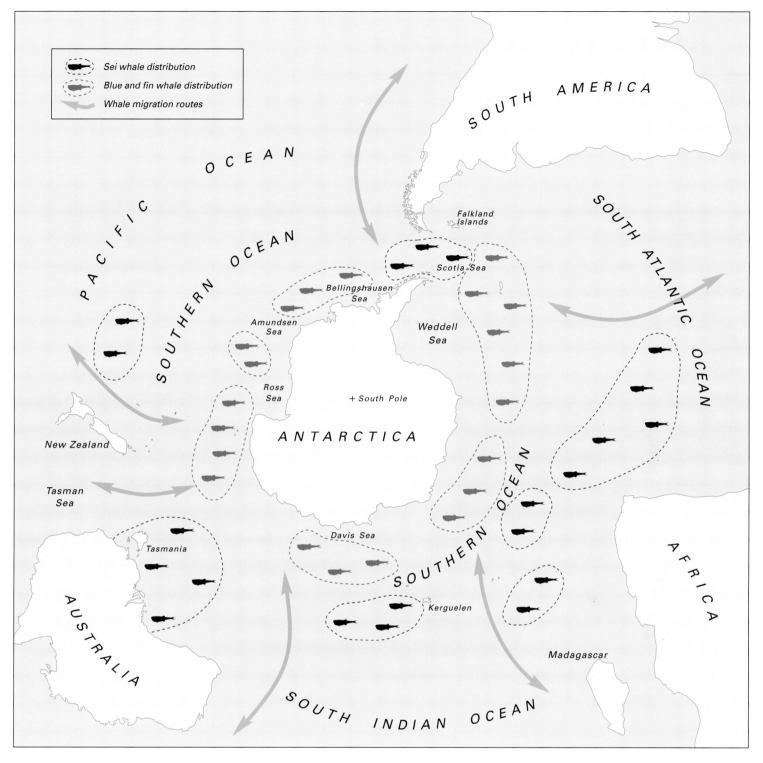

Sei whale distribution

Blue and fin whale distribution

Whale migration routes

SOUTH AMERICA

PACIFIC OCEAN

SOUTHERN OCEAN

SOUTH ATLANTIC OCEAN

Falkland Islands

Scotia Sea

Bellingshausen Sea

Amundsen Sea

Weddell Sea

Ross Sea

+ South Pole

ANTARCTICA

New Zealand

Tasman Sea

SOUTHERN OCEAN

Davis Sea

Tasmania

Kerguelen

AFRICA

AUSTRALIA

Madagascar

SOUTH INDIAN OCEAN

of around 15 million individual crab-eater seals. Although there is some evidence that the elephant seal is declining, all other seal species appear to be increasing in numbers at the present time.

Seven species of filter-feeding whales and eight species of toothed whales inhabit this ocean, although none is confined to this area. The baleen whales, which feed predominantly on krill, undertake annual migrations to warmer, equatorial waters during the southern winter, returning to the Antarctic during the summer when productivity of plankton rises due to the continuous daylight and warmer temperatures.

The Antarctic Treaties

A narrow coastal fringe separates the ice cap from the ocean. On this area of land, which represents less than 2 per cent of the surface of the continent, the majority of the Antarctic wildlife lives and breeds. This is also the area in which more than 70 permanently manned research stations have been established by over 20 nations. Seven nations

have territorial claims to this continent, claims which were recognized but suspended under the 1959 Antarctic Treaty. The treaty also recognizes the interests of non-claimant states in regard to peaceful uses of Antarctica. Biennial meetings between claimant and non-claimant states, the consultative parties, recommend measures designed to further the principles of the treaty and to serve as guidelines for the future actions of the states which claim territorial rights.

In 1978, the Convention for the Conservation of Antarctic Seals came into force to regulate the catch of fur seals and to establish three seal reserves. The more far-reaching Convention on the Conservation of Antarctic Marine Living Resources came into force in 1982, and established a decision-making commission and an advisory scientific body. The Convention on the Regulation of Antarctic Mineral Resource Activities was opened for signature in 1988, but has not yet entered into force. This treaty seeks to regulate future exploitation of mineral resources on the continent.

Deception I

Antarctic Peninsula

● Active
○ Extinct

▲ *Antarctic volcanic* activity has in recent years centred mainly on Deception Island in the South Shetland Islands. The island erupted in 1967 and 1969, throwing pyroclastic material and vapour high into the air and covering the island in ash.

The Southern Ocean Basin

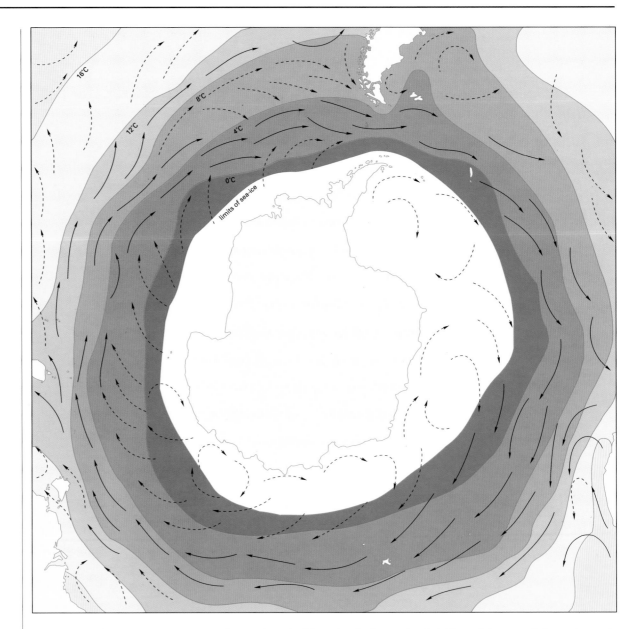

The Antarctic Ocean covers an area of approximately 35 million square kilometres (13.5 million square miles), of which more than half freezes over during the winter. During the summer, around 3 million square kilometres (1.2 million square miles) remain ice covered. There is increasing evidence that interannual variations in ice cover may have profound effects both on the area's ocean and atmospheric circulation patterns. Ice cover affects the exchange of heat and moisture between the ocean and atmosphere, which in turn affects air movements and currents.

Antarctica is the coldest continent, with temperatures in the central region rising briefly to 30°C (86°F) in the two warmest months, and sinking to -65°C (-85°F) in winter. A few coastal locations have less extreme climates, with average temperatures of 1°C (34°F) in December and February dropping to only -10° to -20°C (-4° to -14°F) in July and August. The long daylight hours of the Antarctic summer result in high productivity of phytoplankton.

Bathymetry
The continental shelf of Antarctica is generally much narrower and deeper than that of other continents. It lies at a depth of between 370 and 490 metres (1200 and 1600 feet), and there is a deep depression between the outer edge of the continental shelf and the land. It is believed that this depres-

sion is the result of sinking of the central Antarctic continental landmass under the weight of the overlying ice sheet.

Beyond the continental shelf lies the deep ocean basin, which is bounded to the north by the mid-ocean ridge systems. The ocean basin reaches depths of between 4000 and 5000 metres (13,000 and 16,500 feet) and is subdivided into the Southeast Pacific, southern Indian and Atlantic-Indian basins by ridges which join Antarctica to America, to the Kerguelen Plateau, east of the South Indian Basin, and to Tasmania.

Ocean circulation
North of 60°S, the current flow of the Southern Ocean at all depths is to the east. This Antarctic Circumpolar Current forms the southern boundary to the ocean gyres of the Atlantic, Pacific and Indian oceans. South of 60°S, the flow is to the west, but both current systems contribute to the northward flow of cold water in the surface and deeper layers of the ocean.

In the northern half of the Antarctic Ocean, the southerly movement of water is generally below 1800 metres (6000 feet), but around the Antarctic Convergence Zone it rises sharply to within a hundred metres or so of the surface. The subantarctic waters overlay a column of cold water, which occurs at depths of between 1900 and 3000 metres (6500 and 10,000 feet). This sharp transition zone is known as the polar front, and is generally taken as the limit of the Antarctic Ocean (representing the point at which the cooler, dense Antarctic surface water sinks beneath the warmer subantarctic surface waters). At the surface, this subantarctic convergence zone is recognizable by a sharp rise in temperature of between 1° and 2°C (1° and 3°F) and a change in the composition of the planktonic community.

Some 10° further north is found the less well-defined zone of convergence between the subantarctic and subtropical waters. Although not well-defined as a current boundary, it is recognizable at the surface as a transition zone between the colder waters of southern origin and the warmer, more saline water, characteristic of regions of higher latitude.

The greenhouse effect and the ozone hole

Scientists are now generally agreed that the average surface temperature of the globe has risen by around half a degree over the last 100 years or so. Although there is as yet no firm evidence for significant melting of the Antarctic ice sheet, scientists are monitoring the extent of sea ice – since the melting of only part of the Antarctic land-based ice sheet could have significant impacts on global sea level. In contrast, it has been suggested by some scientists that snowfall in the Antarctic could increase, which would result in an increase in the depth of the ice sheet and an associated lowering of sea level.

Of greater concern at the present time is the hole in the ozone layer, which may have impacts on the phytoplankton and kelp in the region. Both the phytoplankton and kelp form the basis for the Antarctic food web. Depletion of the atmospheric ozone layer allows higher levels of ultra-violet radiation to penetrate to the surface of the ocean which affects plankton productivity. Continued scientific observations and modelling of the Antarctic system is necessary to resolve these uncertainties.

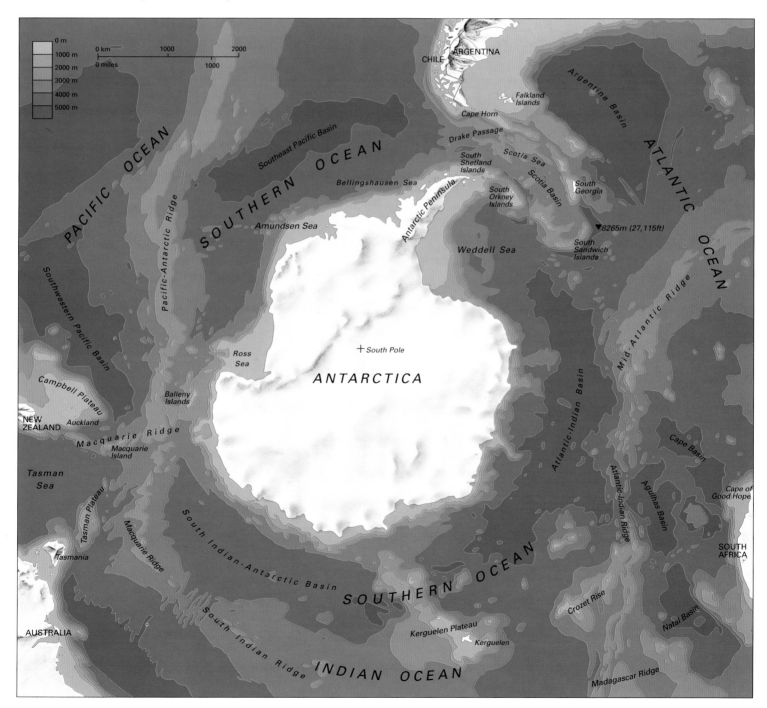

Southern Ocean Resources

The history of resource use in Antarctica is one of short-term perspectives dominated by greed and competition. This is shown in the dramatic decline in numbers of several whale species following over-hunting, and results, in part, from the absence, until recently, of any agreed ownership or international agreement concerning the exploitation of Antarctic resources.

Over the past few years this situation has improved following the negotiation of a variety of international treaties, the aim of which is to regulate the use of Antarctic living marine resources. However, the convention which seeks to regulate mineral extraction in Antarctica has yet to enter into force.

Non-living resources

Iron and coal deposits have been found in the mountain ranges of the Antarctic continent, while the deep-drilling ship, *Glomar Challenger*, found deposits of natural gas in the sediments beneath the Ross Sea. Since Antarctica was once part of the larger landmass of Gondwanaland, it is expected that mineral reserves comparable to those of the other southern continents are likely to occur within the rocks of Antarctica.

At present, the huge economic costs of mineral extraction preclude their exploitation. The inhospitable climate of the Antarctic region, together with the thickness of the ice cap and the problems of exporting the ores once mined, remain the greatest safeguards against unrestrained mineral exploitation.

Whales and whaling

Unlike the mineral resources, the living resources of Antarctica have a history of over-exploitation, of which the best-known example is whaling. The first whales to be exploited were the northern right whales, so called because they could be taken with the primitive equipment of the early whalers. Right whales were taken with hand-held harpoons for their whalebone and oil. By the middle

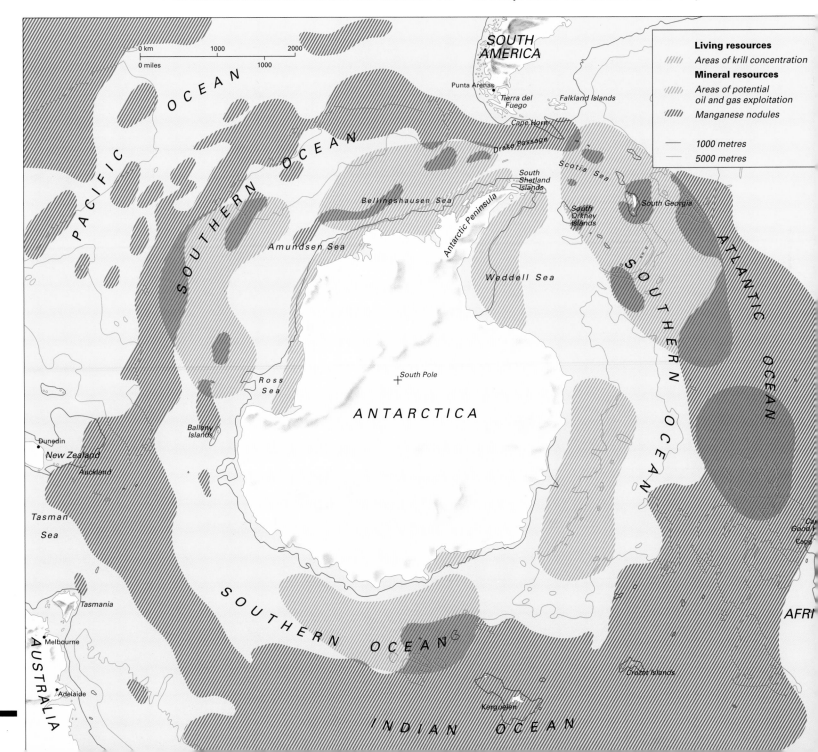

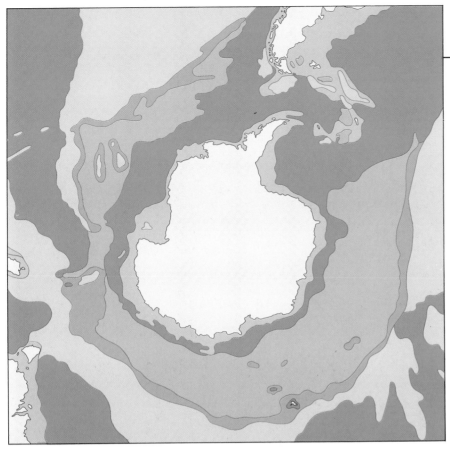

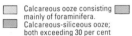

Calcareous ooze consisting mainly of foraminifera.	Siliceous ooze; mainly diatoms, some radiolarians.	Clayey silts and silty clays; red, brown, gray and olive, often with volcanic debris. Calcareous or siliceous near oozes; sandy near land	Shelf and coastal deposits; marine-glacial deposits around Antarctica; calcareous sands and gravels around New Zealand, australia, Africa.
Calcareous-siliceous ooze; both exceeding 30 per cent			

▲ *The giant kelp*
(Macrosystis), reaches the surface from depths of up to 40 m (130 ft) and, floating to the surface, may grow to over 55 m (180 ft) in length. It grows prolifically around Antarctic and subantarctic shores north of 60°S, and around the Falkland Islands. In the future it is hoped that kelp could be used as a source of alginates as well as being used in many food industries.

of the 19th century, both the northern and southern right whales had been reduced in numbers to the point where continued hunting was no longer profitable. The large, sperm whale industry declined at the same time, due to the introduction and increasing use of other oils, the product for which they were initially hunted. At first, the large blue whales and fin whales were too fast to be caught by the early whalers, but following the introduction of steam engines and explosive-tipped harpoons these species quickly became an economically viable catch.

Modern whaling, as described above, commenced in the Antarctic in 1904, when the first whaling station was established on South Georgia. Factory ships moored in sheltered water near land were the next innovation and whales were brought alongside for flensing and processing. By the 1930s, floating factory ships capable of processing whales on board dispensed with the need for land-based factories. Soon these ships were operating throughout the Antarctic during the summer.

In the 1930–31 whaling season, nearly 43,000 whales were killed, and the present whale populations are probably one-sixth to one-tenth of their numbers prior to the commencement of whaling. As individual species of whales became depleted, the industry shifted to more profitable species. For example, between 1927 and 1936, the humpback whale had become unprofitable, and was replaced in the industry by the blue whale. In post-war years, the blue was replaced by the fin whale, and subsequently this species was, in turn, replaced by the smaller sei, bottlenose and minke whales.

It has been estimated that there are at present only about 500 blue whales in the Antarctic, although they may have once numbered 250,000. The number of sei whales is more uncertain due to the fact that their entire range has not been surveyed, but they may have been reduced to a comparable extent. Humpback and right whales, which originally numbered around 100,000, have been reduced to the low thousands.

The status of whale stocks is very uncertain due to the difficulties of surveying these animals, and it will be several decades before firm evidence for a recovery of the most heavily depleted stock becomes available.

Antarctic fur seals

Less well known than the over-exploitation of the whales, but just as damaging to the species involved, is the history of the Antarctic fur seal. This particular species of seal was practically exterminated by the 1820s after several hundred thousand had been slaughtered for their skins. The seals subsequently recovered to some extent, but were again over-exploited in the 1870s, and are now recovering once more.

Elephant seal, which were once exploited for their oil, are no longer captured since the availability of alternative, mineral oils and the costs of labour have made this species uneconomic to exploit.

Krill, squid and fish

The substantial reduction in the numbers of filter-feeding baleen whales must have had a huge impact on the populations of krill, the plankton on which they feed. It has been estimated that before they were exploited, the southern baleen whales alone were consuming around 190 million tonnes of krill each year. These small crustaceans, which filter feed on phytoplankton, can reach lengths of around 5 centimetres (2 inches) when adult. Direct exploitation of krill for processing to produce animal feed is now an important Antarctic fishery.

At present, our knowledge of the biology and ecology of the fish and squid resources of the Antarctic is insufficient to establish whether or not they could sustain long-term harvesting. The history of Northern Hemisphere fisheries, and of the southern cod or orange roughie fishery in subantarctic waters, suggests that such species could support only moderate levels of harvest and may be vulnerable to over-exploitation.

ENCYCLOPEDIA OF MARINE LIFE

Prokaryotes and Eukaryotes

All living organisms fall into two major groups: the eukaryotes, in which the genetic material is arranged in chromosomes contained in a defined nucleus within the cells; and the prokaryotes, which lack a nucleus and defined chromosomes. The blue-green algae, once considered a primitive group of plants are now grouped with the bacteria since they have no organized nucleus or chromosomes and in some, the cell walls are similar to those of bacteria, rather than to those of typical plant or animals cells.

All the other groups of organisms are described as eukaryotes, since they possess a nucleus and chromosomal materials, and in most cases distinct organelles. Multi-cellular organisms often display differentiation of the cellular layers into tissues, groups of cells specialized for reproduction, excretion, co-ordination, locomotion, attachment, or other purposes. The degree of structural complexity in living organisms generally increases in higher groups, with fungi and algae having comparatively simple structures in comparison with the higher plants or angiosperms. The kingdom Protista contains a mix of single-celled organisms some of which display plant-like characters such as the presence of photosynthetic pigments, while others are more animal like in their structure and mode of life.

Cyanobacteria & Bacteria

The cyanobacteria or blue-green algae were once considered a class of algae, the yanophyceae but they are clearly more closely related to the bacteria than they are to the lower plants. Like typical algae, the blue-green algae are autotrophic, they photosynthesize and their photosynthetic pigments give them the colour from which they derive their name. Not all are blue-green however, some are yellow or red and others purple or black depending on the types of pigments they contain. Some species, such as Nostoc, *fix nitrogen from the atmosphere in the same way as the bacteria found in the root nodules of many terrestrial plants. Most marine, blue-green algae are found in shallow waters intertidally or subtidally, living attached to other plants or rocks, while one genus,* Rivularia, *forms an association with a fungus to form the common Lichina, a lichen of inter-tidal rocks. Blue-green algae can reproduce at a remarkably fast rate, forming blooms within a few hours and imparting colour to the water.* Oscillatoria erythrea, *a red-coloured species, gives the Red Sea its name.*

Marine bacteria are most abundant around the coasts, but in the open ocean they occur in association with plankton and many are saprophytes which degrade organic materials including chitin, celluloses, mucins and fats. Like the single-celled algae, bacterial populations explode in numbers under suitable conditions but their tiny size in comparison with even the smallest planktonic organisms means that few animals are capable of feeding directly on bacteria in suspension. One group of Ctenophores, or sea gooseberries however, have evolved specialized feeding nets for filter feeding on bacteria in suspension. Many detrital feeding organisms that ingest sand or mud, feed on the bacteria which they digest from the surface of the sand as it passes through the gut. Decomposing bacteria are as important in the marine environment as on land, while some chemotrophic species form the basis for the food chains of submarine thermal vents and others, the sulphur-reducing bacteria, are important in the anaerobic conditions of intertidal muds.

Kingdom Fungi

Moulds, yeasts and mushrooms, form one of the five kingdoms of living organisms, they are heterotrophic, lacking photosynthetic pigments and differ from bacteria in possessing nuclei and cell organelles. The microbial community in marine sediments can, on occasion, be dominated by fungi, especially in coastal waters subjected to freshwater input containing fungal spores. In the marine environment the dominant fungi are the yeasts, which are saprophytic, feeding on dead organic matter and other species which specialize in attacking organisms, including fish and the shells of molluscs. Ichthyophonus, for example, has been implicated as a significant controlling agent of herring populations.

Plant Kingdom

General Introduction to Plants

Plants form the basis of all life both on land and sea. By converting sunlight to chemical energy through the process of photosynthesis (autotrophy or self feeding), plants make energy available to heterotrophic organisms. The dependence of autotrophic plants on sunlight limits the depth range of these organisms in their environment. In shallow coastal waters, single-celled and multicelled algae may grow attached to the surface of the substrate, while in deeper waters, small planktonic, single-celled organisms form the basis of the food chain.

The algae are a diverse group of organisms which were considered in the past, to include part of the prokaryotes, the cyanobacteria or blue-green algae and, eukaryotic organisms such as the typical seaweeds and the single-celled or colonial algae, many of which are now grouped by zoologists in a separate kingdom, the Protista, and a few specialized angiosperms or higher plants, including the seagrasses.

The cyanobacteria are more closely related to bacteria than they are to lower plants, although they photosynthesize. Most marine blue-green algae are found in shallow waters intertidally or subtidally living attached to other plants or rocks.

The diagram illustrates different types of seaweed. They are not drawn to scale. The seaweeds of the rocky European shores, on the left of the diagram, would not necessarily all occur in the same area. Zonation of seaweed varies with such factors as degree of exposure and boulder height. *Sargassum* is a floating seaweed and *Macrocystis* occurs off the Pacific coast.

SUPER-PHYLUM ALGAE

This division of the plant kingdom is extremely diverse encompassing plants that range from microscopic unicellular organisms that reproduce by simple division to those that are extremely large such as the giant kelps which can attain lengths of up to 50 m (165 ft). The seaweeds are relatively simple plants that have complex life histories and reproductive cycles, and are multi-cellular with differentiated specialist tissues which may include floats, holdfasts for attachment to the substrate, and different male and female organs.

Seaweeds are classified according to their pigments and while all contain the typical green pigment chlorophyll-a, various yellow, brown or red pigments may mask the green coloration completely. The seaweeds of a typical rocky shore are zoned, with the green species occurring high in the intertidal. These are replaced lower down by brown species and lower still by reds. This zonation reflects the penetration of light of different wavelengths with red light penetrating least and blue and green light penetrating to the greatest depths. The complementary colours of red or brown, enable these species to absorb and utilize the blue and green wavelengths needed for photosynthesis under conditions of low light intensity.

Algae are commercially important and the mariculture of seaweeds has a long history in Southeast Asia. Over 70 species are eaten by man, and species are produced commercially as sources of alginin, iodine and various gelling and clarifying agents. In some areas, seaweeds have been used traditionally as agricultural fertilizers, while commercial production of fertilizers from seaweeds has been tried in some areas.

Class Phaeophyceae

The brown seaweeds contain chlorophyll-c and are coloured brown by fucoxanthin pigments. These are the most abundant and widespread seaweeds which store the carbohydrate laminarin, as a food reserve. They range in form from simple filamentous types to the large kelps, *Laminaria* spp. which have long blade-like fronds, a strong stalk and well-developed holdfast for attachment. The simplest brown algae are smaller genera such as *Ectocarpus*, which grows attached to the surface of other algae and *Cutleria*, which has a marked difference in growth form between the sexual and asexual generations. Reproductive cells are formed in special regions found in the plant body, and each cell liberates a single zooid (individual polyp) which forms the sexual phase of the alternating generations.

The most familiar brown seaweeds are the fucoids, common in the intertidal of temperate rocky shores these species have air-filled bladders which support the fronds in the water while the plant as a whole is resistant to exposure to the air, high temperatures and drying. Most of these plants live for between two and four years although some species may grow more slowly and survive for up to 19 years. Re-colonization of a rocky shore denuded of seaweeds may take a considerable period and the grazing action of small molluscs may affect the rate of re-colonisation. *Sargassum*, a brown, floating fucoid, occurs in enormous masses in the Sargasso Sea and forms the habitat for a diverse community of other plants and animals which live attached to the fronds, hide in the mats of material, or feed on the weed itself.

In the subtidal, grow the kelps which may achieve large size and like the *Sargassum* weed form the basis for a specialized community of animals including sea urchins, limpets, small snail-like molluscs, bryozoa, bivalve molluscs, herbivorous fish and worms. Like the bladder wracks of the intertidal, the kelps of the subtidal, represent the asexual generation of these plants which alternates with the microscopic free-living sexual generation. The large kelps of the Pacific coast of North America, *Macrocystis* can grow at the rate of 1.25 metres (4 feet) per day when cut and they are harvested in many areas as sources of mannitol a sweetening agent, oil and alginates.

Class Rhodophyceae

These are the red seaweeds coloured by phycocyanin, phycoerythrin and lutein pigments, and include species of pink, violet, purple, red and brown coloration. Unlike the brown seaweeds these plants only contain chlorophyll-a. Most species live attached to the substrate in the coastal zone and are flattened or filamentous, although a few species of single-celled planktonic forms are also known. A number of different red algae have a calcareous, encrusting growth form (e.g. *Lithothamnion*) and some species are important in consolidating the rubble surface of coral reef flat areas. Others are collected and ground to produce fertilizer, while the flattened *Porphyra* is grown

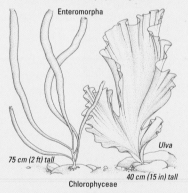

Enteromorpha

75 cm (2 ft) tall

Ulva

40 cm (15 in) tall

Chlorophyceae

commercially and eaten in Japan. However, the most important commercial product from red algae is agar or carrageen, the growth medium used extensively in bacterial culture for a variety of medical purposes all over the world.

Class Chlorophyceae

Some of these green algae are unicellular species with flagella such as *Pyramimonas* and *Tetrahele*, found in the plankton or estuarine and coastal areas, while intertidally are found non-motile forms including the large, multi-cellular genera *Enteromorpha* and *Ulva*. The green seaweeds contain the pigments, chlorophyll-b and lutein and store carbohydrate, as starch. In *Caulerpa*, cell walls are only formed to separate the reproductive parts and numerous nuclei may be found in the remainder of the organism. In contrast *Ulva* has one nucleus per cell. Some of the larger forms are collected for food while economically their importance relates more to their occurrence as fouling organisms on ship hulls. Most species are generally larger than the red seaweeds and may be up to a metre in length but the planktonic forms are members of the picoplankton (between 0.2 and 2 µm in length).

Class Prasinophyceae

These are planktonic green algae, usually with four flagella, sometimes two or one, which differ from most of the Chlorophyceae in having small scales on the flagella and sometimes over the body surface. Some species are colonial and form a colony within a spherical ball of mucus. The separation of this group from the Chlorophyceae is in some doubt since the scales which were originally considered to be a diagnostic feature of this group are now known to occur in some of the Chlorophyceae as well.

Class Bacillariophyceae

The unicellular diatoms are extremely important to the food chains of the open ocean regions where they are important components of the phytoplankton. They form an important food source for filter-feeding animals, but, because they are generally of reasonably large size, they can be fed upon directly by copepods. Diatoms have a rigid siliceous cell wall formed of two overlapping valves making a sculptured little box with a closely fitting lid. Holes in the lid allow the entry of nutrients in solution and appear to be involved in locomotion in some species. The siliceous shell is much denser than seawater and hence many species have spines or hairs projecting from the shell to

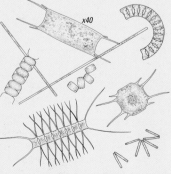

Bicillariophyceae (Diatoms)

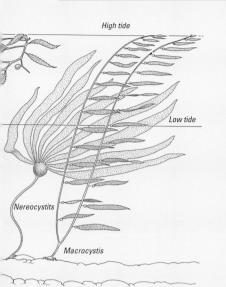

High tide

Low tide

Nereocystis

Macrocystis

reduce the rate of sinking. Production of diatoms is seasonal in many areas, with a spring bloom being followed by decline in numbers as the grazers crop the production and the supply of available nutrients becomes depleted.

ANGIOSPERMS

The higher plants of the marine environment are limited to around some 50 species of monocotyledonous plants, all of which have secondarily invaded the marine environment. They grow in shallow coastal waters and are generally rooted in soft sediments in the intertidal and subtidal zones. Seagrasses often form vast beds and are important in stabilizing the sediment and providing nursery areas for many species of fish and shell-fish of commercial importance. Most species produce flowers and fertilization is by means of floating pollen, although at least one species flowers above water. These plants, like

Zostera 30–60 cm (1–2 ft)

mangroves are viviparous, the seeds germinating on the parent plant and the young propagule growing to relatively large size before becoming detached and assuming an independent existence.

It has been argued that vivipary is an adaptation to the lack of freshwater required in abundance by young, rapidly growing seedlings. Although the centre of diversity of seagrasses is the Indo-West Pacific with meadows of up to 16 species in a single location, seagrasses are also distributed in the Mediterranean where the single species of *Posidonia* are extremely widespread and in temperate Europe, where the typical genus, *Zostera*, has declined in extent in recent decades.

Kingdom Protista

This kingdom includes around 50,000 species of single-celled organisms most of which are motile and heterotrophic. Of the free-living protistans around two-thirds are marine. The protozoa were considered to form a phylum within the Animal Kingdom although some are autotrophic and contain chlorophyll and other pigments making it difficult to decide whether they were animals or plants (algae). The different groups of protozoa, previously considered as classes of the phylum Protozoa are now generally treated as separate phyla within the kingdom Protista.

Protozoa live in the sea, in freshwater or as parasites in other organism. They usually multiply by simple fission (division of the single individual into two or more new cells) although sexual reproduction also occurs in many forms. The different phyla are distinguished by their modes of locomotion; the Sarcomastigophora locomote by means of flagella or pseudopodia; the Ciliophora by means of cilia; and the Sporozoa are parasitic.

PHYLUM SARCOMASTIGOPHORA

Subphylum Mastigophora
The Mastigophora are also known as the flagellata which includes the Euglenida and Volvocida both of which are considered by botanists as belonging to the algal divisions Dinophyta and Euglenophyta. One group, the Chrysophyceae or golden-brown algae including the so-called silicoflagellates, are classified as a distinct class in the algal super-phylum by botanists, but as the order Chrysomonadida, within the Mastigophora by zoologists. These organisms, which are mainly found in freshwater, produce minute urn-shaped siliceous resting spores, while the genera *Dictyota* and *Distephanus* which are widespread in marine phytoplankton have internal skeletons of silica. Siliceous remains of these organisms are abundant in some marine sediments of Tertiary age. The Xanthophyceae or yellow-green algae have green plastids but otherwise resemble the Chrysophyceae, of which *Vaucheria* forms mats of greenish threads of several centimetres length in coastal salt marshes, while *Meringosphaera* is often present in tropical marine nanoplankton.

The Euglenophyceae of botanists are plant-like protistans possessing chlorophyll-a and the pigments xanthophyl and carotene. These are small unicellelar organisms common in estuarine areas and living interstially in sand and muds. Some species are colourless and saprophytic and they locomote using the flagellum as an oar rather than as a propeller. A number of species of Mastigophora are colonial forming large spherical, hollow structures which in the case of Volvox may contain as many as 10,000 individuals. The Cryptophyceae are estuarine flagellates many of which live in symbiotic associations with animals.

The dinoflagellates are motile unicellular organisms with two flagella, one lying in a groove around the body. Cell walls generally consist of cellulose, lined with sporopollenin while some species, such as *Ceratium*, have thickly armoured cell walls. About half the dinoflagellates are colourless and saprophytic, while some marine genera such as *Noctiluca* are phosphorescent when disturbed. Dinoflagellates like diatoms are relatively large in size and are important in the phytoplankton. Some species such as *Gonyaulax* and can produce toxic algal blooms when they occur at high density.

One genus of dinoflagellates, *Symbiodinium*, is extremely important as a symbiont of corals and giant clams. These organisms are found within the tissues of the host from which they derive carbon dioxide and nutrients for growth and photosynthesis. They are implicated in the deposition of calcium carbonate in the skeletons of the host and supply nitrogenous materials for growth of the host animal. Coral and giant clam larvae do not possess the symbionts and must acquire them during the early stages of development, thereafter as the host grows, the population of dinoflagellates increases and the large size achieved by the hermatypic, reef-building corals and giant clams is attributed to the mutual benefits of this association.

SUBPHYLUM SARCODINA
The Sarcodina all possess pseudopodia of one type or another and include a number of important marine groups, the Amoebida, Foraminifera, Heliozoa, Rhizomastigina and Radiolaria. The amoebida includes the familiar amoeba which locomotes and feeds using pseudopodia, a characteristic of the group. Many amoebae are enclosed in non-living shells or tests which are generally vase shaped with a large opening through which the pseudopodia are extended. The shell may be formed of either, mineral particles collected by the animal and cemented together, or of siliceous, or proteinaceous materials.

Foraminifera construct shells of calcium carbonate which differ from the amoeboid tests in being multi-chambered, adding new chambers as they grow. Some of the largest, extinct formainifera had shells up to 5–6 cm (2–2.3 in) across. The shells are penetrated by small holes through which the animal sends out thin branching strands of protoplasm that form a web for catching and digesting diatoms and other small planktonic organisms. Different species construct shells of differing form, ranging from single chambered structures through various complex spirals, to cones of differing shape. The majority of foraminifera are either fixed, or free-living, bottom-dwelling organisms, but some important marine genera such as *Globigerina*, are planktonic. These have long thin spicules from their surface which reduces the rate of sinking.

Radiolarians are all planktonic and differ from the Foraminifera in that their protoplasmic processes are long, thin filaments, which project from a central horny capsule separating the inner mass of the organism from the outer, vacuolated flotation layer. Radiolarians also feed by collecting particles using the protoplasmic threads, while many species have exquisitely formed siliceous skeletons resembling three dimensional snowflakes. Radiolarian skeletons build up to form the radiolarian ooze of deep tropical basins.

The Rhizomastigina are a peculiar group of organisms similar to the Radiolaria but lacking a central capsule and with one or two flagellae. The remaining group, the Heliozoa, also lack a central capsule and also differ from the Radiolaria in having a different arrangement of protoplasmic threads. These animals are carnivorous and are capable of subduing copepods and even nematodes. Some are semi-sessile being fixed on a long stalk bearing the "head" from which the axopods protrude.

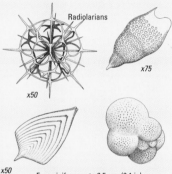

Radiolarians

x50

x75

x50

Foraminifera *up to 2.5 mm (0.1 in)*

PHYLUM SPOROZOA
Divided into the Apicomplexa and Microspora this group is composed of parasitic species living within or between the cells of the host organism. Many species are found in the gut of invertebrates or the gall bladder and other tissues of vertebrates including fish. Some sporozoans are multi-nucleate and may pass through a multicellular stage during the life cycle. Sporozoan infections may be common and comparatively harmless to the host under natural conditions, but under the crowded conditions of hatcheries and shellfish beds they may become a serious economic problem.

PHYLUM CILIOPHORA
The ciliate protozoans number more than 7000 species which move and also feed by means of cilia. Marine ciliates occur in the plankton, interstially and as benthic forms. They differ from other protozoans in having two types of nuclei, while the body form is generally constant and asymetrical. Simple ciliates are covered with rows of cilia arranged over the surface and around the mouth, while one group, the suctoria has lost the cilia and developed poisonous tentacles. These animals live like tiny sea anemones, usually on the exoskeleton of crustacea, where they feed on other ciliates. The Tintinnids a group of around 900 species have lost the cilia and developed a crown of membranelles around the mouth.

The Animal Kingdom

General introduction to animals

There is considerable diversity of animals in the marine environment although, in terms of the absolute number of species, they are less numerous than on land, where the enormous numbers of insect species dwarf diversity in the other taxa. In contrast however, there are many phyla of marine organisms which have no terrestrial or freshwater representatives and hence the diversity of the marine fauna when measured in terms of higher taxa or genetic diversity is greater than on land.

The chemical and physical properties of seawater have dominated the evolution of the marine animals such that large marine animals have achieved sizes which could not be reached successfully on land, their large mass being buoyed up by seawater. Because the plants of the sea which form the basis of the food chains are all small in size, the nature of marine food chains differs considerably from those of terrestrial environments. Small phytoplankton are fed on by small herbivorous zooplankton which in turn are eaten by slightly larger predatory zooplankton before being consumed by small fishes. Food chains in the marine environment tend therefore, to have more steps than those on land and, as a consequence, the top carnivores of the marine ecosystems tend to range over wide areas and occur at relatively low density.

Another feature of marine ecosystems is the standing stock or biomass of organisms present at any one time. On land, the standing stock is dominated by the plants, while most animals in contrast are of smaller size and occur at lower densities. In contrast the phytoplankton, although of small size have a rapid turnover time with new individuals being produced every few hours or days. This production is rapidly consumed by slower-growing animals such that at any one time a lower biomass of phytoplankton is present than the biomass of zooplankton. Measured over a year, however, phytoplankton production exceeds that of zooplankton.

PHYLUM PORIFERA

The sponges are sessile animals, with a simple body form consisting of two different cell types, on the inner and outer surfaces. They filter feed by means of cilia, which generate a current of water that flows into the animal through numerous pores over its surface. Food particles are filtered out of this flow of water which passes out of the animal via one or more excurrent openings. The ciliated inner cells are responsible for absorbing the food from the water current. Sponges show varying grades of organization with the simplest asconoid type consisting of a cup-shaped structure which lacks folds in the body wall, and in which the inner wall is lined completely with flagellated cells. This grade of organization is characteristic of many of the smaller, glass sponges, while the syconoid grade, where the wall is folded and the flagellated cells are restricted to pockets in the body wall can grow to larger size. The largest and most complex sponges are of the leuconoid grade of organization where the central chamber of the sponge has been completely filled by tissue and a complex series of flagellated chambers line the passages leading from the incurrent canals to the excurrent openings.

Although, unlike other animals, the sponges are multi-cellular and display a level of cellular differentiation they are not as co-ordinated as are higher animals and resemble in some respects, aggregations or colonies of individuals, rather than a complex multi-cellular organism. They are sometimes placed in a separate subkingdom, the Parazoa, to distinguish them from the Metazoa where the cells and tissues are more complex and in most cases much better co-ordinated. Single sponge cells can continue to exist in the absence of others and in some instances, if the cells of a sponge are separated, they can re-aggregate to form the sponge once more. Four classes of sponges are recognized based in part on the nature of the skeleton.

CLASS HEXACTINELLIDA

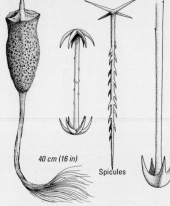

40 cm (16 in)

Spicules

Hexactinellid

The glass sponges are the most primitive of all the living sponges, whose cells are assembled around a trellis work of siliceous spicules with four, five or six branches. In some species, such as *Euplectella*, the Venus' flower basket, the spicules are fused to form a delicate basket-like structure. Most of the living species are deep-sea forms although, during the Jurassic and Cretaceous period these were the dominant sponges of all depths. The extant species live anchored in deep-sea oozes and other similar soft sediments and have developed basal spicules of considerable length to ensure they remain in an upright position in the sediment.

CLASS CALCAREA

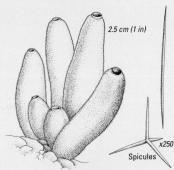

2.5 cm (1 in)

x250

Spicules

Calcareous sponge

These are generally small, mainly littoral sponges, which often encrust rocks and are built on a skeleton of two, three or four rayed calcareous spicules. These sponges are more complex in structure than the glass sponges since the inner cells are generally confined to pockets in the body of the sponge rather than lining a single large central chamber.

Class Desmospongia

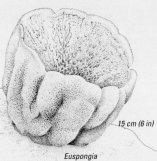

15 cm (6 in)

Euspongia

This is by far the largest class of living sponges, characterized by a much more complex canal system and a wider diversity of cell types separating the inner and outer layers. When present, the skeleton is constructed either, of siliceous spicules or, of spongin fibres or, of a mixture of both. The traditional bath sponges *Hippospongia* and *Euspongia* have only spongin filaments and are collected commercially. One of the largest species is *Hircinea gigantea* which may weigh up to 50 kg (110 lbs) and measure a metre (3 ft) in diameter.

Many of the desmosponges are brightly coloured and they are often inhabited by a wide range of small commensal crustaceans, brittle stars and molluscs, which take food from the current of water drawn in by the sponge itself.

Class Sclerospongia

There are only a few species in this class of sponge which, in addition to an internal skeleton of similar construction to that of the desmospongia, have an outer casing of calcium carbonate.

PHYLUM CNIDARIA

In common with the phylum Ctenophora, the cnidarians display radial symmetry and a more complex level of organization than the simple sponges. Like the sponges, however, their body plan is based on two layers of cells separated by a non-cellular layer of fibres, and proteins, the mesogloea. The diagnostic feature of this group of animals is the presence of the stinging cells, nematocysts or cnidocytes, which vary in form and function but are primarily used in feeding. They range from pentrating forms containing toxic solutions, to types which extrude threads that curl around the bristles and small appendages of the prey. The most toxic cnidarians are capable of killing humans although the nematocysts, which are grouped on the tentacles are primarily used to immobilize and hold prey, which

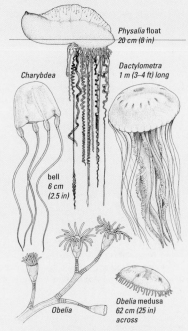

Physalia float 20 cm (8 in)

Charybdea

Dactylometra 1 m (3–4 ft) long

bell 6 cm (2.5 in)

Obelia

Obelia medusa 62 cm (25 in) across

are then transferred to the mouth, located in the centre of the ring of tentacles. The unique nematocysts are produced by amoeboid cells and transported to the surface of the tentacles by the movement of these cells through the tissues of the animal. They discharge in response to chemical and tactile stimuli from the prey organisms and the cnidarians have some measure of control over this discharge. Following a meal, anemones for example, require much stronger stimuli to discharge their nematocysts then when they are hungry. Some animals such as nudibranchs or shell-less molluscs, feed on cnidarians and ingest the nematocysts whole. These are then stored in special organs found on the upper surface of the animal and which serve as protection for the mollusc.

All the cnidarians have the same basic plan, with a mouth surrounded by tentacles, that opens into a central body cavity. The mouth also functions as an anus and the undigested material such as molluscan shells and crustacean exoskeletons are voided via this central opening. The body plan of cnidarians is of two basic types, the polyp, a sedentary tubular animal with the mouth facing upwards, and the pelagic or free-swimming medusae, in which the mouth generally faces downwards. These two body plans may alternate in the two generations of the same species, as in siphonophores, for example or, they may occur together in colonial forms.

All the 9000–10,000 species of cnidaria are aquatic and the vast majority of species are marine, including the jellyfish, hydroids, sea anemones and corals. The latter group of animals are important in forming coral reefs, geological structures characteristic of the clear nutrient poor waters of tropical areas, where the water temperatures range between 20 and 30°C (68 and 86°F). The hermatypic or reef-building corals, are notable for the presence of symbiotic algae which aid the animal in skeleton formation and supplement the diet of small plankton. There are four classes of Cnidaria of which only one the Hydrozoa contains freshwater species, the remainder being marine.

Class Hydrozoa

The hydrozoans are the simplest of the cnidarians with simple nerve nets and non-cellular mesogloea, and number around 2000 species. Muscle tissue is generally poorly developed and the majority of hydrozoan polyps are generally of small size, although the genus *Branchiocerianthus* found in depths of several thousand metres may reach 2 m (7 ft) in length. The hydromedusae, or free-swimming forms, are generally simple and throughout this group there has been a tendency to reduce the medusae, which remain attached to the parent polyp in genera such as *Campanularia* or, are completely lacking in genera such as *Hydra*. An exception to this trend is *Trachylina* in which the polyp is suppressed and the medusa is the only body plan.

In many colonial hydrozoans specialization of the polyps occur, with some adapted for feeding and others for reproduction as in the case of *Obelia*, which also secretes a perisarc, or protective covering of horny material. The greatest degree of specialisation is found in Siphonophores such as *Physalia*, the Portugese man-'o-war. In these animals three different types of polyp are found. The first is a large and highly specialized polyp which secretes gas and serves as a float, keeping the animal at the surface. The polyps beneath are grouped in long tentacles and include, specialized dactylozooids with high densities of poisonous nematocysts and gonozoids, which produce medusoid individuals that remain attached to the colony. Amongst the siphonophores there has been an evolutionary tendancy to reduce the number of tentacles, while at the same time increasing their length and hence efficiency in catching the epipelagic fish on which they feed. One species of fish (*Nomeus gronovii*) is found living among the tentacles of *Physalia*. Apparently, it fails to cause the nematocysts to discharge and thus is protected from predators by the poisonous tentacles of its host.

Class Scyphozoa

The Scyphozoa number around 250 species and include some of the largest known cnidarians such as *Cyanea*, with a bell diameter of 2–3 m (7–10 ft) and up to 800 or more tentacles, each up to 60 m (200 ft) in length. Most members of this group are free-swimming, medusoid forms, although the Stauromedusae, such as *Haliclystus*,

the stalked jellyfish are bottom-dwelling, sessile animals.

Generally the Scyphozoa are more complex than the hydrozoa but as a group they display much less variation in body form. The most familiar scyphozoans are the small jellyfish such as *Aurelia* and *Cyanea* which swim by muscular contractions of the bell. The vast majority of these forms are suspension feeders feeding on small planktonic organisms which become entrapped in the mucus covering the ciliated undersurface of the animal. Trapped particles are swept towards the mouth by four oral arms. In the deep-water, coronate forms, the bell has a ring of leaf-shaped structures, each bearing a single tentacle around the rim. In contrast, the Rhizostomae have no marginal tentacles and the oral arms surrounding the mouth are fused into a tube-like structure. These animals occur in shallow waters and are common in the tropics and sub-tropics.

As their name suggests, the cubomedusae or box jellyfish, are basically cuboidal in form with the bell having four flattened sides. At each corner of the box are single, extended tentacles carrying extremely toxic nematocysts. Small sense organs are present in the centre of each side consisting of statocysts or balancing organs and six small ocelli or light-sensitive organs. These enable the animal to adjust its position in the water column, which it does by muscular contractions of the bell that force water out of the central body cavity. The peculiarities of this group have resulted in their being classified by some taxonomists as the separate class, Cubozoa

Class Anthozoa

This is the largest class of Cnidaria encompassing somewhere in the region of 6000–8000 species of sea anemone and corals. The anthozoa all have a basic, polyp body form in which the central gastric cavity is divided by septae that increase the surface area for digestion. They have an upwardly facing mouth surrounded by a ring of tentacles; eight branched tentacles in the case of the Octocoralia, but six, or multiples of six, in sea anemones and true corals.

The anthozoa are divided into two subclasses, the Octocoralia and the Zoantheria. The latter group is itself divided into the main order of sea anemones (Actinaria) and the true stony corals (Madreporia or Scleractinia).

The sea anemones are rather sluggish sessile animals although some species such as *Stomphia* are capable of movement by vigorous lashing of the body column from side to side. An interesting example of commensalism is found amongst the sea anemones with *Calliactis parasitica* being found in association with a hermit crab. While both partners can survive apart, they are commonly found together and a similar association occurs between the anemone, *Adamsia palliata*, which is found associated with the hermit crab, *Eupagurus prideauxi*. In this case, the crab has been observed feeding the anemone by placing food in its mouth. Large anemones such as *Stoichactis* are also host to the brightly coloured anemone fish, which shelter amongst the tentacles and hide from larger predatory fish.

The most conspicuous and important members of this group are the stony, hermatypic, reef-building corals, whose calcium carbonate skeletons form massive geological structures such as the Australian Great Barrier Reef. The individual coral polyps resemble small sea anemones, and these are grouped in colonies, connected by lateral folds of the body wall. Each polyp sits within a cup in the coral skeleton, into which it can withdraw when threatened by predators. Some, such as *Fungia* are solitary animals, the single polyp reaching up to 25 cm (10 in) in diameter, while others such as *Acropora*, grow as branched colonies. Some genera, such as *Porites*, form large boulders or micro-atolls in the lagoons of reef systems. The striking variation in skeletal growth form reflects differences in the energy of waves, with the more fragile forms occurring in largely sheltered areas, and the more compact, brain corals being exposed to surf and wave action in the more exposed areas.

The Antipatheria, or black corals, have their polyps arranged around a black and horny skeleton which is much sought after for the manufacture of jewellery, and as a result, fetches a high price. The octocoralia include the sea pens, sea fans, whips and pipe corals in which the polyps are rather small, growing as a colony supported by a central, rod-like skeleton of gorgonin. In the precious coral, *Corallium*, the skeleton is composed of fused calcareous spicules. Calcareous spicules, are also present in the soft corals, while the organ pipe coral has a red, tube-like skeleton which is sometimes used in traditional medicine in various different parts of Asia.

PHYLUM CTENOPHORA

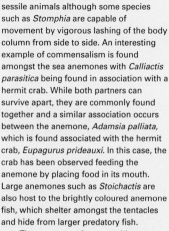
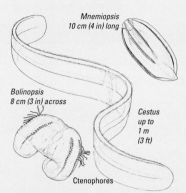

Mnemiopsis 10 cm (4 in) long

Bolinopsis 8 cm (3 in) across

Cestus up to 1 m (3 ft)

Ctenophores

The transparent and delicate sea gooseberries are a remarkable group of around 50 species of animals, many of which luminesce at night. Like the cnidaria, many ctenophores are apparently radially symmetrical, although on closer inspection this is seen to be biradial symmetry. They also possess cells, the colloblasts, with similarities to the cnidarian nematocysts. Unlike the cnidaria, however, they possess eight rows of fused cilia forming a comb-like structure which beats to provide the power for swimming.

Ctenophores do not show the wide range of body form characteristic of the phylum cnidaria, but like the Cnidaria they are composed of two layers of cells, the outer epidermis and the inner gastrodermis, separated by a mesogloea.

Most species are pelagic and the class Tentaculata possess two pairs of elongate tentacles which trail behind the animal as it swims. A few species in this group have become adapted to a creeping, bottom-dwelling mode of life. The class Nuda are without tentacles and one order in this group has become specialized for feeding on other ctenophores. *Pleurobranchia*, the typical sea gooseberry, has a spherical shape and is considered to be a primitive member of the phylum. In contrast, *Cestum*, the Venus' girdle, has an elongate and flattened body and can swim by lateral undulations of the body, in addition to the beating of the comb rows. Cestum includes some of the largest ctenophores which may reach a metre (3 ft) in length

PHYLUM PLATYHELMINTHES

The 12,000–15,000 species of flatworms are not true worms at all, but flattened, bilaterally symmetrical, unsegmented animals which lack a true body cavity or coelom. The external epidermis is covered with cilia and the gut, if present, lacks an anus. They are generally hermaphroditic, although cross fertilization generally occurs leading to the production of shelled eggs which hatch, to yield a ciliated larva or, a miniature adult. Amongst the flatworms there are both parasitic and free-living forms and the more advanced, parasitic forms show remarkable adaptations to their mode of life. These include specialized structures for attachment to the host, resistance to the digestive enzymes of the host and elaborate lifecycles involving several hosts at different stages of growth. In some of the monogenean flukes infestation by one parasite may convey

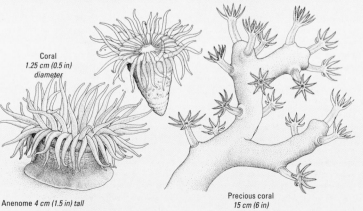

Coral 1.25 cm (0.5 in) diameter

Anenome 4 cm (1.5 in) tall

Precious coral 15 cm (6 in)

immunity to subsequent attack, while different groups of fishes are parasitised by different numbers of fluke species. Twelve families of fluke are known to infect teleost or bony fish; five, attack elasmobranchs; four, the ratfishes or holocephali; and, only two are found in chondrostean fishes. This suggests that the bony fish were the first group of hosts for these animals in the marine environment.

There are four classes of platyhelminthes: the Turbellaria, or free-living flatworms; the Cestoda, or tapeworms; and, the Monogenea and Digenea, or flukes, grouped by some authors into a single class, the Trematoda.

Class Turbellaria

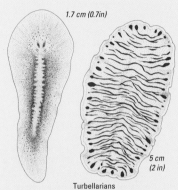

1.7 cm (0.7in)

5 cm (2 in)

Turbellarians

The turbellaria are chiefly marine, with a few freshwater and a few terrestrial species, and are exclusively free-living, although one genus, *Bdelloura*, lives commensally in the gills of horse-shoe crabs, and a few species are found in association with crabs and molluscs. The majority of species are generally small in size being less than 1 cm (0.4 in) in length. Some species achieve a size of 2 cm (1 in). These animals are carnivorous, gliding over the substrate using their cilia and a film of mucus produced by glands in the epidermis. The simplest group, the Acoela, have no gut but merely a muscular pharynx, while the more advanced polyclads and triclads have a well-developed gut with many branches in the polyclads and three branches in the triclads. Some species are often brightly coloured, but most are dull brown or black and one species, *Convoluta*, which contains algal symbionts, is green and may sometimes colour the surface of sandy areas at low water.

Class Digenea

The digenea are parasitic flukes, who usually have at least two, and sometimes as many as four, different hosts during the various stages of the lifecycle. A free-swimming, ciliated larva enters the primary molluscan host and multiplies until thousands of tailed, swimming larvae emerge. These, in turn, may penetrate other hosts, or the final vertebrate host directly, where the adult flukes mature and the cycle recommences.

Digenea have only two suckers when they are fully developed, a resistant outer cuticle, and a bi-lobed gut, although most, if not all, of the internal space of the animals is occupied by the reproductive organs.

Class Monogenea

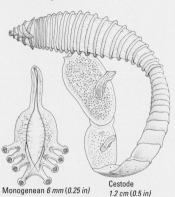

Monogenean 6 mm (0.25 in)

Cestode 1.2 cm (0.5 in)

These are also parasitic flukes but, unlike the digenea, they infect only a single species of vertebrate host and the free-swimming larvae infect the host directly. The adults are usually external parasites of fish, feeding on blood and skin tissue with a specialized organ, the opisthaptor, at the rear of the animal, for attachment to the gills of the host. A few species are internal parasites, one being found in the hind gut of the ratfishes.

Class Cestoda

Adult tapeworms live inside the gut of vertebrate hosts and are attached by suckers and hooks on the worm's anterior end, or scolex. The body is divided into sections, or proglottids, and the gut is absent, since the bulk of the body is occupied by the uterus. There are normally two or three hosts of which the final vertebrate host is usually specific, so much so, that different species of tapeworms have been used to distinguish between two species of thresher shark. The subclass Cestodaria are a peculiar group of gut parasites of fishes which resemble flukes in their general body form but, like the typical tapeworms lack a digestive system.

PHYLUM NEMERTINEA

15–50 cm (6–20 in)

average 4.5 m (15 ft)

Nemertines

This small phylum of some 600 species of ribbon worms are, like the flatworms, creeping, burrowing animals with a ciliated epidermis. They are effective predators catching and consuming their prey by means of a long proboscis armed with sharply pointed stylets in the class Enopla and lacking such structures in the Anopla. Most species are small, a few millimetres or centimetres in length, although the typical bootlace worm, *Lineus longissimus*, of temperate shores can achieve lengths of up to 5 m (16.5 ft). Most species are

marine, bottom-dwelling, littoral organisms but a few species burrow, some are found at abyssal depths and a few, are pelagic.

PHYLUM MESOZOA

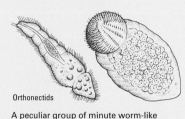

Orthonectids

A peculiar group of minute worm-like animals which includes around 50 known species. They are all internal parasites of invertebrates including flatworms, nemerteans, annelids, molluscs and echinoderms and are characterized by an unusual, solid, two-layered body, and complex lifecycles similar to those of the parasitic protistans. The simplest order, Orthonectida have separate sexes and the adults when mature leave the host. The eggs are fertilized in the body of the female, subsequently developing into a ciliated larva which infects a new host. The order Dicyemida are restricted to the kidneys of cuttlefish, squid and octopus. This phylum is regarded by some authors as a degenerate class of flatworms, while others consider them an early, metazoan offshoot of the Protista.

PHYLUM GNATHOSTOMULIDA

Gnathostomula

A phylum of around 80 species of small – 0.5–1 mm (0.02–0.04 in) – acoelomate worms which live between sand grains. They have an elongate body with a ciliated epidermis and feed on bacteria and fungi which are seized by a pair of toothed jaw-like structures. They lack excretory and respiratory organs and are hermaphroditic with fertilization occurring when the male penetrates the female body wall with a copulatory organ. The eggs are shed by rupture of the female's body wall.

THE PSEUDOCOELOMATE PHYLA AND SUPER-PHYLUM ASCHELMINTHES

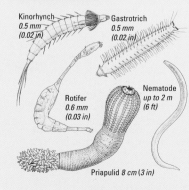

Kinorhynch 0.5 mm (0.02 in)

Gastrotrich 0.5 mm (0.02 in)

Rotifer 0.6 mm (0.03 in)

Nematode up to 2 m (6 ft)

Priapulid 8 cm (3 in)

Six invertebrate phyla lack a true coelom or body cavity but possess a pseudocoelom

which lacks a membranous lining, being formed by splitting of the mesoderm or middle-body layer during development. This grouping is of no evolutionary significance although in the past, five of these phyla were considered classes of the phylum Aschelminthes. Most authors now recognize that the differences between these five groups warrant their separation into individual phyla within the super phylum Aschelminthes. A sixth pseudocoelomate phylum, the spiny-headed worms or Acanthocephala are considered to show important differences from the rest and hence are not included in this super-phylum.

The super-phylum Aschelminthes are worm-like animals which lack definite heads, but display some degree of radial symmetry at their anterior end. They have a complete digestive system with mouth and anus, a simple reproductive system, but both respiratory and circulatory systems are absent. Another characteristic of the group is the reduced number of cells making up the body and the presence, in many groups, of a well-developed scleroprotein cuticle lying on the surface of the epidermis.

PHYLUM NEMATODA

This is a vast phylum and although only some 10,000 species have been described it has been estimated that as many as 500,000 species may exist. They are, however, small and inconspicuous, and in addition to the vast array of species there are immense numbers of individuals, one rotting apple may hold an estimated 40,000 individuals for example. The majority are free-living and these animals are distributed from the tropics to the poles and from the tops of mountains to the abyssal depths of the ocean.

The phylum is unique in lacking either flagella, or cilia, and in general the number of cells making up an individual is small, less than 500 in the body wall. The body is almost perfectly circular in cross section and covered in a multi-layered cuticle containing keratin, elastic tissues and a lattice-work of collagen fibres. The cuticle is tough, elastic and permeable to both gas and water, forming a remarkable protective covering. The phylum is divided into two classes the Aphasmida, which lack chemoreceptive organs and the Phasmida which have a pair of chemoreceptive organs on either side of the tail. The former include the majority of marine and freshwater species, while the latter includes the majority of parasitic forms.

PHYLUM ROTIFERA

These tiny animals number around 1500 species and are comparable in size to the ciliated protistans with which they often live and compete. They derive their common name of "wheel animalcules", from the corona, or crown of cilia, which, when it is in motion resembles a spinning wheel. Only a few species are marine but like the nematodes they are cosmopolitan in distribution. Most species feed on protista which are filtered from the water as the animals stand upright, with the foot attached to the substrate. These animals may reproduce both sexually and parthenogenically, females producing eggs that hatch to produce more females, enabling populations to expand rapidly when conditions are suitable.

PHYLUM GASTROTRICHA

This phylum contains only about 150 species of mainly freshwater animals, with a small number of species occurring interstially, in intertidal sands. Although they have a superficial resemblance to rotifers they lack the corona and are generally scaly or spiny on the dorsal surface, with cilia for locomotion beneath. Most species feed on organic debris, algae, protista and bacteria, which are swept into the terminal mouth by four groups of cilia around the anterior end of the animal. These animals generally have a short lifespan of between 3 and 21 days and in contrast to most other pseudocoelomates the animals are hermaphroditic. Four or five eggs are produced at any one time and these are of two types; thick-shelled ones, used to avoid unfavourable conditions and thin-shelled ones which hatch in a few days, the animals maturing in two days.

PHYLUM KINORHYNCHA

Like the rotifers and gastrotrichs the Kinorhynchs are microscopic, less than a millimetre in total length. The body is generally elongate, consisting of a head, neck and 11 body segments. The majority of the 100 known species are found in ocean muds in shallow waters. The cuticle is divided into plates, with the dorsal surface bearing spines, while the head has 5–6 circles of spines and an oral cone of stylets. Detrital material is sucked in by contractions of the muscular pharynx and the head can be retracted into the body. Like the gastrotrichs, the Kinorhynchs have a pair of posterior adhesive organs allowing them to attach temporarily to the surface of the substrate.

PHYLUM NEMATOMORPHA

The horsehair or hair worms are a small phylum of some 230 species of extremely slender worms which are brown or black in colour. Most species are terrestrial, but a single genus, *Nectonema*, occurs in the marine environment where it parasitizes crabs in its juvenile stages. They may reach 1 m (3 ft) in length, but are usually less than a millimetre (0.04 in) in diameter. The juveniles are parasitic in arthropods while the adults, which are normally free living, do not feed. The sexes are separate, males entwine around the females to fertilize the eggs which hatch as larvae that enter the host, either by being eaten or, by penetration.

PHYLUM ACANTHOCEPHALA

Although the phylum Acanthocephala or spiny-headed worms possess a pseudocoelom like the members of the super-phylum Aschelminthes, this develops in a different manner and hence this phylum is considered distinct from the other pseudocoelomate groups. The 800 or so known species are all internal parasites

of vertebrate guts, and have an intermediate, arthropod host. Fish are the most common vertebrate host and the majority of these worms are around 2 cm

Acanthocephala *up to 2.5 cm (1 in)*

(0.8 in) in length. They may be present in large numbers and over a thousand have been reported from a single seal.

As their name suggests these animals are armed with a spiny proboscis used for attachment to the digestive tract of the host. The sexes are separate and fertilization occurs within the female. The eggs develop into an acanthor larva surrounded by a protective shell which is shed from the female's body and passes out of the host with the faeces. The resistant egg can survive for several months and when ingested by a crustacean, intermediate host, it hatches and the larva bores through the gut wall using its hook-bearing rostellum. After passing through two more larval stages in the body cavity of the intermediate host, the animal only completes its development when the intermediate host is eaten by the vertebrate primary host. *Corynosoma* has two intermediate hosts, developing first in amphipods, before undergoing further development in a fish, finally completing its lifecycle in the gut of seabirds, or seals.

THE COELOMATE ANIMALS AND METAMERIC SEGMENTATION

All the remaining members of the animal kingdom possess, at least during some stage of their life history, a true body cavity or coelom. There are five major phyla (Annelida, Arthropoda, Mollusca, Echinodermata and Chordata) with several minor phyla. In addition to the possession of a body cavity all of these animal groups show considerable specialization in comparison with the acoelomate and pseudocoelomate groups.

An important structural adaptation seen in the coelomate animals is the development of segmented body form, although this may be so modified in arthropods and chordates as to be unrecognisable. Segmentation involves the replication of all structures in each segment along the length of the body and the coelom is divided between the segments allowing individual segments to be stretched or shortened, important for locomotion and in particular burrowing.

PHYLUM ANNELIDA

This is an important phylum of some 9000 species of worms, recognisable by the ring

like divisions of the body which represent the individual segments. The general body plan of these animals resembles an elongate tube with terminal mouth and anus and a straight digestive tract between. The body cavity is lined with a membranous peritoneum and the coelom of each segment is separated from its neighbours by an extension of the membrane forming a septa. The tubular gut is supported in the centre of this cavity by dorsal and ventral mesenteries. Both the septa and mesenteries may be perforated by small holes allowing some circulation of the fluid between the individual cavities of the coelom. The fluid filled cavities of the body form a hydraulic skeleton against which the muscles of the body wall can act. Contraction of the longitudinal muscle fibres causes shortening and thickening of the segment and hence of the animal as a whole, while contraction of the antagonistic circular muscles causes elongation and narrowing of the segment. By alternating contractions of these muscles in different segments the animal can pull itself forward in a burrow, or, by means of its appendages and bristles, move over the surface of the substrate.

There are three classes of annelids: the polychaetes, which include some 5500 species of marine worms, each segment bearing paired appendage and bristles; the predominantly freshwater and terrestrial oligochaetes, the familiar earthworms with fewer bristles; and the Hirudinea, with 500 species of marine, freshwater and terrestrial leeches.

Class Polychaeta

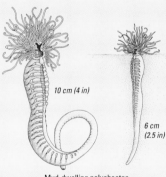

Mud-dwelling polychaetes

This class contains over 60 per cent of all annelids and almost all species are marine. Most are relatively small animals inhabiting coastal areas, either free-living under stones, in crevices and temporary burrows, or, living a sedentary existence in permanent burrows in sand and mud, or secreted tubes.

Sometimes these tubes are cemented together forming massive colonies but more usually they are solitary such as *Spirobis* or *Pectinaria*. The body form and habits of polychaetes are quite diverse and most modifications of the head relate to the mode of feeding. Free-living worms are generally active carnivores having an extendible toothed proboscis, up to six large eyes and sensory tentacles on the head, to detect the prey. The nereid ragworms are of this type and *Neanthes* may attain lengths of up to 0.6 m (2 ft). Some free-living species have fewer body segments and have adopted a short fat

body form such as *Aphrodite*, the sea mouse for example, whose dorsal covering of hair-like structures obscures the animal's segmentation.

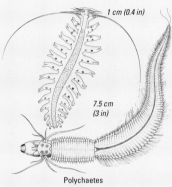

Polychaetes

Scale worms generally have peculiar plates on the dorsal surface which may be shed when the animal is handled. In contrast the syllids have beautifully coloured, thread-like bodies and elongate dorsal cirri. Several polychaete families have adopted a pelagic mode of life and although they retain the same basic body plan as surface-dwelling species many are transparent. Burrowing has lead to a variety of adaptations exemplified by the lugworm *Arenicola*. Many burrowers build tubes as linings to their burrows, while others construct calcareous tubes on the surface of the substrate. Terebellids cement sand grains together to form a tube, sabellids use mud, and serpulids secrete a hard calcareous substance.

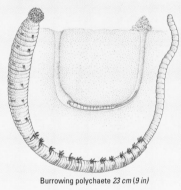

Burrowing polychaete *23 cm (9 in)*

Some of the burrowing species feed by ingesting the substrate in the same way as earthworms, while many of the sedentary forms are filter feeders drawing a current of water into the burrow, or, as in the case of fan worms, filtering suspended material from the water column by means of a crown of pinnate processes.

Class Myzostomaria

This group of small, flattened, disk-shaped worms are normally parasites of echinoderms and are usually found only on the crinoids. They are considered by some authors to be members of the class polychaeta. Their body shape is designed for easy attachment to the feather star arms and the host responds by enclosing them in cysts, which affect the skeletal arrangement of plates on the crinoid's arms. Similar deformities in the arms of fossil feather stars suggest, that the Myzostomaria are far from modern creatures, but rather ancient parasites of this group of echinoderms.

Class Oligochaeta

These animals are found in freshwater and terrestrial environments. However, one family, the enchytraeidae, has managed to invade the marine environment, and it is confined to living beneath stones and amongst seaweed in the upper littoral zones. None have managed to descend below the tidemark and invade the true ocean realm, and the group is insignificant in the sea, though abundant in freshwater. Unlike the polychaetes they lack the appendages, or parapodia on their segments, and have simple heads without the ornate and complex structures used for filter feeding by many polychaetes.

Class Hirudinea

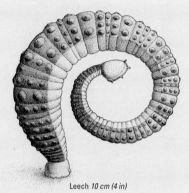

Leech 10 cm (4 in)

The leeches are a relatively uniform group of highly specialized annelids, which include some 500 described species. Most are freshwater but some are external parasites of both elasmobranchs and bony fish, remaining on a single host and feeding on blood which is digested by bacteria living in specialized pockets in the gut. To prevent the blood from clotting they produce anticoagulants and remain attached to the host by means of suckers at both ends of the body. The sexes are separate and fish leeches lay a single egg in cocoons.

PHYLUM ARCHIANNELIDA

The archiannelids were once thought to be primitive members of the Annelid phylum but are now believed to represent a group of unrelated families. They are generally interstitial and highly aberrant, lacking bristles, parapodia and any signs of external segmentation. The members of this group live in surface mud and in the splash zone and have external ciliation and simply arranged heads. Some at least, may be survivors of an archaic group ancestral to the annelids.

PHYLUM ARTHROPODA

The phylum Arthropoda is the largest group in the animal kingdom with an estimated million species and over 800,000 described species at the present time. Although they are extremely numerous, both in terms of diversity and individual species numbers, their total biomass is not as large as one might expect since many species are quite small. Representatives of the phylum are found in every known habitat and many species are of economic importance, either as food, or as pests. The

arthropods display some affinity with the annelids since both groups possess metameric segmentation, although in the arthropods, the process of tagmatization or specialization of regions of the body for particular purposes, is a major characteristic of the group. Together with the molluscs these two groups display protostomy, in which the mouth develops from the larval blastopore, both show specialization of the nervous and circulatory systems, and paired, coelomic cavities comparable with those of annelids are seen in young arthropods.

The main features distinguishing arthropods, however, are the tough semi-rigid exoskeleton consisting basically of chitin, a polysaccharide and protein mixture, with a thin, non-wettable external cuticle. In marine arthropods such as crabs the cuticle is thick and calcified. The external exoskeleton allowed the development of jointed appendages, which can be moved independently of one another and of the body, by internally inserted muscle blocks. It has jointed limbs which gives the phylum its name. The group is divided into four subphyla, one of which, the Trilobomorpha, is an entirely fossil group of primitive animals, the trilobites. The remaining three subphyla include the Chelicerata, which lack antennae and have feeding appendages called chelicerae, the Crustacea and the Uniramia, distinguished on the basis of their limb anatomy.

SUBPHYLUM CHELICERATA

King crab 53 cm (1.75 ft) long

This group includes the well-known spiders, scorpions and ticks and some less well-known aquatic groups, the Merostomata or king crabs and the Pycnogonida or sea spiders. Of the better-known terrestrial groups only a few species of spider have managed to invade the intertidal while some species of ticks are found on the external surface of seasnakes of the genus *Laticauda*, which also serve as hosts to lung mites.

Class Merostomata

There are four living species of horseshoe or king crabs in three genera, of which the North American species, *Limulus polyphemus* is best known. It is a bottom-dwelling animal which ploughs through soft mud in search of molluscs and worms that form its prey. The food is caught and brought to the gnathobases between the front limbs where it is broken up before being passed to the mouth and ingested. The body is dorsoventrally flattened and consists of two major sections, the

horseshoe-shaped prosoma, which covers the smaller opisthosoma, and a terminal, caudal spine. The animals mate in spring and the young larva resembles a trilobite taking approximately 13 months to achieve the adult body form and three years to reach sexual maturity.

Class Pycnogonida

Pycnogonid 1 cm (0.4 in)

This is a group of 500 rather aberrant marine species known as sea spiders. Most species are small, inhabiting the lower intertidal and subtidal but some of the abyssal species may have a leg span of 75 cm (30 in). The body is narrow and separated into three regions an anterior proboscis, a cephalothorax with four pairs of eight-jointed, walking legs and a posterior abdomen, consisting of a single segment. *Nymphon gracile* a common intertidal species of the northern temperate zone has both chelicerae and pedipalps, whereas, *Pycnogonum littorale*, has neither. In some species there is an additional pair of ovigerous, ten-jointed legs, held beneath the body and used by males to carry cemented masses of eggs. These animals are carnivorous, normally sucking juices and small particles from soft bodied sponges, hydroids, and sea anemones.

SUBPHYLUM CRUSTACEA

In contrast to the insects which are predominantly terrestrial, all but a few species of crustacea are aquatic, with the vast majority being found in the ocean realm. Marine crustaceans are extremely important as major components of the grazing levels of the pelagic food chains; as human food resources; and, as important components of all marine ecosystems. A few species are found above the high-tide level, but most terrestrial species are restricted to damp environments. The crustacea are distinguished by the presence of biramous appendages, a pair of mandibles and gills, associated with the thoracic appendages. About 31,300 species of crustacea are presently known to science and these are generally divided into seven classes of which, the Malocostraca is the most advanced and numerous containing some 20,000 species.

Class Branchiopoda

The branchiopods show considerable variation in body form but all posses flattened, leaf-like appendages and flattened gills. The limbs are important, not only in locomotion and respiration but also in filter feeding, being armed with brush-like setae for trapping suspended food

particles. Most species are of small size less than 10 cm (4 in) in length and two basic body forms are common in the group. Of these *Daphnia* is typical with a short, stocky body and a well-developed carapace which covers the limbs. The head protrudes anteriorly and bears a pair of large and powerful antennae used to row the animal upwards in the water. The remainder of the body is visible through the carapace and bears five pairs of trunk appendages. Many species reproduce parthenogenically and eggs may be retained in a brood pouch until the young animals emerge. *Chirocephalus* the fairy shrimp is an example of the elongated body form found in many branchiopods. It lacks a carapace and hence the segmentation is clearly visible. It swims upside down using its limbs, which also function in feeding and respiration.

Class Ostracoda

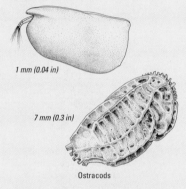

1 mm (0.04 in)

7 mm (0.3 in)

Ostracods

The ostracods are a group of around 2000 species with a wide distribution in marine and freshwater habitats. They are small, generally around 1 cm (0.4 in) long, and the majority are bottom dwellers, although a few are pelagic. The ostracod body form is short, enclosed in a bivalved case with the two halves hinged by a ligament.

Class Copepoda

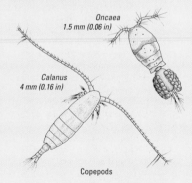

Oncaea
1.5 mm (0.06 in)

Calanus
4 mm (0.16 in)

Copepods

The copepods, with some 7500 species, are the most highly successful and abundant group of small crustaceans, feeding on microscopic organisms and providing food for a number of carnivores. They are the most important link between the marine phytoplankton and higher trophic levels and at certain times of the year *Calanus inmarchus* may form the main food of economically important fish species, such as herring. Several groups of copepods are parasitic with *Penella* for example, being parasitic on fish and whales and others parasitizing polychaete worms. The free-living forms may be either planktonic or benthic and the body form resembles that of a crayfish. Like the ostracods and branchiopods most copepods swim by means of the enlarged antennae and the abdomen of many species bears no appendages. Females of many copepods may be recognized by the pair of large egg sacs attached to the first abdomenal segment. The eggs hatch into nauplius larvae between 12 hours and five days after fertilization and a subsequent batch is immediately produced. Five or six nauplian larval stages precede the copepoid larval form which, although it resembles an adult, undergoes a further five instars before the final adult form is assumed.

Class Cirripedia

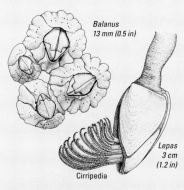

Balanus
13 mm (0.5 in)

Lepas
3 cm
(1.2 in)

Cirripedia

The barnacles are another world-wide group and are the only non-parasitic crustaceans to have adopted a sessile mode of life. Their structure is so unusual with the exoskeleton reduced to a series of calcareous plates, that they were originally described as molluscs and it was not until their larvae, which resemble ostracods were described that their true affinities were realised. Agassiz, the French zoologist, aptly described a barnacle "as nothing more than a little shrimp-like animal standing on its head in a limestone house and kicking food into its mouth".

Two basic body forms are found in the group, both of which filter feed from a current of water created by the fringed thoracic limbs which protrude from the top of the plates. Typical stalked barnacles such as Lepas, the goose barnacle, are amongst the largest barnacles, having five translucent plates around the body. Attached forms without a stalk, such as Balanus and Cthalamus, the typical encrusting barnacles of rocky shores, have six plates arranged symmetrically around the body and a further four which form a valve that can be closed over the withdrawn cirri at low tide.

Class Mystacocarida

This is a minor class, related to the copepods and barnacles, and including a single genus, Derocheilus, first discovered in 1943. The animals are small, less than 5 mm (0.2 in) in length and live between sand grains in the littoral zone where they are thought to feed on detritus. They have elongate bodies and long mouthparts fringed with setae.

Class Branchiura

The branchiurans are a group of around 75 species of ctoparasitic, blood suckers, living on fish and some amphibians. They attach to the skin or gill chamber of the host and adaptations for attachment include, the modification of the first pair of appendages as claws, or of the bases of the first pair of appendages as suckers. They have well-developed thoracic appendages which are used for swimming when they are detached from the host.

Class Malacostraca

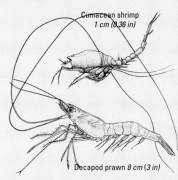

Cumacean shrimp
1 cm (0.36 in)

Decapod prawn 8 cm (3 in)

About 75 per cent of the lower crustaceans including all the largest and most successful forms are members of this class the majority of which are entirely aquatic and marine. Typical sand hoppers such as the genus Gammarus have laterally compressed bodies and the posterior three segments are modified for jumping. Among the dorsoventrally flattened isopods are found a number of terrestrial species while Ligia is a common isopod of the supralittoral splash zone. There are several small orders of Malacostraca including the primitive, burrowing Nebalia; the stomatopods, such as Squilla, the mantis shrimps, so-called because of their resemblance to the praying mantis; and the mysids, swimming forms with long abdomens and a thin carapace, found in colder temperate waters.

The most successful group is the super-order Eucarida, characterized by a well-developed carapace and including such well-known marine organisms as the krill, which form the food of whales in the Antarctic. The decapods are specialized either as swimmers, the shrimps and prawns, or crawlers, the bottom-dwelling crabs, crayfish and lobsters. Swimming forms generally have a lighter carapace. The prawns and shrimps may be distinguished on the basis of the structure of the rostrum, which is well developed in prawns and greatly reduced in shrimps. In contrast, the crabs, lobsters and crayfish have heavy skeletons and are not laterally flattened like prawns and shrimps. They are considerably more diverse and three divisions are recognized: the Macrura of true lobsters and crayfish; the Anomura the squat lobsters and hermit crabs; and, the Brachyura or true crabs. The Macrura typically have an elongate abdomen with terminal fan of swimming appendages while in the Anomura the abdomen is reduced and the swimming fan is absent. In the true crabs, the Brachyura, the abdomen is reduced and bent forwards beneath the animal and the carapace is typically flattened and rounded. Crabs range considerably in size, from minute pea crabs such as Pinnotheres, to the solid, rounded, edible crab, Cancer. The largest living arthropod is the Japanese spider crab which reaches a length of over 3 m (10 ft) with its chelipeds outstretched.

SUBPHYLUM UNIRAMIA

This subphylum contains five classes, of which, only the Insecta contain marine representatives and these number only a few hundred from the 750,000 described species. The myriapods (centipedes, millipedes, pauropoda and symphyla) are all terrestrial and number in total some 10,500 species characterized by division of the body into two tagmata or regions.

Marine insects include oceanic seaskaters and a few saltmarsh, littoral and mangrove species, including the seaweed flies Coelopidae which use the seaweed heaps at the strand line, for food and shelter. Some anopluran and mallophagan lice may be considered at least partially marine being ectoparasites of seals and seabirds, but in general the diversity of insects in the marine environment is extremely limited.

PHYLUM PRIAPULOIDEA

Priapulus

This phylum contains nine species of small burrowing animals which live in mud or sand in cool- or cold-water areas, down to depths in excess of 7000 m (23,000 ft). They range in size from the microscopic Tubiluchus which is only 0.5 mm (0.2 in) in length, to some species of Priapulus which can reach 20 cm (8 in). The body is somewhat barrel shaped with an eversible proboscis armed with spines used to catch slowly moving worms and other soft-bodied animals. The sexes are separate and the larva inhabit a secreted case, buried in the mud where they undergo a series of moults before assuming the adult body form, and lifestyle.

PHYLUM SIPUNCULOIDEA

This is a widespread group of around 350 species of active burrowers such as Sipunculus nudus, which is found in the North Sea and Atlantic, usually in the middle shore or downwards. Other species of the phylum inhabit either the abandoned shells of other animals, or bore into coral reefs. Animals of this phylum range in size from 2–720 mm (0.08–29 in) and are cylindrical, with no external segmentation. They have a short protrusible proboscis which ends in an oral disc of frilly tentacles. The mouth lies on the end of the proboscis, and the anus is situated on the top at the front. The sipuncculoids often have a chequered appearance due to bands of transverse and longitudinal muscles which can be seen through the skin.

Unlike the priapulids, the larval forms of sipunculids are planktonic, (aiding dispersal) and include a typical trochophore larva resembling that of annelid worms.

PHYLUM ECHIUROIDEA

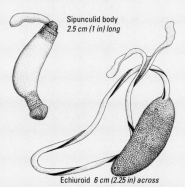

Sipunculid body
2.5 cm (1 in) long

Echiuroid 6 cm (2.25 in) across

The 100 or so species show some affinity to the sipunculid worms and like them, are sausage-shaped animals living buried in mud or sand. They are readily recognizable by the extremely long proboscis which is non-retractable. In several genera, such as Bonellia and Echiurus, the minute males are parasitic, living permanently attached to the surface of the female. One small species, Thelasemma melita inhabits sand dollars, although generally echiuroids are sedentary, inhabiting permanent burrows. Like the sipunculid worms these animals have a trochophore larva and are considered to be related to the annelids.

PHYLUM POGONOPHORA

The pogonophora inhabit marine environments below depths of around 100 m (330 ft) and are thread-like animals, up to 35 cm (1.2 ft) in length, but usually no more than 1 mm (0.04 in) in diameter. Only discovered in 1900, the 100 or so species are now known to be widely distributed. Although some species have a single anterior tentacle most have a tuft of tightly packed tentacles surrounding the anterior end. They have no mouth or digestive tract and are believed to digest their food externally, absorbing the products through the tentacles. The elongate body ends in an opisthoma, which bears bristle-like chaetae and serves to anchor the animals in their burrows.

PHYLUM TARDIGRADA

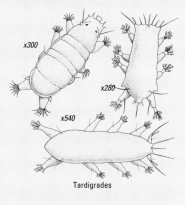

x300

x280

x540

Tardigrades

The water bears, are a peculiar group of cosmopolitan animals generally less than 0.5 mm (0.02 in) in length. Their cylindrical bodies are equipped with four pairs of short, stubby legs, which end in bunches of four or eight hooks. The external surface is covered with a cuticle which is moulted periodically and mating only occurs during the time of the moult. As in the case of rotifers, some species produce either thick- or thin-shelled eggs in response to external environmental conditions. Tardigrades display similarities to both gastrotrichs and rotifers and it is believed that they may be related to the Aschelminthes. Some species are marine, living amongst sand particles, while others occur in freshwater, but the majority are semi-aquatic living in the water film associated with terrestrial vegetation such as liverworts and lichens.

PHYLUM PENTASTOMIDA

The pentastomids are a group of around 90 species of blood-sucking parasites which are believed to be related to the brachyuran crustacean fish parasites. As adults they inhabit the respiratory tract of reptiles, mammals and birds. The elongate body has five protuberances at the anterior end, four of which are legs bearing claws, while the fifth bears the mouth. Common intermediate hosts of pentastomids include fish and small mammals.

PHYLUM MOLLUSCA

This is the second largest phylum of the animal kingdom with around 100,000 living species and some 35,000 described fossils. The group is diverse and includes well-known animals such as land snails, oysters and mussels, squid and octopus. The vast majority of species are marine and the group evolved in the ocean realm, having only a few gastropods (snails) and bivalves in freshwater and even fewer gastropods on land.

Molluscs are a group of bilaterally symmetrical animals with a reduced coelom in which the well-developed head, with associated sense organs is continuous with a muscular structure, the foot. The body organs form a visceral mass on the dorsal surface and the foot is modified in different groups for creeping, burrowing and digging. In molluscs the epidermis covering the visceral mass is referred to as the mantle, or pallium, folds of which extend ventrally like a skirt around the head and foot, to form the mantle cavity between it and the body wall. Gills are found in the mantle cavity and these may be modified as lungs in terrestrial molluscs such as slugs. A well-known characteristic of the molluscs is the well-developed and often beautifully coloured shell which is secreted by the mantle and consists of a protein matrix reinforced by crystalline calcium carbonate in the form of calcite or aragonite. Most shells are lined by mother of pearl or nacre which consists of layers of aragonite blocks. Pearls are also formed of nacre and are secreted around particles such as sand grains which become lodged in the mantle cavity. The enormous diversity in body form is reflected in the five classes of molluscs each of which is based on a somewhat different body plan. The Monoplacophora, Aplacophora, and Scaphopoda are represented by only a few living species. The Polyplacophora or chitons, have retained a somewhat primitive body form, while the Bivalvia

have a shell of two halves hinged together. In contrast the Gastropoda have a single shell or have lost it altogether and the Cephalopods are soft bodied with a generally active mode of life and well-developed sense organs.

Class Monoplacophora

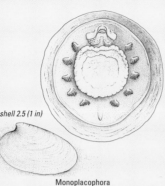

shell 2.5 (1 in)

Monoplacophora

Long considered to be extinct, ten specimens of a small, 3–20 mm (0.1–0.8 in) long mollusc with a symmetrical shell were recovered from a deep ocean trench off Costa Rica in 1952. Subsequently other species in the genus *Neopilina* have been discovered from abyssal depths of between 2000 and 7000 m (6600 and 23,000 ft). These animals have a single shell covering a broad flat foot, a simple head, and five or six pairs of gills. *Neopilina* apparently feeds on diatoms, forams and sponges and it possesses a simple radula of short teeth which can be protruded and retracted using a complex series of retractor and protractor muscles.

Class Polyplacophora

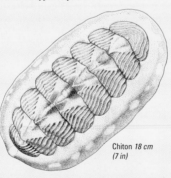

Chiton *18 cm*
(7 in)

A far better known group of animals are the chitons or mail-shells which are frequently found in the intertidal zone and are easily recognizable by their shell. Its unique shell is formed of eight separate plates which are embedded in, and partially cover, the dorsal surface of the animal. Like many of the other molluscs, chitons feed by using the radula to scrape algae from the surface of the rocks to which they cling strongly with their muscular foot. Specialized mantle cells with sensory nerve endings are grouped in minute canals actually in the shell and are believed by many to be photoreceptors. These are modified in one family (Chitonidae) as what are termed single "eyes". Chitons have two separate sexes with external fertilisation. The larval stages of this animal can last from anywhere between five and eight days.

Class Gastropoda

15 cm (6 in)

2 cm (0.8 in)

Conus 9 cm (3.5 in)

Gastropods

This is the largest molluscan class containing some 75,000 living species, the name means "stomach-foot" and the foot is highly modified in different species for different forms of locomotion although the basic form of movement is a creeping movement using a combination of ripples of muscular contraction and ciliary action. Gastropods undergo torsion, such that the visceral mass is twisted through 180° and the anus and opening for the mantle cavity now lie behind the head rather than facing the rear of the animal. The shell, which is often coiled, provides protection and the animal can withdraw its soft parts completely into the shell.

Of the three gastropod subclasses the Prosobranchia are marine having an anterior mantle cavity and gills, the Opisthobranchia show detorsion and have reduced the shell and mantle cavity and the Pulmonata have no gills and the mantle cavity has become modified as a lung. The least modified prosobranchs, the archaeogastropoda, browse on algae and this group includes the limpets and ormers which have flattened or conical shells often with slits that allow the passage of water out of the mantle cavity. Members of the most advanced group, the neogastropods such as whelks, have a typical spiral or coiled shell which can be closed by means of an operculum, a horny plate, which fits tightly into the shell aperture. More advanced, neogastropods have specialized siphons for the intake of water, formed from a rolled fold of the mantle these can be extended away from the substrate and prevent the intake of mud and silt which would foul the delicate surfaces of the gills.

Mesogastropods are recognizable by the nacreous lining to the shell and include the common *Littorina* of the intertidal, and the unusual *Xanthina exigua* which is pelagic, floating at the surface by means of a bubble raft and preying on jellyfish. Amongst the predators, animals such as the oyster drill, *Ocenebra erinacea* have modified the radula to form a drill used to pierce the shells of its bivalve prey. The Opisthobranchia are an entirely marine group which show a trend towards reduction or loss of the shell, and detorsion. In addition many have developed bilateral symmetry and lost the original gills which are replaced by specialized folds of the mantle on the dorsal surface. Many species have become adapted for a free-swimming or pelagic mode of life, while animals such as *Aplysia*, the sea hare, can crawl or swim using modified flaps of tissue, parapodia, developed from the foot.

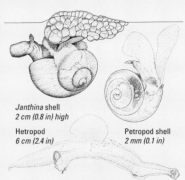

Janthina shell
2 cm (0.8 in) high

Hetropod
6 cm (2.4 in)

Petropod shell
2 mm (0.1 in)

Pelagic molluscs

The pelagic pteropods may be shelled or unshelled and are suspension feeders, trapping food in mucus on the parapodia. The creeping nudibranchs are brightly coloured, lacking a shell and mantle cavity they have achieved total bilateral symmetry and the "naked" secondary gills are external often being grouped around the anus. The opisthobranchs include one parasitic order of worm-like animals which are found in sea cucumbers. The pulmonata are the highly successful land snails and slugs but also include a number of brackish and freshwater species, the few marine representatives are confined to estuaries or intertidal waters.

Land snails lack an operculum and have modified the mantle cavity to form a lung, while slugs have modified the body form and reduced the shell to a small horny, or partly calcified plate, buried in the dorsal surface. Many pulmonates serve as the intermediate hosts for parasitic nematodes, tapeworms and flukes which infect vertebrates.

Class Bivalvia

As their name suggests this group of molluscs is easily recognizable by the shell which is composed of two valves hinged at the dorsal edge. The foot is also flattened and the crawling surface is not developed although the foot is important in many burrowing bivalves, being used to draw the animal deeper into the substrate. In bivalves such as the oysters, the foot is lost altogether and the left valve of the shell is cemented to a solid object such as a rock or mangrove root. The other common mode of attachment of sessile bivalves is by means of byssus threads. These are secreted by a special gland located on the foot and protruded between the ventral edges of the shell. By shortening the foot using the retractor muscles the animal can tighten the byssal threads and clamp more securely to the surface. Originally divided into three subclasses this group is considered by palaeontologists to consist of five subclasses, a classification which probably represents more accurately the relationships between the groups. Many of the most primitive forms live buried in mud or silt and feed by selectively picking up small ostracoda , organic debris or other animals using labial palps and tentacle-like structures. The foot is usually well developed and in addition to aiding in burrowing can be used to move in a series of leaps across the bottom. More advanced bivalves are all filter feeders drawing a current of water into the mantle cavity by means of cilia and removing suspended materials from the incoming water. Advanced burrowing forms may extend a siphon to the surface ensuring that the water supply does not include too many inorganic particles which would foul the surface of the gills. A few species, such as *Pecten*, are active swimmers, moving around on the bottom by opening and closing the valves and forcing a stream of water out through modified flaps of tissue at the edge of the mantle. The highly specialized ship worms, *Toredo*, bore into submerged wood and are elongate and worm like, with the shell reduced to two small valves which are used as a drill. The mantle, which covers the body behind, secretes a hard calcareous lining for the burrow, and the animal's digestive system contains bacteria capable of digesting cellulose from the sawdust produced by its burrowing activities. The most advanced bivalves are predators, feeding on small crustaceans which they suck into the mantle cavity using a specially modified muscular pumping system. Once inside, the small animals are moved by the labial palps to the mouth.

Class Scaphopoda
The tusk shells are a small group of tubular molluscs which generally live buried in sand or mud, with the tip of the shell protruding above the surface. The head is reduced and proboscis like and the foot is modified for burrowing. These animals are detritus feeders using thread-like feeding tentacles to pick up their organic food.

Class Aplacophora
A group of peculiar, shell-less, worm-shaped molluscs which live amongst corals and hydroids on which they feed. They have lost all their molluscan characters except for the radula and style sac.

Class Cephalopoda

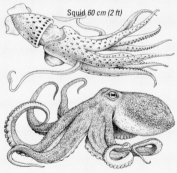

Squid, 60 cm (2 ft)

Cephalopods   Octopus 1 m (3 ft)

These are the most specialized and highly organized molluscs having evolved to become rapid, free-swimming and active predators, with well-developed sense organs including eyes. There are around 650 species, all of which share certain major features. The front of the foot is modified to form tentacles or arms which surround the mouth, the shell is reduced to a simple internal structure in most forms and in place of free-swimming larval stages, development is direct, from large yolked eggs. All modern cephalopods belong to two subclasses the Coleoidea which include the cuttlefish, squid and octopi and the Nautiloidea which includes the living species *Nautilus pompilio*. In the

nautiloids the shell is composed entirely of nacre and is coiled in a flat spiral the rear chambers of which serve as a buoyancy regulator into which gas can be secreted. *Nautilus* also lacks suckers on its numerous tentacles and the two living species are found only at depths of around 200 m (660 ft) in the Pacific.

Squid are actively swimming, torpedo-shaped animals, with the triangular fins restricted to the posterior end. The animals move rapidly by jet propulsion, expelling water from the mantle cavity via a modified siphon. Most squid are active predators of crustaceans living in mid-water but the giant squid, *Architeuthis* lives in deep water and can reach a length of 6 m (20 ft) when fully extended. Both squid and cuttlefish have a pair of elongate tentacles for catching prey and the radula has been modified to a horny beak-like structure used to tear up food. Unlike the squid, cuttlefish have a continuous finfold surrounding the dorsoventrally flattened body. The octopi have abandoned a mid-water swimming mode of life in favour of a benthic existence, clambering over the sea bottom using their eight long arms or lurking in crevices in rocks and coral reefs, where they wait for passing snails, crustaceans or small fish. With the exception of *Argonauta* the paper nautilus, octopods have lost all traces of a shell while that of cuttlefish is reduced to an internal cuttlebone of calcium carbonate and protein which serves as a buoyancy organ into the cavities of which, gas or liquid, can be secreted to alter its density.

THE LOPHOPHORATE PHYLA
The bryozoa, brachiopods and phoronids are three minor phyla of coelomate animals which all possess a lophophore, a cluster of tentacles used for gathering food. They are sometimes grouped into a single super-phylum the Lophophorata. All are sessile, with the body divided into three sections, the middle section of which is modified to form the crown of tentacles, and all possess a U-shaped digestive tract such that the anus opens on the upper surface.

PHYLUM BRYOZOA

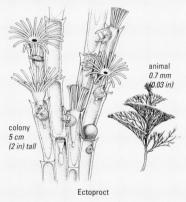

animal 0.7 mm (0.03 in)

colony 5 cm (2 in) tall

Ectoproct

The sea mosses number some 4000 species and this is the most abundant and widespread lophophorate phylum occurring on virtually any solid substrate along the shore and fouling marine structures and ship hulls. One class, of 50 species, the Phylactolaemata is confined to

freshwater and one, the Stenolaemata includes the majority of the fossil species although a few extant representatives of this group occur in marine environments. The majority of living species are included in the class Gymnolaemata, which is predominantly marine. These animals form colonies of individuals normally less than 1 cm (0.4 in) in height although colonies of the larger genus, *Flustra*, which are often washed up on temperate shorelines might be mistaken for small seaweeds. Individuals grow inside a cuticle of chitin and calcium carbonate through which the food gathering tentacles or lophophore are protruded. Individuals within the colony are polymorphic with some being adapted for defence, others for feeding, reproduction, attachment, or creating water currents. The tissues of individuals are connected through pores in the cuticle. The lophophore or crown of tentacles is used for feeding and cilia lining the tentacles beat downwards driving suspended material towards the mouth. Bryozoa show different forms of larval development with some species brooding their eggs and larvae and others producing free-swimming larvae which feed and survive in the plankton for several months.

PHYLUM BRACHIOPODA

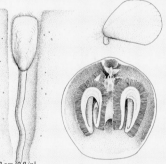

2 cm (0.8 in) across   Brachiopods   2.5 cm (1 in) across

The lampshells are an entirely marine group of lophophorates which superficially resemble bivalve molluscs since they have a shell comprised of two horny valves impregnated with calcium carbonate. The shell encloses the animal from top to bottom rather than from side to side and the ventral valve, which is generally larger, is normally attached to the substrate. The surviving 280 living species, represent a small selection of the 30,000 or so known extinct species, and this group of animals has a fossil record extending back as far as the Cambrian.

The group is divided into two classes the Articulata, in which the valves are connected by a hinge, as well as muscles, and the Inarticulata, in which the valves are connected only by muscles. *Lingula*, a living brachiopod found on tropical shorelines is one of the oldest genera of living animals, having persisted almost unchanged since the Ordovician period 500–440 million years ago. The animal lives in vertical burrows and is anchored in place by a long, muscular pedicel with root-like extensions at the lower end.

The two valves of the shell, which enclose the main body of the animal, are held slightly open, just below the surface when the animal is feeding. When

disturbed the pedicel retracts, drawing the animal into the relative safety of the burrow. When the animals is feeding, the water current is drawn in at the sides of the valves and passes out of the lophophore through the centre.

Eggs are brooded by some species of the phylum in the mantle cavity and free-swimming larvae are produced, which closely resemble the adult form. In some species of the phylum, such as *Lingula*, the free-living larval phase is extremely brief before the animal assumes the adult form.

PHYLUM PHORONIDA

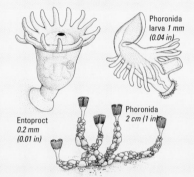

Phoronida larva 1 mm (0.04 in)

Entoproct 0.2 mm (0.01 in)

Phoronida 2 cm (1 in)

The phoronida or horseshoe worms comprise one of the smallest phyla, of around 10 marine species which live in chitinous tubes buried in sand in shallow water. The body structure of these animals is simpler than that of the other lophophorates consisting of a tubular, cylindrical body with no differentiation or appendages apart from the horseshoe-shaped lophophore. The digestive tract is U-shaped and the anus opens outside the crown of feeding tentacles. Like the brachiopods and bryozoans there is no respiratory system but a simple blood vascular system with contractile vessels, as is a simple nerve net with a nerve ring at the base of the lophophore. Phoronids are hermaphroditic with fertilization taking place outside the body to produce an egg which hatches as a long, ciliated, free-living planktonic larvae which finally settle to the bottom and secrete a tube, to assume the adult lifestyle.

PHYLUM ECHINODERMATA
This phylum contains approximately 6000 living species of exclusively marine animals. The group has an extensive evolutionary history extending back to the Cambrian and including over 20,000 described fossil species. As a group they are immediately recognizable by their pentaradial (five-rayed) symmetry although some of the starfishes for example, have numerous arms in multiples of five, while the five-rayed symmetry of some burrowing sea urchins is difficult to see at first glance. In sea urchins and starfish, the mouth generally faces downwards, in others such as the feather stars and sea lillies, the mouth faces upwards.

The name echinoderm (spiny-skinned) derives from the characteristic spines which are present in most representatives of the group. Some species of urchins such as the tropical *Diadema* and *Echinothrix* have long, fine spines armed with backwardly pointing barbs and are hollow, containing toxic secretions. One distinctive feature of the echinoderms is their complex water vascular system, a network of fluid-filled tubes which connect the tube feet that protrude outside the calcium carbonate skeleton and which are used for locomotion, burrowing and in some groups feeding. The skeleton consists of a "test", or box of plates, joined edge to edge and perforated by a series of pores through which the tube feet and respiratory papullae are extended. The surface of the test is covered by the animals epidermis and specialized structures, the pedicellariae may be used to remove settling larvae from the surface of slowly moving sea urchins and some starfish.

There are five classes of echinoderms: the Crinoidea, which includes the sea lillies, most of which occur in deep waters and the feather stars, which are abundant in shallow tropical waters; the Asteroidea or familiar sea stars, which are widely distributed in shallow water environments; the Ophiuroidea or brittle stars, another widely distributed group; the Echinoidea or sea urchins; and finally the Holothuroidea or sea cucumbers.

Class Crinoidea

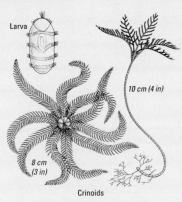

Crinoids

Although there are only around 80 living species of sea lillies and feather stars the group has a long fossil history and they are generally regarded as the most primitive group of echinoderms. Two basic body forms are seen, the sessile sea lillies which have a long basal stalk up to 1 m (3 ft) in length which live in deep water below 100 m (330 ft) and the more numerous generally sedentary but free-swimming forms, the feather stars, which lack a basal stalk but possess jointed cirri to attach to rocks in shallow waters. The crinoid body, or calyx, consists of a small, cup-shaped structure, with the mouth facing upwards and supporting five long, feather-like arms. The arms are formed of small skeletal units like vertebrae which in turn support the smaller plates of the pinnules. Along the inner surface of the pinnules and arms are tube feet which secrete mucus, used for trapping food particles in suspension. These mucus strands with the trapped food are then passed back down the arms by the tube feet to the mouth at the centre of the calyx. Feather stars swim by alternating up-and-down movements of the arms and use this method of locomotion to move from place to place depending on the availability of food in the neighbouring water.

Class Asteroidea

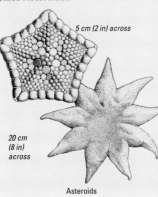

Asteroids

Many starfish display obvious, five-rayed symmetry, having five well-developed arms radiating from a central, barely distinguishable body. Some species such as *Porania*, however, have reduced the arms and resemble cushions, others such as *Luidia*, *Solaster* and *Crossaster* may have 7 or 13 arms. In general these animals are slowly moving, benthic forms with the mouth facing downwards and the under surface of the arms bearing numerous tube feet which can be extended and attached to the substrate, then shortened, to drag the animal along. The tube feet are arranged in two rows alongside a groove on the undersurface of the arms and are protected by short, stout spines.

The digestive system is vertical with the mouth below and the anus on the upper surface. Extensions from the central digestive cavity extend along each arm, which also contains the reproductive organs of these animals. Some species of starfish are predators of bivalve molluscs and *Asterias* feeds on mussels by grasping the two halves of the shell with the tube feet of opposing arms and forcing apart the valves. The stomach is then extruded into the shell and enzymes commence digestion, producing a partly digested mixture, which is taken in by the starfish.

Sea stars like other echinoderms are broadcast spawners, when ripe, a female discharges millions of eggs into the water which stimulates the production of eggs and sperm by neighbouring individuals. Fertilized eggs produce a dipleurula larva which passes through a series of larval stages of development in the plankton. The larval stages may last several weeks and the larvae feed on organisms such as diatoms before sinking to the bottom to undergo the final stages of metamorphosis and assume the adult body form.

Class Ophiuroidea

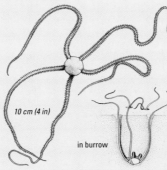

Ophiuroids

The brittle stars have distinct arms, which radiate from a central, disc-shaped body. The arms are much more flexible than is the case with sea stars and this results from the jointed ossicles which are hinged together and operated by muscles in much the same way as the vertebral column of the higher animals. These are abundant animals of the shallow coastal seas and over 2000 living species are known. Most are small, agile, rapidly moving animals which hide under stones but in tropical areas the massive, conspicuous basket stars with numerous arms are obvious, nocturnal inhabitants of lagoon areas around coral reefs.

Most ophiuroids are scavengers, feeding on organic debris on the surface of the substrate which is trapped with sheets of mucus and rolled into balls by the tube feet, before being passed down the underside of the arms to the centrally located mouth. Some species will hold themselves in position with two arms while extending the other three to feed. Reproduction generally involves planktonic larvae of quite distinct form from those of other echinoderms, although some species are known to brood their larvae in special pockets or bursae.

Class Echinoidea

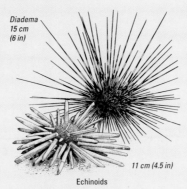

Echinoids

The sea urchins lack arms and are generally less active than the sea stars and brittle stars. Typical sea urchins are rounded although many of the burrowing species, such as sand dollars, are considerably flattened, while other burrowing species such as the heart urchin *Echinocardium*, have assumed secondary, bilateral symmetry. Most urchins move by means of the tube feet although the long spines may be used both in moving and in holding the animal in place in rocky retreats. The urchins have a special feeding apparatus consisting of ossicles, supporting five teeth, surrounding the mouth. These teeth can be protruded and withdrawn to scrape algae from the surface of the substrate or graze on larger seaweeds and seagrasses. Since the majority of urchins are herbivores the gut tends to be extremely long and is coiled inside the circular test of the animals with the anus opening in the centre of the upper surface. Given their comparatively large volume-to-surface ratio urchins have numerous respiratory papulae which extrude through the test allowing gas exchange between the internal body fluids and the surrounding water. Gas exchange also takes place through the tube feet which in many surface dwelling forms have sucker-like discs at their ends for attachment to the surface.

The sexes are generally separate and five large gonads occupy much of the space inside the animals testes when they are ripe. The eggs are shed and fertilized externally and the larvae pass rapidly through several stages before settling, frequently in the vicinity of other, adult urchins.

Class Holothuroidea

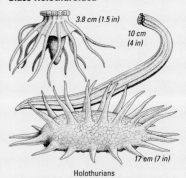

Holothurians

The sea cucumbers or beche-de-mer comprise one of the smaller echinoderm classes with only about 900 species which have, like the burrowing urchins, assumed a secondary bilateral symmetry. In these animals the test is considerably reduced, being present only as a series of spicules embedded in the body wall of the animal. They are greatly elongated and the under surface is lined with three rows of suckered tube feet while the tube feet of the dorsal surface are greatly reduced in number, lack suckers and have generally assumed a rather warty appearance. Some sea cucumbers have lost the tube feet altogether and move in a worm-like manner using the hook-shaped spicules for attachment. Some species are burrowers, some slowly creeping, bottom scavengers and some have adopted a suspension mode of feeding extending their anterior ends from rock crevices or holes in the surface of coral reefs. All have well-developed buccal tentacles surrounding the mouth and these have a variety of shapes depending on the food source. In some they have short foliose ends used for picking up debris from the surface, in others they are more extensible and feathery being used for trapping particles in suspension.

Since many of these animals occur in areas of low oxygen concentration they have specially modified respiratory structures associated with the terminal cloaca. Water is pumped into and out of the respiratory trees by means of muscular contractions of the cloaca and some species may obtain more than half of their total oxygen requirements via the respiratory trees. A number of species such as *Borhadsia argus* possess a remarkable defense system, the Cuverian organ, which consists of a series of long, blind-ending tubules attached to the ducts of the respiratory trees. When attacked, these are everted, like turning the fingers of a rubber glove inside out, and the sticky surface adheres to the would be predator allowing the cucumber to make good its escape. Some species will also discharge their gonads and entire digestive system which are later regenerated. Only a single gonad is present and, as in most other echinoderms, fertilization is external. The larvae undergo most of their metamorphosis in the plankton, finally settling out as young sea cucumbers.

PHYLUM CHAETOGNATHA

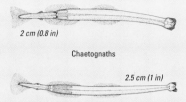

2 cm (0.8 in)

Chaetognaths

2.5 cm (1 in)

This is a group of 50 species of marine organisms which are extremely common in the oceanic plankton, all of which display a rather uniform body plan. They are recognizable by their elongate, arrow-shaped body, paired lateral fins and a single tail fin. The anterior mouth has strong grasping spines and these animals are among the most important carnivores of the plankton community, feeding particularly on copepods. Some, such as *Sagitta bipunctata*, are recorded as feeding on larval fish. The prey are seized with the spines and bitten with the spine-like teeth, while some species are believed to secrete venom, to kill and immobilize the prey. They are generally small, around 2 cm (0.8 in) in length and the head is comparatively small, most of the animal consisting of a trunk section, with paired lateral fins which terminate just in front of the single tail fin. Fertilization is internal and some species carry their eggs, others release them into the plankton, while the benthic genus *Spadella* attaches them to rocks and seaweeds in the littoral zone.

PHYLUM HEMICHORDATA

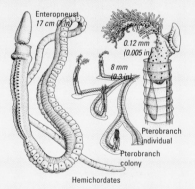

Enteropneust
17 cm (7 in)

0.12 mm (0.005 in)

8 mm (0.3 in)

Pterobranch individual

Pterobranch colony

Hemichordates

This phylum of around 100 species displays two distinct body forms: the class Enteropneusta or acorn worms are elongate worm-like animals, while the Pterobranchs are sessile, tube-dwelling, deep-water forms of peculiar body form. The presence of gill slits suggests a relationship between these animals and the chordates but they are sufficiently different to be considered a distinct phylum. They are recognizably distinct from the preceding animal phyla in having the body clearly divided into three regions: an anterior protosome, modified into a short proboscis in the acorn worms and as a shield-shaped structure which secretes the tube in pterobranchs; a mesosome, or collar into which the acorn worm can retract the proboscis and which in

pterobranchs is strikingly modified into a series of hollow arms bearing ciliated tentacles used in feeding; and, a metasome or trunk which is elongate in the acorn worms and short and barrel-shaped in pterobranchs.

PHYLUM CHORDATA

The phylum chordata, which includes the vertebrates, is the largest deuterostome group of animals and the youngest in evolutionary terms. While the fossil record for invertebrate phyla extends back some 1600 million years, fossils of the earliest vertebrates are only 500 million years old. Chordates are however, one of the major animal phyla dominating land, air and water, although they do not exceed the arthropods either in terms of numbers of species or individuals but far outstrip them in terms of total biomass and ecological importance. Three chordate characters are considered diagnostic of this phylum. The first is the presence, at least during some stages of their life history, of an axial skeleton either in the form of a stiffened rod, the notochord, or in higher forms a jointed bony structure the vertebral column. Secondly, the presence of clefts or gill slits, and thirdly the possession of a dorsal, hollow nerve tube, expanded at the anterior end to form a brain, often protected by a box-like structure the cranium. In addition most chordates possess a post-anal tail which is flexible and muscular and in most aquatic forms is used for locomotion. The circulatory system consists of a closed tubular network with a muscular pump, the heart.

SUBPHYLUM UROCHORDATA

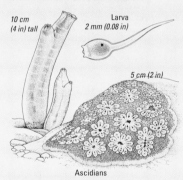

10 cm (4 in) tall

Larva 2 mm (0.08 in)

5 cm (2 in)

Ascidians

Urochordates or tunicates are common marine animals numbering some 1300 species which as adults bear little resemblance to other chordates and are placed by some authors in a separate phylum, Tunicata. The most common and widespread group, the class Ascidiacea, or sea squirts, are adapted to a sessile mode of life, resembling a barrel, bearing two siphons through which water is drawn for both respiration and feeding. The water is passed through a net and food particles are trapped in mucus and rolled towards the opening of the digestive tract. Water, which has been filtered, passes through the net and out of the body of the animal via the atrial siphon. Only during development do these animals display the chordate features of a notochord or axial skeleton, a dorsal hollow nerve cord and gill clefts. Fertilization is external and the eggs hatch

to produce a minute, tailed, tadpole-like larvae which swims by means of side-to-side movements of the tail. Following a period in the plankton the larva settles and undergoes extensive metamorphosis during which the tail is reabsorbed, the nerve cord becomes reduced to a small ganglion lying between the siphons of the adult and the gill slits become modified to form the filtering, feeding net of the adult.

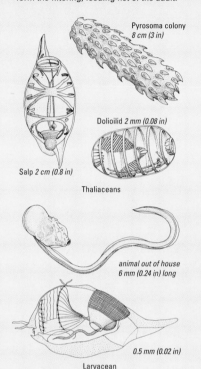

Pyrosoma colony 8 cm (3 in)

Dolioliid 2 mm (0.08 in)

Salp 2 cm (0.8 in)

Thaliaceans

animal out of house 6 mm (0.24 in) long

0.5 mm (0.02 in)

Larvacean

Two other classes of ascidean, the Thalacea and the Larvacea are both modified for a pelagic existence. The thalaceans, a small group of only six genera, resemble small barrels with one siphon at each end, so that water passes through the length of the animal. A typical filtering network is present and the feeding current also serves for respiration and locomotion. Circular muscle bands passing around the body force water out of the central chamber when contracted. In contrast the larvaceans retain the tadpole-like body form throughout life, the tail is long and extends at right angles from the dorsal surface. The filtering net of larvaceans is adapted for feeding on fine particles, less than 1 μm in diameter, which are not normally available to other planktonic filter feeders. These animals may themselves be important components of larval fish diets.

SUBPHYLUM CEPHALOCHORDATA

Only two genera comprise this subphylum, *Branchiostoma* and *Assymmetron* distinguished by the fact that in *Assymmetron* the gonads are restricted to the right side of the body and in *Branchiostoma* they are paired. Adults are bottom-dwelling, filter-feeding animals, using a net of similar construction to that of tunicates, for trapping fine, suspended particles. These animals are generally small, around 4–8 cm (1.6–3 in) in length, and the characteristic notochord and dorsal hollow nerve cord of the chordates are

present throughout life, as is the post anal tail. The muscles, nerves and blood vessels of the body wall are segmentally arranged and the alternate contractions of the myotomes, or muscle blocks, on each side of the body cause lateral movements, allowing the animal to swim in a fish-like manner. Unlike the higher vertebrates the head is less distinct in the cephalochordates although the nerve cord is expanded to form a brain-like ganglion and simple light and chemical receptors are present on the anterior end of the animal. Although most authors consider these animals to be primitive members of the phylum Chordata some consider them a separate phylum, Acrania.

SUBPHYLUM VERTEBRATA

The classification of vertebrates varies, with some authors recognizing the Agnatha, or jawless forms as a separate subphylum or superclass, distinct from the Gnathostomata or jawed animals. The most primitive vertebrates were jawless, fish-like animals which appear in the fossil record during the Ordovician and are related to the extant hagfish and lampreys. The old grouping of animals into the class Pisces encompassing sharks and rays together with the bony fish is now recognized as being somewhat artificial in that the elasmobranchs (sharks and rays) have probably had a lengthy period of separate evolutionary history from the bony fishes. Modern classifications now recognize three major groupings of extant fishes and fish-like animals: the Agnatha, hagfish and lampreys; the Elasmobranchiomorphi (Chondrichthyes) sharks, rays and ratfishes, with a cartilaginous skeleton; and the Osteichthyes or bony fishes. In addition an extinct fossil group, the Placodermi, includes early, heavily armoured, bottom-dwelling forms considered either as a subclass of the Elasmobranchiomorphi or as a separate class.

Class Agnatha

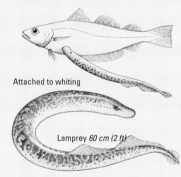

Attached to whiting

Lamprey 60 cm (2 ft)

The lampreys and hagfish lack a vertebral column and paired fins, but have a notochord like the cephalochordates. They also lack the jaws of the higher vertebrates and the mouth is circular, terminal and armed with hook-like structures. Lampreys have a muscular rasping tongue which is used to scrape the surface of the fish to which they attach, feeding on blood and tissue. In contrast the hagfishes seem to be scavenging animals feeding on dead and dying animals. The gills of lampreys are

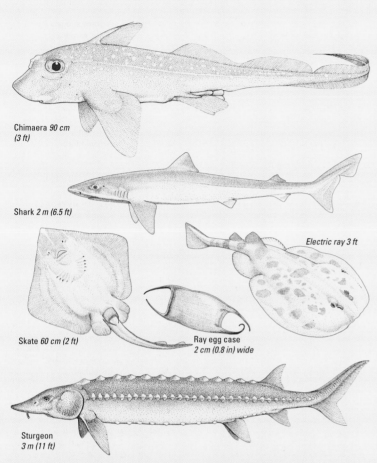

Chimaera *90 cm (3 ft)*

Shark *2 m (6.5 ft)*

Skate *60 cm (2 ft)*

Electric ray *3 ft*

Ray egg case *2 cm (0.8 in) wide*

Sturgeon *3 m (11 ft)*

unusual in that water passes into and out of the gill pouches through the same opening rather than entering through the mouth and exiting via the external gill slits. Sea lampreys may attach to large basking sharks and whales but unlike the hagfish which are entirely marine, lampreys return to freshwater to breed, where the larvae, which resemble cephalochordates undergo a lengthy period of development. The relationships between the living agnathans and the four extinct agnathan orders, collectively known as placoderms, is obscure, although it is clear that the living species are highly specialized for different modes of life in which they have few competitors.

Class Elasmobranchiomorphii (Chondrichthyes)

This group of animals is distinguished from the higher fishes by the presence of a cartilaginous, rather than a bony skeleton and is divided into two major groupings, the Holocephali or ratfishes, and the Elasmobranchii, which includes over 2000 species of sharks, skates and rays. The ratfishes are a small group of peculiar fishes, in which the gill slits open externally through a single opening. The tail is long and whip like and these animals swim by means of enlarged pectoral fins. The head is large and the jaws have strong, plate-like teeth for crushing the molluscs on which they feed.

Modern elasmobranchs have well-developed jaws generally armed with sharp teeth and capable of powerful jaw closure although generally without the capacity to chew. The great white shark, *Carcharodon carcharias*, has serrated teeth and is a voracious predator of fish,

including other sharks and large vertebrates such as seals, sealions and dolphins. Most sharks are solitary, although the thresher shark, *Alopias* hunts in packs using its long whip like tail to drive the herring, mackerel or pilchards into a compact shoal. A number of species, including the familiar dogfish, *Scyliorhinus*, are sluggish bottom dwellers, feeding on molluscs and crustaceans, while the frilled shark, *Chlamydoselachus* is a deep water form, living between 200 and 1000 m (660 and 3300 ft) and feeding on cephalopods.

In contrast the basking shark, *Cetorhinus maximus*, which grows to a length of 10.5 m (35 ft) and the whale shark, *Rhincodon*, which grows to a length of 18 m (60 ft), feed by swimming around with their mouths open, filtering small plankton from the surrounding water.

The skates and rays are specialized for a bottom-dwelling existence being flattened forms with enlarged, lobe-like pectoral fins, which form the typical wings. The tail fin is not generally used in locomotion, and in many species it is reduced to a long, whip-like structure, armed with spines which in some species, such as the sting ray, *Dasyatus*, have an associated poison gland. Most species have a grinding dentition and *Myliobatis*, the eagle ray, feeds almost exclusively on clams and oysters which are crushed between a grinding mill of flattened bar-like teeth.

Class Teleostomii (Osteichthyes)

Two major groupings of bony fish are recognized the Sarcopterygii or lobe-finned fishes with only seven living species and the Actinopterygii or ray-finned fishes which number in excess of 30,000 species. The lob-finned fish are important as the ancestral group from which the land vertebrates evolved and the living *Latimeria*, or coelocanth of the Indian Ocean and the freshwater lungfishes of tropical regions are the only living representatives. *Latimeria* is a large fish reaching over 90 kg (200 lbs) in weight and feeding on other fish which it usually swallows whole.

Three grades of organization are recognized amongst the ray-finned fish of which the group Chondrostei is represented only by the marine sturgeons which migrate to freshwater to breed. Although the internal skeleton is largely cartilaginous, bony plates are present in the skin which also contains denticle-like structures similar to those of the sharks. The flattened snout is used to stir up the muddy bottom and reveal small invertebrates on which these animals feed. The Holostean fishes are another small group including the freshwater garpike, *Lepisosteus* and the bowfin *Amia calva*.

Most of the 30,000 living marine and freshwater fishes are members of the Teleostei, of which the herring-like fishes are the most primitive, having an elongate fusiform body with pectoral and pelvic fins placed towards each end of the animal, a median dorsal and anal fin which act as stabilizers during swimming and a large caudal or tail-fin which provides the main propulsive force during swimming.

Alternating contraction of the muscle blocks on each side of the animal sweep the tail from side to side and in more advanced teleosts this lateral movement involves only the tail fin itself, the remainder of the body being held relatively rigid. The diversity of body form in this group of fishes is staggeringly large, ranging from long, eel-like burrowing forms, globular slow-swimming forms, and flattened, bottom-dwelling forms, to high-speed swimmers such as tuna with a torpedo-shaped body and greatly enlarged pectoral fins. Tuna also increase their muscular efficiency by retaining heat in the core of the body musculature and by storing oxygen in the muscle tissue. Many coral reef fish are laterally compressed allowing them to slip between the coral colonies, while pipe fishes and sea horses have assumed an elongate and upright body form allowing them to hide amongst the seaweeds they inhabit.

Part of the success of the teleosts is based on their ability to regulate their buoyancy by means of an internal gas bladder into which gas can be secreted from the blood stream, or from which gas can be reabsorbed. This allows the animal to alter its density according to the external water pressure and density, thus remaining in one position without sinking or floating to the surface. Many sedentary fishes have neutral buoyancy and locomote by fine movements of the paired fins, rather than broad sweeps of the tail. Quite delicate movements allow the animals to remain in one position, either to feed or to hide amongst rocks or weeds in their shallow water habitats.

Class Reptilia

Leatherback turtle *3 m (10 ft)*

Although this group of animals dominated the marine environment during the Mesozoic era they are now restricted to seven species of marine turtle, several hundred species of sea snakes, one species of marine iguana and a few estuarine crocodilian species.

The turtles have horny beaks and modified front limbs which serve as flippers and are the main source of power for locomotion, the rear flippers being used only for steering. In contrast to their land-based relatives the tortoises, sea turtles have reduced the weight of the skeleton and assumed a more flattened streamlined shape. They come ashore to breed, laying eggs in excavated nests at the head of beaches. Many of the young are lost to predators on their journey from the nest to the sea and a more recent problem has been the inland movement of hatchlings attracted to the lights of tourist hotels which are a stronger stimulus than the moonlight ocean.

Herring *30 cm (12 in)*

Coelacanth *1.2 m (4 ft)*

The marine lizards are few in number, although some of the squamate lizards will actively forage between the tide lines. The marine iguana, *Amblyrhynchus*, of the Galapagos Islands is really only an amphibious lizard spending a considerable portion of its time on land. In general body form they are very like their terrestrial relatives, although they swim and dive, and feed on seaweeds. Like seabirds, these animals have salt-excreting glands in the nostrils, in contrast to turtles which excrete excess salt through ducts in the eye socket.

Sea snakes are a diverse group of highly venomous, marine animals, the vast majority of which occur in shallow waters of the Southwest Pacific Ocean, between Malaysia and Australia. Sea snakes are related to the cobras and kraits on land. The tail is laterally compressed, forming a paddle and the animals swim by lateral flexure of the body. Pelamis is completely pelagic, and feeds on small fish which it catches by means of a lateral strike. Many of the smaller-headed species of sea snake feed on fish eggs, while the genus *Laticauda*, the sea kraits, feed on moray eels on coral reefs. Most sea snakes are ovo-viviparous, the eggs hatching inside the females reproductive tract and the three to eight young emerging as fully formed juvenile snakes. Birth takes place at sea and with the exception of the Laticaudine sea snakes no species comes on land. Sea snakes, other than the sea kraits are well adapted to their marine environment. Unlike terrestrial snakes, the belly scales of sea snakes, are much smaller, similar in size to the dorsal scales. Sea snakes will often migrate great distances and in great numbers.

Class Aves

King penguins *1m (3 ft)*

Seabirds range from wading species found around the fringes of the world's oceans to truly oceanic species, such as petrels and albatrosses which come on land only to nest. All seabirds feed on marine organisms with the waders probing mud and silt to extract small worms and crustaceans; oyster catchers using the chisel-shaped beak to hammer bivalves off rocks and open them; and, diving birds which feed on fish in surface waters. Perhaps one of the most specialized groups are the penguins which have lost the ability to fly and have modified the wings to form flippers for swimming underwater in pursuit of their fish prey. Other diving birds, such as the true divers and cormorants, use the feet for underwater swimming.

Most of the seabirds which hunt fish from the air are white or pale grey below, presumed to provide camouflage when viewed from below against the bright sky.

Like the turtles, seabirds must nest on land, and seabird colonies are generally concentrated in suitable coastal areas. Kittiwakes and gannets nest on rocky cliffs, while many gulls and terns nest on the ground in sand dune areas, boobies and albatross in trees on isolated oceanic islands, and penguins on ice free areas of the Antarctic and other southern continents.

Class Mammalia

Marine mammals belong to a number of different groups with the polar bear and sea otters being representatives of the predominantly land-based order, the Carnivora. As might be expected of animals which have recently returned to the marine environment they display fewer adaptations to a marine existence than do the more specialized seals, the Pinnepedia, to which they are related. Three groups of true marine mammals display adaptations for a marine existence: the Pinnipedia or seals; the Sirenia or sea cows; and, the Cetacea, the whales and dolphins.

The dugong and manatees are the smallest of these groups with only four living species although a fifth, Steller's sea cow was hunted to extinction in the last century. Distantly related to elephants these animals have lost the hind limbs completely and modified the fore-limbs to form strong paddles. The tail is flattened like the flukes of whales and the animals are herbivorous being confined to shallow water areas, estuaries and large river systems of the tropics and sub tropics where they graze on sea grasses.

The seals, sealions and walruses are a more diverse group, of mainly fish-eating

species, divided into three sub-groups: the earless or true seals; the eared seals or sealions; and, the walrus. All possess streamlined bodies with reduced or absent hair cover and insulation provided by a layer of fat beneath the skin. The true seals cannot move the hind limbs forward under the body when on land, but the sea lions can rotate the short hind limbs and move with a galloping gait on land. The walrus, with its elongate downwardly pointing tusks, is an Arctic animal adapted to feeding on molluscs.

The most highly adapted marine mammals are the whales and dolphins with their streamlined body, no visible neck, well-developed front flippers and flattened and expanded tail flipper or flukes. Like the seals they are hairless and posses a thick layer of fat or blubber below the skin for insulation. Whales cannot move at all on land and the skin must be wet at all times. They give birth in the water and the baby whales are born tail first and nudged to the surface to take their first breath. There are two groups of whales, the toothed whales, which include the smaller sperm whales, narwhals and dolphins are active predators, pursuing fish, squid and in the case of killer whales, seals and penguins.

The second group, the baleen whales, lack teeth in place of which they have plates of baleen, set in the upper jaw which act as sieves, straining out the krill as the animals swim through the plankton with their mouths agape. These animals reach a large size, the smallest being 5 m (17 ft) when fully grown and the largest, the blue whale, which can reach over 33 m (108 ft) in length, is the largest mammal to have ever lived on Earth.

Tables

Continental shelf areas *(seabed not deeper than 200 m/656 ft)*

ATLANTIC OCEAN		
Ocean/Subarea	Square kilometers	Square miles
Northwest Atlantic	**1,260,000**	**486,360**
W. Greenland	85,000	32,800
Labrador	100,000	38,600
Newfoundland	490,000	189,140
Nova Scotia & St. Lawrence	310,000	119,660
New England	185,000	71,410
Middle Atlantic	90,000	34,740
Northeast Atlantic	**3,011,000**	**1,162,250**
Iceland	142,000	54,810
Barents Sea	1,300,000	501,800
North Sea	550,000	212,300
Skagerrak & Kattegat	40,000	15,440
British Isles	449,000	173,310
Baltic	420,000	162,120
Spain – Portugal – France	60,000	23,160
Southwest Atlantic	**1,950,000**	**752,700**
Guyana	160,000	61,760
Brazil	610,000	235,460
Uruguay	150,000	57,900
Argentina	1,030,000	397,580
Southeast Atlantic	no data	
Eastern Central Atlantic	**480,000**	**185,280**
Morocco coastal	65,000	25,090
Sahara "	65,000	25,090
C. Verde "	110,000	42,460
C. Sherbo "	70,000	27,020
W. Gulf Guinea	50,000	19,300
C. Gulf Guinea	65,000	25,090
S. Gulf Guinea	55,000	21,230
Western Central Atlantic	**1,280,000**	**494,080**
US east coast	110,000	42,460
Bahamas – Cuba	120,000	42,360
Gulf of Mexico	600,000	231,600
Caribbean	250,000	96,500
S. America	200,000	77,200
Total Atlantic	**7,981,000**	**3,080,670**

PACIFIC OCEAN		
Ocean/Subarea	Square kilometers	Square miles
Northwest Pacific	**995,00**	**384,070**
Okhotsk Sea	610,000	235,460
Japan Sea	202,000	77,970
N.W. Pacific areas	183,000	70,638
Northeast Pacific	**1,276,000**	**492,540**
Bering Sea	1,000,000	386,000
Oregon – S.E. Alaska	96,000	37,060
Gulf of Alaska	100,000	38,600
Alaska Peninsula	80,000	30,880
Southwest Pacific	**930,000**	**358,980**
New Zealand	200,000	77,200
Australia	730,000	281,780
Southeast Pacific	**177,000**	**68,320**
Peru	87,000	33,580
Chile	90,000	34,740
Western Central Pacific	**4,610,000**	**1,779,460**
Yellow Sea – E. China Sea	950,000	366,700
Formosa Strait	280,000	108,080
Gulf of Tongking	200,000	77,200
Gulf of Thailand	305,000	117,730
S. China Sea	970,000	374,420
Java Sea	580,000	223,880
Gulf of Carpentaria	960,000	370,560
Islands	365,000	140,890
Eastern Central Pacific	**450,000**	**173,700**
Total Pacific	**8,438,000**	**3,257,070**

INDIAN OCEAN		
Ocean/Subarea	Square kilometers	Square miles
East Africa	390,000	150,540
Arabian Sea	400,000	154,400
Bay of Bengal	610,000	235,460
Indonesia	130,000	50,180
Western Australia	380,000	146,680
South Australian coast	260,000	100,360
Red Sea	180,000	69,480
Persian Gulf	240,000	92,640
Madagascar	210,000	81,060
Oceanic Islands	200,000	77,200
Total Indian	**3,000,000**	**1,158,000**

Production by fishery commodity and content *(thousand metric tonnes)*

FISH: FRESH, CHILLED OR FROZEN

Continent/Area	1982	1983	1984	1985	1986	1987	1988	1989	1990	1991
Africa	322	317	343	396	407	427	404	374	380	376
N. America	810	755	741	785	872	917	986	1,057	1,410	1,295
S. America	486	486	563	566	563	578	604	693	774	665
Asia	5,698	5,702	6,313	6,319	6,849	6,649	7,267	7,174	7,106	7,057
Europe	1,878	1,977	2,056	2,050	2,214	2,266	2,433	2,459	2,459	2,639
Australia and Oceania	100	124	122	120	151	133	191	173	158	202
Former USSR	3,174	2,982	3,194	3,291	3,431	3,272	3,412	3,406	3,133	2,990
Total	12,470	12,344	13,333	13,526	14,488	14,242	15,297	15,336	15,420	15,224

FISH: DRIED, SALTED OR SMOKED

Continent/Area	1982	1983	1984	1985	1986	1987	1988	1989	1990	1991
Africa	238	242	250	271	291	328	328	343	382	384
N. America	117	96	88	10	74	96	100	105	116	110
S. America	56	61	66	60	63	65	77	73	73	73
Asia	2,302	2,346	2,301	2,383	2,387	2,440	2,554	2,289	2,499	2,527
Europe	515	471	501	509	510	526	531	515	437	422
Australia and Oceania	2	1	0.5	2	3	2	6	10	8	6
Former USSR	773	793	794	850	860	841	799	769	730	723
Total	4,007	4,011	4,002	4,174	4,188	4,298	4,396	4,106	4,245	4,245

FISH: AIRTIGHT CONTAINERS

Continent/Area	1982	1983	1984	1985	1986	1987	1988	1989	1990	1991
Africa	145	145	157	124	121	141	108	143	182	217
N. America	345	390	331	378	335	409	424	487	499	488
S. America	104	90	90	96	94	116	116	126	124	132
Asia	1,188	1,337	1,420	1,664	1,735	1,756	1,901	1,923	1,965	1,934
Europe	562	584	600	573	614	676	711	685	727	732
Australia and Oceania	5	6	7	9	8	8	32	36	37	37
Former USSR	427	486	517	547	617	690	743	796	820	887
Total	2,776	3,038	3,122	3,391	3,524	3,796	4,035	4,196	4,354	4,427

CRUSTACEANS & MOLLUSCS: FRESH, FROZEN, DRIED, SALTED, ETC.

Continent/Area	1982	1983	1984	1985	1986	1987	1988	1989	1990	1991
Africa	77	102	100	106	135	134	142	141	134	147
N. America	375	355	367	399	438	442	457	445	487	532
S. America	106	117	121	100	98	132	131	135	138	228
Asia	802	846	819	851	855	1,040	977	1,226	1,136	1,209
Europe	220	267	293	291	296	354	267	290	264	246
Australia and Oceania	68	66	75	67	57	63	67	101	54	46
Former USSR	-	-	-	-	-	-	-	-	-	-
Total	1,648	1,751	1,775	1,813	1,879	2,165	2,041	2,339	2,213	2,407

CRUSTACEANS & MOLLUSCS: AIRTIGHT CONTAINERS (CANNED)

Continent/Area	1982	1983	1984	1985	1986	1987	1988	1989	1990	1991
Africa	1	1	1	✳	✳	✳	✳	✳	✳	✳
N. America	62	65	66	63	61	67	71	70	73	77
S. America	3	3	3	2	3	2	3	2	2	3
Asia	58	49	60	60	58	66	57	73	69	61
Europe	27	29	29	30	36	39	34	38	40	42
Australia and Oceania	-	1	1	2	2	2	2	2	2	2
Former USSR	6	6	5	5	5	4	3	2	2	2
Total	157	154	165	162	165	180	170	187	188	187
Grand total	21,058	21,298	22,397	23,066	24,244	24,681	25,939	25,977	26,420	26,490

FISH OIL AND FATS

Continent/Area	1982	1983	1984	1985	1986	1987	1988	1989	1990	1991
Africa	37	48	31	39	34	92	41	31	18	21
N. America	178	199	189	181	188	175	139	146	167	155
S. America	352	79	353	341	506	297	414	597	395	436
Asia	321	361	432	429	498	483	517	485	447	344
Europe	342	380	422	419	348	294	331	251	253	297
New Zealand and Oceania	-	-	1	2	0.8	0.7	0.9	1	1	1
Former USSR	67	73	93	85	95	111	112	119	127	120
Total	1,300	1,141	1,520	1,496	1,671	1,453	1,556	1,631	1,408	1,373

FISH MEAL AND SOLUBLES

Continent/Area	1982	1983	1984	1985	1986	1987	1988	1989	1990	1991
Africa	172	213	153	145	170	309	287	177	146	136
N. America	665	651	606	674	582	677	566	588	583	569
S. America	1,635	1,163	1,795	2,137	2,519	2,082	2,462	2,711	2,378	2,648
Asia	1,353	1,549	1,708	1,619	1,651	1,636	1,623	1,594	1,634	1,436
Europe	965	1,112	1,167	1,076	1,018	1,013	1,126	1,048	932	937
Australia and Oceania	7	7	5	2	4	4	4	4	4	9
Former USSR	600	605	674	657	747	767	769	752	695	632
Total	5,392	5,293	6,107	6,311	6,689	6,487	6,837	6,875	6,372	6,367
Grand total	6,692	6,434	7,627	7,807	8,360	7,940	8,393	8,506	7,780	7,740

Landings of fish by oceans *(metric tonnes)*

ATLANTIC OCEAN LANDINGS

Diadromous Fish	1985	1986	1987	1988	1989	1990	1991
Sturgeon, etc.	43	21	21	18	5	3	18
River eels	5,073	5,315	4,371	5,466	4,685	4,180	4,557
Salmon, trout, etc.	73,539	98,896	113,130	162,894	216,172	225,449	246,678
Shad, milkfish, etc.	27,614	21,104	26,780	17,453	21,768	19,286	19,085
Miscellaneous	1,108	25	61	65	11	25	20
Total	107,377	125,361	144,363	185,896	242,687	248,943	270,358

Marine Fish	1985	1986	1987	1988	1989	1990	1991
Flounder, sole, etc.	632,617	614,186	637,937	615,001	608,442	603,040	629,359
Cod, hake, etc.	5,493,621	5,686,568	5,703,464	5,458,793	5,091,574	4,479,555	4,175,652
Redfish, bass, etc.	2,062,466	2,438,853	2,358,301	2,394,299	2,430,905	1,796,050	2,238,714
Jack, mullet, etc.	3,452,402	2,786,166	2,630,601	2,830,029	2,509,832	2,430,435	2,693,318
Herring, sardines, etc.	5,122,859	4,593,582	6,163,598	5,045,408	5,795,624	5,590,124	5,233,312
Tuna, bonito, etc.	517,968	498,086	485,501	532,143	541,509	571,594	580,834
Mackerel, snoek, etc.	947,123	1,046,327	992,791	1,187,597	1,087,882	996,229	989,688
Sharks, rays, etc.	223,021	217,067	239,896	240,620	230,419	248,336	243,186
Miscellaneous	961,724	991,877	1,046,296	960,110	951,138	1,047,103	903,003
Total	19,413,801	18,872,712	20,258,385	19,264,000	19,247,325	17,762,466	17,687,066

Crustaceans	1985	1986	1987	1988	1989	1990	1991
Crabs, etc.	220,407	206,810	201,845	215,711	199,592	195,818	220,461
Lobsters, etc.	160,145	159,203	162,671	164,897	163,671	168,498	172,819
Shrimps, prawns, etc.	539,546	613,695	481,821	510,588	552,647	566,621	579,637
Krill, etc.	80,807	425,871	346,504	364,173	395,643	344,445	231,337
Miscellaneous	9,230	9,825	14,984	19,159	7,799	6,239	4,532,
Total	1,010,135	1,407,565	1,194,828	1,274,528	1,313,542	1,277,372	1,206,245

Molluscs	1985	1986	1987	1988	1989	1990	1991
Winkles, conchs, etc.	20,635	21,828	23,086	25,188	29,076	25,537	21,620
Oysters	20,635	406,087	369,879	332,574	332,844	313,067	328,268
Mussels	550,398	642,908	555,585	530,145	495,513	485,480	498,439
Scallops, etc.	286,778	213,064	371,093	381,783	323,535	288,307	261,000
Clams, cockles, etc.	494,260	489,100	497,170	472,556	504,468	513,792	503,365
Squid, cuttlefish, etc.	521,723	628,928	1,018,712	915,800	1,062,497	848,211	1,009,208
Miscellaneous	33,275	23,139	30,510	28,425	32,555	25,682	18,465
Total	1,927,704	2,425,054	2,866,035	2,686,471	2,780,488	2,500,076	2,640,365

Grand total							
	22,459,017	22,830,692	24,463,611	23,410,895	23,584,042	21,788,857	21,804,034

MEDITERRANEAN SEA LANDINGS

Diadromous Fish	1985	1986	1987	1988	1989	1990	1991
Sturgeon, etc.	1,327	1,427	1,363	567	602	997	994
River eels	3,687	3,457	3,176	3,759	2,879	4,627	4,360
Shad, milkfish, etc.	134,287	102,843	95,039	38,070	40,750	5,632	28,073
Total	139,301	107,727	99,578	42,396	44,231	11,256	33,427

Marine Fish	1985	1986	1987	1988	1989	1990	1991
Flounder, sole, etc.	13,585	15,181	15,563	13,836	14,939	13,309	629,359
Cod, hake, etc.	83,225	75,253	90,965	89,572	74,927	67,615	76,001
Redfish, bass, etc.	156,965	150,203	150,839	146,097	162,487	165,210	163,686
Jack, mullet, etc.	245,713	208,165	196,666	195,587	198,155	164,276	123,238
Herring, sardines, etc.	832,102	935,128	842,018	989,402	646,010	502,600	448,713
Tuna, bonito, etc.	66,617	61,684	66,849	77,675	60,640	66,616	67,286
Mackerel, snoek, etc.	41,385	48,430	57,097	51,540	44,059	36,967	32,725
Sharks, rays, etc.	25,589	23,913	21,726	22,962	220,041	17,552	18,978
Miscellaneous	114,225	119,823	121,737	118,555	110,083	103,699	95,835
Total	1,565,821	1,622,599	1,547,897	1,691,390	1,516,402	1,124,535	1,655,822

Crustaceans	1985	1986	1987	1988	1989	1990	1991
Crabs, etc.	1,467	1,592	2,601	3,002	2,003	1,320	1,492
Lobsters, etc.	7,503	8,394	8,128	10,594	7,727	8,426	8,891
Shrimps, prawns, etc.	35,682	31,836	33,218	33,266	31,377	38,102	30,627
Miscellaneous	7,149	9,095	10,000	8,649	8,706	7,806	11,691
Total	51,801	50,917	53,947	55,511	50,812	55,654	52,701

Molluscs	1985	1986	1987	1988	1989	1990	1991
Winkles, conchs, etc.	346	–	602	–	–	–	–
Oysters	11,596	18,304	14,131	12,043	17,233	15,2707	14,398
Mussels	90,591	93,325	105,361	120,346	125,480	136,358	126,567
Scallops, etc.	2	4	4	4	1	1	0
Clams, cockles, etc.	27,095	28,724	45,154	39,324	47,954	47,104	57,931
Squid, cuttlefish, etc.	65,723	68,403	67,688	83,321	76,920	71,884	66,745
Miscellaneous	13,867	16,949	16,089	23,089	22,779	17,253	14,965
Total	208,874	225,709	228,427	278,127	290,367	425,307	280,606

Grand total							
	1,965,797	2,006,952	1,929,849	2,067,424	1,901,812	1,616,752	2,032,556

Landings of fish by oceans *(metric tonnes)*

PACIFIC OCEAN LANDINGS

Diadromous Fish	1985	1986	1987	1988	1989	1990	1991
Sturgeon, etc.	8	6	5	5	10	9	3
River eels	1,503	1,158	1,075	1,197	1,232	1,309	786
Salmon, trout, etc.	784,358	659,791	625,549	618,626	853,426	833,572	1,463,165
Shad, milkfish, etc.	35,511	45,406	33,578	31,322	34,437	42,355	47,838
Miscellaneous	38,145	37,596	35,977	39,472	37,185	103,893	57,999
Total	*859,525*	*743,957*	*696,184*	*690,622*	*926,290*	*981,138*	*1,569,791*

Marine Fish	1985	1986	1987	1988	1989	1990	1991
Flounder, sole, etc.	684,000	658,194	614,656	693,734	543,255	567,540	426,787
Cod, hake, etc.	6,883,344	7,805,909	7,989,025	8,082,376	7,737,285	7,278,285	6,214,203
Redfish, bass, etc.	2,304,896	2,631,509	2,489063	2,389,218	2,430,953	2,549,502	2,383,777
Jack, mullet, etc.	4,183,064	4,088,625	3,720,076	5,575,717	6,056,509	6,545,104	6,634,726
Herring, sardines, etc.	14,550,415	17,458,301	14,751,933	16,731,022	17,683,129	15,313,322	14,949,106
Tuna, bonito, etc.	2,061,174	2,336,403	2,396,827	2,655,205	2,714,967	2,949,263	3,064,180
Mackerel, snoek, etc.	2,490,442	2,625,379	2,302,867	2,352,319	2,284,029	2,146,756	2,117,472
Sharks, rays, etc.	241,266	247,354	254,924	251,857	268,551	276,705	275,029
Miscellaneous	5,908,561	6,370,211	6,676,345	4,720,032	6,929,877	7,240,355	7,387,160
Total	*39,307,162*	*44,221,885*	*41,195,716*	*43,451,480*	*46,648,555*	*44,866,832*	*43,452,440*

Crustaceans	1985	1986	1987	1988	1989	1990	1991
Crabs, etc.	652,353	673,656	752,388	817,756	944,016	928,078	1,107,986
Lobsters, etc.	12,820	14,487	13,658	11,143	11,908	12,187	12,076
Shrimps, prawns, etc.	1,056,166	1,200,750	2,103924	1,420423	1,357,769	1,368,386	1,410,976
Krill, etc.	4,721	3,892	394	–	–	658	3
Miscellaneous	16,126	15,410	4,693	6,541	6,879	8,564	7,398
Total	*1,742,186*	*1,908,195*	*2,875,057*	*2,255,863*	*2,320,572*	*2,317,873*	*2,317,873*

Molluscs	1985	1986	1987	1988	1989	1990	1991
Winkles, conchs, etc.	53,646	55,637	70,638	64,443	52,696	40,701	41,791
Oysters	656,444	656,795	726,422	745,614	688,332	675,302	1,338,640
Mussels	316,502	352,982	465,570	622,400	673,548	689,898	697,419
Scallops, etc.	302,668	305,221	358,559	479,208	512,993	587,126	546,862
Clams, cockles, etc.	1,345,682	831,618	875,711	881,786	834,296	838,935	901,265
Squid, cuttlefish, etc.	1,144,498	991,331	1,832,208	1,061,783	1,458240	1,330,247	1,404,369
Miscellaneous	256,158	376,511	464,657	525,021	429,921	460,994	459,040
Total	4,075,598	3,570,095	4,793,765	43,80,255	4,650,026	4,623,203	5,389,386

Grand total							
	46,843,996	*53,096,284*	*50,256,906*	*50,778,220*	*55,471,733*	*53,770,184*	*54,519,847*

INDIAN OCEAN LANDINGS

Diadromous Fish	1985	1986	1987	1988	1989	1990	1991
Shad, milkfish, etc.	101,130	116,928	139,717	138,378	144,937	145,711	152,137
Miscellaneous	5,838	5,590	6,374	6,656	8,889	6,968	8,683
Total	*106,968*	*122,518*	*146,091*	*145,034*	*153,826*	*152,679*	*160,820*

Marine Fish	1985	1986	1987	1988	1989	1990	1991
Flounder, sole, etc.	20,749	26,846	23,493	18,945	37,721	38,725	42,100
Cod, hake, etc.	1,848	3,125	2,514	4,851	1,205	1,736	1,378
Redfish, bass, etc.	683,155	734,311	710,729	771,871	888,860	896,531	936,345
Jack, mullet, etc.	360,374	362,276	421,403	486,025	536,423	533,937	570,127
Herring, sardines, etc.	582,648	596,036	613,393	634,015	672,672	774,323	772,880
Tuna, bonito, etc.	662,078	706,980	825,622	970,494	764,634	965,598	989,476
Mackerel, snoek, etc.	344,051	287,073	291,550	270,129	355,051	325,237	339,757
Sharks, rays, etc.	133,054	141,757	150,899	173,669	158,665	147,799	161,056
Miscellaneous	1,513,115	1,631,108	1,669,617	1,680,576	1,897,666	1,826,469	1,914,939
Total	*4,301,072*	*4,489,512*	*4,709,220*	*5,010,575*	*5,312,897*	*5,510,355*	*5,728,058*

Crustaceans	1985	1986	1987	1988	1989	1990	1991
Crabs, etc.	12,284	15,084	15,186	15,725	17,572	15,556	20,454
Lobsters, etc.	21,513	21,301	21,673	23,016	22,708	21,106	20,454
Shrimps, prawns, etc.	393,987	387,133	387,386	407,393	416,321	436,994	433,962
Krill, etc.	4,721	3,892	394	–	–	658	3
Miscellaneous	28,601	30,958	42,029	39,801	38,295	45,001	45,811
Total	*461,106*	*458,368*	*466,668*	*485,935*	*494,896*	*519,315*	*520,684*

Molluscs	1985	1986	1987	1988	1989	1990	1991
Winkles, conchs, etc.	7,049	6,500	6,210	6,345	5,212	4,901	4,875
Oysters	2,703	1,453	1,698	1,456	2,372	1,274	4,875
Mussels	3,221	4,886	7,337	11,441	10,617	7,245	7,825
Scallops, etc.	14,717	12,400	8,900	7,050	1,791	1,976	7,826
Clams, cockles, etc.	76,371	73,171	70,394	69,513	70,020	62,257	75,929
Squid, cuttlefish, etc.	55,341	57,636	78,254	94,170	124,485	77,408	79,622
Miscellaneous	3,599	7,648	9,988	4,595	4,499	2,996	4,221
Total	*163,001*	*163,694*	*182,781*	*194,570*	*218,996*	*158,057*	*185,173*

Grand total							
	5,032,147	*5,234,092*	*5,504,760*	*5,691,080*	*6,179,468*	*6,341,553*	*6,594,735*

Glossary

abyssal Lying in the deep ocean.

abyssal plain The ocean floor between 4000 and 6000 metres (13,000 and 20,000 feet).

annular Ring shaped.

anticline A fold of rock in the form of an arch

archipelago A group of islands. Often used to denote an island chain.

asthenosphere A semimolten layer of the Earth's upper mantle on which the crustal plates move.

barrel A unit of volume in the measurement of liquids. The unit is different in size for different liquids. 1 barrel of petroleum = 35 gallons (UK) = 42 gallons (US).

bathymetry The measurement of depths of water in the oceans.

bathypelagic Living in deep water below the level of light penetration, but above the abyssal regions.

bedding planes A plane in a sedimentary rock parallel to the original surface of deposition. A sedimentary rock usually splits along such a plane.

bedrock Unweathered rock lying beneath the soil or surface sediment.

benthic Occurring at the bottom of the sea.

cap rock The rock formation above a salt dome or oil and gas trap that prevents the escape of fluids.

continental rise The gently sloping part of the seabed leading down from the continental slope into the abyssal plain.

continental shelf The part of the seabed, normally not deeper than 200 metres (660 feet), adjacent to a landmass and forming a submerged part of the continent itself.

continental slope The steeply sloping part of the seabed leading down from the continental shelf.

crust The outermost large-scale division of the Earth's structure. It is between 16 and 40 kilometres (10 and 25 miles) thick, the thinnest parts lying beneath the oceans, the thickest beneath mountain ranges.

cuticle A superficial covering layer on either a plant or an animal.

demersal fish Fish living on or near the seabed.

detritus A mass of loose rock fragments.

diadromous fish Fish that migrate between fresh and seawater.

diapir The intrusion of relatively less dense material through the lower layers of more dense overlying rocks causing doming at the surface.

dolomite Limestone containing more than 15 per cent magnesium carbonate.

dory A small, high-sided, flat-bottomed boat.

drogue A device towed behind an underwater instrument to regulate its speed or its direction of movement.

Earth's axis The line about which the Earth rotates.

ecosystem The relationship between a community of plants and animals and its environment.

eddy A small circular current.

epeirogenic movements The uplift or depression of large areas of land.

epipelagic Description for those organisms living in the upper, lighted layer of seawater.

euphotic zone The upper layer of seawater having sufficient light to support plant life.

eustatic changes Changes in sea level.

evaporite A sedimentary rock; produced by the evaporation of salt water.

fault A break in a rock along which displacement has occurred.

thrust fault A fault in which one block of rock is forced over another.

transform fault A fault occurring between crustal plates where the mid-ocean ridge is offset.

graben A rift consisting of a downthrown block between parallel faults.

guano A deposit of seabird excrement.

guyot A flat-topped mountain occurring on the abyssal plain.

gyre Circular or spiral movement.

horst An upthrown block between parallel faults.

hot spot A postulated source of volcanic heat located in the Earth's mantle. Such a spot tends to remain stationary while the plates move over it, giving rise to chains of volcanic islands.

hydrothermal Concerning natural hot springs.

isostatic movement The movement of a landmass to attain equilibrium.

kinetic energy Energy associated with movement.

lagoon A shallow lake at the edge of a deeper body of water, often separated from it by a reef.

lithosphere The outer rigid skin of the Earth consisting of the crust and the topmost part of the mantle. This constitutes the "crustal plates", which move on the more fluid asthenosphere.

littoral zone The area of coast found between high and low tide.

magma Molten igneous material.

magnetic poles The opposite ends, or poles, of the Earth's magnetic field.

mantle (biology) The material covering the body of a mollusc lying immediately inside the shell.

mantle (geology) The large-scale division of the Earth's structure lying between the surface crust and the innermost core.

medusa Jellyfish and the free-swimming stages of some other marine organisms.

meridian A line of longitude.

metric tonne One thousand kilograms.

nanoplankton Minute planktonic organisms, too small to be caught in a plankton net.

notochord The axial skeletal support in chordates.

oolith A spherical rock particle formed by the accretion of calcite around a nucleus.

orogeny The process of mountain-building.

ovoviviparous Producing eggs that are hatched within the maternal body.

pahoehoe Lava having a ropey appearance.

pedicellaria Pincer-like spines occurring on the surface of sea urchins and starfish.

pelagic fish Free-swimming fish inhabiting the open sea, independently of the seabed.

placer deposits Deposits of economic minerals eroded from a landmass and accumulated by waves, currents or tides underwater.

plate tectonics The modern study of the large-scale features of the Earth's surface involving the creation of "crustal" plates at mid-ocean ridges and their destruction in the deep-sea trenches.

polyp The sedentary form of Cnidaria (corals, sea anemones, etc.), some of which form colonies.

potential energy Energy latent in a body by virtue of its position or relation to other bodies.

pyroclastic Formed of igneous rock blown into the air by a volcanic explosion.

rift valley An elongated, subsided block bounded by parallel faults in the form of a valley.

seismic Concerning earthquake waves and the artificial production of similar shock waves.

sessile Attached by the base. Normally refers to organisms fixed to the ground or seabed.

setae Bristle-like structures that occur on some invertebrate animals.

sonograph A picture of the seafloor produced by means of sound waves and having the appearance of a photograph.

spicules Spines or needles of calcium carbonate occurring in the tissues of certain organisms, such as sponges.

subaerial On the Earth's surface, i.e. below the air, as opposed to *submarine* – below the sea.

subduction zone The area at the margin of a crustal plate where one plate is being destroyed beneath another. This coincides with a deep-sea trench.

tectonics The study of the large-scale features of the Earth's surface and their evolution.

thermocline The layer in a stratified body of water in which the temperature changes most rapidly with changes in depth.

trench An elongated depression in the seabed produced as one crustal plate slides beneath the other. The trenches are the deepest parts of the ocean.

troll To fish with hook and line drawn through the water, often behind a boat.

turbidity current The movement of a slurry of suspended material, such as silt or clay, in water.

unconformity A break in a sequence of sedimentary rocks. It is produced when the depositional process that laid down the sediments is halted and begins again after a pause. Bedding planes above and below the unconformity frequently lie at different angles.

Conversion factors

1 metre	=	3.281 feet		
1 centimetre	=	0.394 inches		
1 micron (μm)				
(1 millionth of a meter)	=	0.039 thousandths of an inch		
1 foot	=	0.305 metres		
1 kilometre	=	1000 metres	= 0.621 miles	
1 mile	=	1.609 kilometres		
1 gram	=	0.035 ounces (dry)		
1 ounce (dry)	=	28.35 grams		

1 kilogram	=	1000 grams	=	2.205 pounds
1 pound	=	0.454 kilograms		
1 metric tonne	=	0.984 long tons (UK)	=	1.102 short tons (US)
1 long ton UK)	=	1.016 metric tonnes	=	1.120 short tons (US)
1 short ton (US)	=	0.907 metric tonnes	=	0.893 long tons (UK)
1 litre	=	0.220 gallons (UK)	=	0.264 gallons (US)
1 gallon (UK)	=	4.546 litres	=	1.201 gallons (US)
1 gallon (U.S)	=	3.785 litres	=	0.833 gallons (UK)
1 barrel (oil)	=	0.132 metric tonnes	=	0.134 long tons (UK)
	=	0.150 short tons (US)	=	159 litres
	=	35 gallons (UK)	=	42 gallons (US)

Threatened Marine Species

All Oceans

Mammals
E Blue whale (*Balaenoptera musculus*)
V Fin whale (*Balaenoptera physalus*)
V Humpback whale (*Megaptera novaeangliae*)

Arctic

Mammals
V Bowhead whale (*Balaena mysticetus*)
V Polar bear (*Ursus maritimus*)

North Atlantic

Mammals
E Northern right whale (*Eubalaena glacialis*)
V Bowhead whale (*Balaena mysticetus*)
V Northern bottlenose whale (*Hyperoodon ampullatus*)
Others
V Starlet sea anemone (*Nematostella vectensis*)

Mediterranean

Mammals
E Mediterranean monk seal (*Monachus monachus*)
Birds
E Freira (*Pterodroma madeira*)
E Dalmatian pelican (*Pelecanus crispus*)
Reptiles
E Green turtle (*Chelonia mydas*)
V Loggerhead turtle (*Caretta caretta*)

Wider Caribbean

Mammals
V West Indian (Caribbean) manatee (*Trichechus manatus*)
V Amazonian manatee (*Trichechus inunguis*)

Birds
E Cahow (Bermuda petrel) (*Pterodroma cahow*)

Reptiles
E Green turtle (*Chelonia mydas*)
E Olive Ridley turtle (*Lepidochelys olivacea*)
E Leatherback turtle (*Dermochelus coriacea*)
E Hawksbill turtle (*Eretmochelys imbricata*)
E Kemp's Ridley turtle (*Lepidochelys Kempii*)
V Loggerhead turtle (*Caretta caretta*)

Southwest Atlantic

Mammals
V Southern right whale (*Eubalaena australis*)
V La Plata otter (*Lutra longicaudis*)

Reptiles
E Green turtle (*Chelonia mydas*)
E Olive Ridley turtle (*Lepidochelys olivacea*)
E Leatherback turtle (*Dermochelus coriacea*)
E Hawksbill turtle (*Eretmochelys imbricata*)
V Loggerhead turtle (*Caretta caretta*)

West Africa

Mammals
West African manatee (*Trichechus senegalensis*)

Reptiles
E Green turtle (*Chelonia mydas*)
E Olive Ridley turtle (*Lepidochelys olivacea*)
E Leatherback turtle (*Dermochelus coriacea*)
E Hawksbill turtle (*Eretmochelys imbricata*)
V Loggerhead turtle (*Caretta caretta*)

Indian Ocean

Mammals
V Southern right whale (*Eubalaena australis*)
V Dugong (*Dugong dugon*)

Birds
E Amsterdam albatross (*Diomedea amsterdamensis*)
V Kerguelen tern (*Sterna virgata*)
E Madagascar fish eagle (*Haliaeetus vociferoides*)

Reptiles
E Green turtle (*Chelonia mydas*)
E Olive Ridley turtle (*Lepidochelys olivacea*)
E Leatherback turtle (*Dermochelus coriacea*)
E Hawksbill turtle (*Eretmochelys imbricata*)
V Loggerhead turtle (*Caretta caretta*)

Others
V Coelacanth (*Latimeria chalumnae*)

East Asian Seas

Mammals
V Dugong (*Dugong dugon*)

Birds
E Abbott's booby (*Sula abbotti*)
E Christmas frigatebird (*Fregata andrewsi*)

Reptiles
V Estuarine crocodile (*Crocodylus porosus*)
E Green turtle (*Chelonia mydas*)
E Olive Ridley turtle (*Lepidochelys olivacea*)
E Leatherback turtle (*Dermochelus coriacea*)
E Hawksbill turtle (*Eretmochelys imbricata*)
V Loggerhead turtle (*Caretta caretta*)

Molluscs
V Southern giant clam (*Tridacna derasa*)
V Giant clam (*Tridacna gigas*)

South Pacific

Mammals
V Southern right whale (*Eubalaena australis*)
V Hector's dolphin (*Cephalorynchus hectori*)
V Dugong (*Dugong dugon*)
E Hawaiian monk seal (*Monachus schauinslandi*)

Birds
V Pycroft's petrel (*Pterodroma pycrofti*)
V Chatham Island petrel (*Pterodroma axillaris*)

V Newell's shearwater (*Puffinus newelli*)
V Black petrel (*Procellaria parkinsoni*)
E Chatham Island oystercatcher (*Haematopus chathamnsis*)
V Black-fronted tern (*Sterna albostriata*)
V Yellow-eyed penguin (*Megadyptes antipodes*)

Reptiles
V Estuarine crocodile (*Crocodylus porosus*)
E Green turtle (*Chelonia mydas*)
E Olive Ridley turtle (*Lepidochelys olivacea*)
E Leatherback turtle (*Dermochelus coriacea*)
E Hawksbill turtle (*Eretmochelys imbricata*)
V Loggerhead turtle (*Caretta caretta*)

Molluscs
V Southern giant clam (*Tridacna derasa*)
V Giant clam (*Tridacna gigas*)

North Pacific

Mammals
E Northern right whale (*Eubalaena glacialis*)
V Bowhead whale (*Balaena mysticetus*)
V Guadeloupe fur seal (*Arctocephalus townsendi*)
E Vaquita (*Phocoena sinus*)

Birds
E Townsend's shearwater (*Puffinus auriculus*)

Others
V Starlet sea anemone (*Nematostella vectensis*)

Southeast Pacific

Mammals
V Southern right whale (*Eubalaena australis*)
V Juan Fernandez fur seal (*Arctocephalus philippii*)
V Marine otter (*Lutra felina*)

Birds
V Pink-footed shearwater (*Puffinus creatopus*)
V Defilippe's petrel (*Pterodroma defilippiana*)

Reptiles
E Green turtle (*Chelonia mydas*)
E Olive Ridley turtle (*Lepidochelys olivacea*)
E Leatherback turtle (*Dermochelus coriacea*)
E Hawksbill turtle (*Eretmochelys imbricata*)

Antarctica and Southern Oceans

Mammals
V Southern right whale (*Eubalaena australis*)

E=Endangered
V=Vulnerable

Index

Index

Index

Index

Index

Index

Index

Index for the *Encyclopedia of Marine Life*

Index for the *Encyclopedia of Marine Life*

ACKNOWLEDGEMENTS

Illustration Credits

Credits read from left to right in descending order on each page
12 Arka graphics; Chris Forsey; Chris Forsey. **13** Chris Forsey. **14** Arka Graphics, Sidney Woods, Arka Graphics. **15** Sidney Woods. **16** Chris Forsey. **17** Chris Forsey. **18** Chris Forsey. **23** Keith Williams. **24** Arka Graphics; Mike Saunders; Brian Delf. **25** Brian Delf; Mike Saunders; Brian Delf. **26** Chris Forsey. **27** Arka Graphics. **28** Colin Rose. **29** Richard Lewis. **31** Mike Saunders. **32** Chris Forsey. **33** Mike Saunders. **34** Ed Stuart **36** Arka Graphics. **37** Ed Stuart. Arka Graphics. **38** Mike Saunders. **39** Mike Saunders. **41** Brian Delf; Mike Saunders; Mike Saunders; Mike Saunders. **42** Keith Williams. **47** Keith Williams; **53** Chris Forsey. **55** Ray Grinaway; Alan Suttie. **64** Arka Graphics; Arka Graphics; Sidney Woods. **65** Arka Graphics; Sidney Woods. **66** Sidney Woods. **67** Sidney Woods. **68** Brian Delf. **74** Les Smith. **75** Sandra Doyle. **76** Brian Delf. **77** Ray Grinaway; Ray Grinaway; Brian Delf; Ray Grinaway. **79** Michael Woods. **80** Michael Woods; Michael Woods; Michael Woods; Michael Woods; Ray Grinaway. **81** Keith Williams. **82** Chris Forsey. **84** Sidney Woods. **85** Keith Williams. **90** Chris Forsey. **91** Chris Forsey. **96** John Davis. **97** John Davis. **99** Keith Williams. **104** Andrew Thompson. **105** Brian Delf. **106** Andrew Thompson. **108** Terry Collins, Chris Forsey; Mike Saunders. **109** Alan Suttie. **110** Mike Saunders. **111** Mike Saunders. **118** Pierre Charron. **120** Sidney Woods; Arka Graphics. **121** Arka Graphics. **122** Andrew Thompson/Marian Steele. **123** Andrew Thompson; Arka Graphics. **124** Gordon Miles; Arka Graphics. **125** Andrew Thompson/Marian Steele. **126** Sidney Woods; Mike Saunders; Andrew Thompson. **127** Andrew Thompson/Marian Steele; Mike Saunders. **128** Andrew Thompson/Marian Steele. **129** Andrew Thompson. **130** Sidney Woods; Arka Graphics; Mike Saunders. **131** Andrew Thompson/ Marian Steele; Mike Saunders. **132** Andrew Thompson/Marian Steele. **134** Arka Graphics; Arka Graphics; Andrew Thompson; Arka Graphics; Arka Graphics. **135** Andrew Thompson/Marian Steele; Arka Graphics; Arka Graphics. **136** Arka Graphics. **137** Andrew Thompson. **138** Sidney Woods; Andrew Thompson/Marian Steele. **139** Andrew Thompson; Arka Graphics. **141** Andrew Thompson. **142** Sidney Woods. **144** Arka Graphics; Andrew Thompson. **145** Andrew Thompson; Andrew Thompson/Marian Steele. **147** Andrew Thompson. **148** Sidney Woods; Trianon. **149** Andrew Thompson/Marian Steele; Andrew Thompson. **150** Andrew Thompson. **151** Trianon; Keith Williams. **153** Andrew Thompson/Marian Steele. **155** Andrew Thompson. **156** Sidney Woods; Arka Graphics. **157** Mike Saunders. **158** Andrew Thompson. **159** Andrew Thompson/ Marian Steele; Mike Saunders. **160** Andrew Thompson. **161** Gordon Miles. **164** Sidney Woods, Arka Graphics. **165** Andrew Thompson; Mike Saunders. **166** Andrew Thompson. **167** Andrew Thompson/ Marian Steele. **168** Andrew Thompson. **169** Mike Saunders. **172 – 175** Michael Woods. **176** Michael Woods; Raymond Turvey; Michael Woods; Raymond Turvey; Michael Woods; Michael Woods. **177 – 178** Michael Woods. **179** Michael Woods; Michael Woods; Michael Woods; Raymond Turvey; Michael Woods. **180 – 185** Michael Woods.

Photographic Credits
Heather Angel/Biofotos 54 *top,* **68** *bottom,* **72** *centre left,* **90** *bottom left,* **91** *top right,* **91** *bottom left;* **Arbeitsgemeinschaft Gewinnbare Rohstoffe, Frankfurt 27** *top right;* **Alan Archer 124, 147** *left;* **Bridgeman Art Library** /National Maritime Museum **49** *bottom,* /Royal Geographical Society, London **48-49; Bruce Coleman Limited** /Erwin & Peggy Bauer **79** *top,* /S.C. Bisserot **90** *bottom right,* /Massimo Borchi **147** *right,* /Jane Burton **68** *top,* **91** *bottom right,* /Francisco Erize **113** *right,* /Inigo Everson *half-title page,* /Frithfoto **79** *bottom,* /Fraser Hall **8-9,** /Dr Rocco Longo **140** *top,* /William S. Paton **83** *bottom,* /Dr Eckhard Pott **102,** /G. Ziesler **67; Bruce Coleman Inc. New York** /R.N. Mariscal **91** *top left;* **ESA 53, 138** *centre;* **Explorer** /C. Delu **110-111; Mary Evans Picture Library 48** *top, centre and bottom left;* **Fishing News 100** *bottom;* **FLPA** /Christiana Carvalho **78** *bottom;* **Greenpeace** /Hewetson **112** *bottom;* **Robert Harding Picture Library** /Robert Frerck **141,** /Photri **162; HMSO 98** *centre;* **Michael Holford Photographs** /National Maritime Museum **50** *centre left;* **Hutchison Library 154; Institut für Meereskunde** /D. Carlsen **52** *bottom left;* **Institute of Oceanographic Sciences Deacon Laboratory 24** *centre,* **25, 51, 52** *bottom right;* **Jacana 63** *bottom right,* /C. Carré **75** *centre left;* **Paolo Koch 133** *top left;* **The Motorship Magazine** /Quadrant, Asturias S.L. **98-99; NASA 40** *top,* **123** *left,* /Johnson Space Center, Houston **126** *right;* **Natural History Museum, London 27** *left;* **Natural Science Photos** /M.E. Bacchus **84** *top right,* /Nat Fain **75** *right;* **NHPA** /Matt Bain **17** *top,* /Anthony Bannister **105** *centre,* /G.I. Bernard **50** *bottom left,* /Peter Johnson **66** *left,* /B. Jones & M. Shimlock **73** *top,* /Tsuneo Nakamura **66** *right,* /Lady Philippa Scott **97** *bottom,* /Roy Waller **71** *bottom right,* /David Woodfall **19** *top;* **Oxford Scientific Films 27** *top left,* **88** *top left,* **88-99, 89** *top and bottom right,* /Bruce Coleman **72** *centre right,* /Jeff Foott Okapia **69** *right,* /Michael Pitts **22** *left;* **Photo Researchers Inc. 101** *top right;* **Planet Earth Pictures** /Kurt Amsler **92** *left,* /Gary Bell **172-173,** /Richard Chesher **40** *bottom,* /Dick Clarke **57** *top,* /P.David **63** *right centre,* /Alex Double **94-95,** /J. Duncan **98** *bottom,* /Pieter Folkens **69** *top,* /Jim Greenfield **71** *bottom left,* /Robert Hessler **27** *bottom,* /Ken Lucas **70,** /John Lythgoe **34, 35, 85, 90** *top right,* /Richard Matthews **83** *top,* /Michael McKinnon **153** *bottom,* /Bora Merdsoy **33,** /Doug Perrine **81** *right,* /Linda Pitkin **93** *bottom right,* **115** *top,* /Mike Potts **84** *bottom right,* /Peter Scoones *back cover right,* **6, 93** *bottom left,* **116-117, 164,** /Marty Snyderman **73** *bottom,* /Ken Vaughan **54** *bottom,* /Warren Williams **104; Quadrant Picture Library** /Jan Strømme **109; Science Photo Library** /Alex Bartel **17** *bottom,* /David Campione **21** *top,* /Earth Satellite Corporation **18,** /Gene Feldman, NASA GSFC **171** *top,* /Geospace **36** *top,* /Jan Hinsch **27** *centre left,* /Institute of Oceanographic Sciences, NERC **107** *bottom right,* /NASA **43** *top,* **149** *bottom,* /David Parker **22-23,** /Sam Pierson Jr. **38,** /University of Cambridge Collection of Air Photographs **21** *bottom,* /Tom Van Sant, Geosphere Project, Santa Monica **11, 121, 143, 157** *top;* **The Sea Library** /P. Capen **72** *bottom centre,* /H. Genthe **69** *bottom,* **72** *bottom left;* **Seismograph Service Ltd. 108** *bottom;* **South American Pictures** /Tony Morrison **43** *bottom,* **101** *top,* **163** *top;* **Still Pictures** /Greth Arnaud **153** *centre,* /Mark Edwards **46,** /Pierre Gleizes **133** *bottom,* /Al Grillo **112-113,** /Michel Gunther **19** *bottom,* **140** *bottom,* /Yves Lefevre **75** *top left,* /Andre Maslenniker **31,** /Edward Parker **97** *top;* **Tony Stone Images** /Paul Chesley **92-93; Survival Anglia Limited Photographic Library** /Michael Pitts **56, 144-145; Haroun Tazieff 148** *bottom right;* **John de Visser 133** *top right;* **D.P. Wilson 62** *top, centre and bottom left,* **63** *right,* **68** *centre,* **72** *bottom right;* **Woods Hole Oceanographic Institution 56; Zefa Picture Library (U.K.) Ltd. 20, 28-29, 71** *top,* **103** *top,* **114, 129** *bottom,* **136,** /APL *front cover,* /B. Crader **103** *bottom,* /Darodents *title page,* /Janoud **101** *bottom,* /Lanting **60-61,** /Jim Watt **115** *bottom.*